Chemistry

OCR and Heinemann are working together to provide better support for you

Heinemann is an imprint of Pearson Education Limited, a company incorporated in England and Wales, having its registered office at Edinburgh Gate, Harlow, Essex CM20 2JE. Registered company number: 872828

www.heinemann.co.uk

Heinemann is a registered trademark of Pearson Education Limited

Text © Pearson Education Limited 2008

First published 2008

12 11 10
10 9 8 7 6

British Library Cataloguing in Publication Data
A catalogue record for this book is available from the British Library

ISBN 978 0 435691 98 1

Edited by Tony Clappison
Index compiled by Wendy Simpson
Designed by Kamae Design
Project managed and typeset by Wearset Ltd, Boldon, Tyne and Wear
Original illustrations © Pearson Education Limited 2008
Illustrated by Wearset Ltd, Boldon, Tyne and Wear
Picture research by Wearset Ltd, Boldon, Tyne and Wear
Cover photo shows the growth of a fluorapatite (calcium fluoride phosphate) crystal from a hexagonal rod to a sphere, composite scanning electron micrographs, © Science Photo Library.
Printed in China (GCC/06)

Acknowledgements
We would like to thank the following for their invaluable help in the development and trialling of this course: Andrew Gelling, Peter Haigh, Amanda Hawkins, Dave Keble, Maggie Perry, Michael Taylor and Chris Wood.

The authors and publisher would like to thank the following for permission to reproduce photographs:

p3 Laguna Design/Science Photo Library; p4 Cordelia Molloy/Science Photo Library; p5 T Alan L. Detrick/Science Photo Library; p5 M Charles D. Winters/Science Photo Library; p6 T Science Photo Library; p6 B Getty Images; p8 Clive Freeman, The Royal Institution/Science Photo Library; p11 TL Phillipe Psalia/Science Photo Library; p11 TM Gustoimages/Science Photo Library; p11 TR Kevin In Canada; p11 M Garry D. McMichael/Science Photo Library; p15 T Dave Gent; p15 M Jim Zipp/Science Photo Library; p16 Martyn F. Chillmaid/Science Photo Library; p17 T Science Source/Science Photo Library; p17 B Rob Ritchie; p17 B Rob Ritchie, Collectstamps; p18 Martyn F. Chillmaid/Science Photo Library; p19 T Cordelia Molloy/Science Photo Library; p19 M Rob Ritchie; p19 BM Rosal/Science Photo Library; p19 B Tek Image/Science Photo Library; p21 M David Gubernick/AgstockUSA/Science Photo Library; p21 B Jiri Loun/Science Photo Library; p23 M Dave Gent; p23 BL Adey Bryant; p23 BR GoGo Images Corporation/Alamy; p24 Dave Gent; p25 Dave Gent; p26 Dave Gent; p27 T Dave Gent; p27 B Andrew Lambert Photography/Science Photo Library; p28 TL David Munns/Science Photo Library; p28 TR Image Source/Getty Images; p28 TM Maximilian Stock Ltd/Science Photo Library; p28 B Charles D. Winters/Science Photo Library; p31 T George & Judy Manna/Science Photo Library; p31 B John Heseltine/Science Photo Library; p32 T Maximilian Stock Ltd/Science Photo Library; p32 M Sinclair Stammers/Science Photo Library; p32 B Erika Craddock/Science Photo Library; p33 Bildagentur-Online/TH Foto/Science Photo Library; p34 Martin Bond/Science Photo Library; p35 T Bsip Vem/Science Photo Library; p35 B Mr. Nut/Alamy; p36 T Gustoimages/Science Photo Library; p36 M Daniel Sambraus/Science Photo Library; p38 Andrew Lambert Photography/Science Photo Library; p39 BL Edward Rozzo/Corbis; p39 BM Burn; p47 Alex Bartel/Science Photo Library; p48 Steve Gschmeier/Science Photo Library; p51 Dorling Kindersley; p53 SleeperService on Flickr; p54 Dave Gent; p55 Huhtamaki; p56 Peter Menzel/Science Photo Library; p57 T Getty Images; p57 B Cordelia Molloy/Science Photo Library; p61 T Peter Menzel/Science Photo Library; p61 B Sainsburys; p62 Estate of Francis Bello/Science Photo Library; p66 Cordelia Molloy/Science Photo Library; p67 Martyn F. Chillmaid/Science Photo Library; p75 Mason Morfit, Peter Arnold Inc./Science Photo Library; p76 Maximilian Stock Ltd/Science Photo Library; p79 Mehau Kuylk/Science Photo Library; p82 Dr Jurgen Scriba/Science Photo Library; p83 NASA/Science Photo Library; p85 Will Davey; p94 Colin Cuthbert/Science Photo Library; p98 Blend Images/Alamy; p99 T Erich Schrempp/Science Photo Library; p99 B Zephyr/Science Photo Library; p111 David Munns/Science Photo Library; p114 M Rob Ritchie; p114 B Rob Ritchie; p119 Gianni Tortoli/Science Photo Library; p121 Martyn F. Chillmaid/Science Photo Library; p131 Charles D. Winters/Science Photo Library; p134 Science Photo Library; p135 T Science Photo Library; p135 M Science Photo Library; p137 T Martyn F. Chillmaid/Science Photo Library; p137 B Rob Ritchie; p139 Charles D. Winters/Science Photo Library; p140 Science Photo Library; p142 BL Science Photo Library; p142 BR Thomas Deerinck, NCMIR/Science Photo Library; p143 David Munns/Science Photo Library; p144 Charles D. Winters/Science Photo Library; p146 Edward Kinsman/Science Photo Library; p147 T The Great Shadow; p147 M Dave Gent; p148 CCI Archives/Science Photo Library; p151 Mary Evans Picture Library/Alamy; p153 JUPITERIMAGES/Comstock Images/Alamy; p156 Rob Ritchie; p165 Dr Mark J. Winter/Science Photo Library; p174 Rob Ritchie; p179 T Rob Ritchie; p179 B Martyn F. Chillmaid/Science Photo Library; p180 Science, Industry & Business Library/New York Public Library/Science Photo Library; p184 Rob Ritchie; p188 E.R. Degginger/Science Photo Library; p190 Rob Ritchie; p191 Nikon/Scott Andrews/NASA/Science Photo Library; p192 M Martin Bond/Science Photo Library; p192 B Vanessa Vick/Science Photo Library; p193 Justin Sullivan/Getty Images; p201 Mauro Fermariello/Science Photo Library; p204 TL Tysto Nicogag; p204 ML Steve Fairman; p204 MR James King-Holmes/Science Photo Library; p204 B Dave Gent; p206 T Astrid & Hanns-Frieder Michler/Science Photo Library; p206 M Dick Luria/Science Photo Library; p207 T Charles D. Winters/Science Photo Library; p207 M Dave Gent; p207 B Dave Gent; p210 Science Photo Library; p214 Dave Gent; p215 T Dave Gent; p215 M Dave Gent; p216 T Andrew Syred/Science Photo Library; p216 M John Bavosi/Science Photo Library; p218 Dave Gent; p219 Dave Gent; p221 Dave Gent; p222 Dave Gent; p223 Justin Kase Ztwoz/Alamy

Every effort has been made to contact copyright holders of material reproduced in this book. Any omissions will be rectified in subsequent printings if notice is given to the publisher.

Websites
There are links to websites relevant to this book. In order to ensure that the links are up-to-date, that the links work, and that links are not inadvertently made to sites that could be considered offensive, we have made the links available on the Heinemann website at www.heinemann.co.uk/hotlinks. When you access the site, the express code is 1981P.

Exam Café student CD-ROM
© Pearson Education Limited 2008

Original illustrations, screen designs and animation by Michael Heald
Photographs © iStock Ltd
Developed by Elektra Media Ltd

Technical problems
If you encounter technical problems whilst running this software, please contact the Customer Support team on 01865 888108 or email software.enquiries@heinemann.co.uk

OCR Chemistry

A2

OCR and Heinemann are working together to provide better support for you

Dave Gent and Rob Ritchie
Series editor: Rob Ritchie

www.heinemann.co.uk
✓ Free online support
✓ Useful weblinks
✓ 24 hour online ordering

01865 888080

Official Publisher Partnership

Contents

Contents

Introduction

How to use this book

In this book you will find a number of features planned to help you.

- **Module opener pages** – these carry an introductory paragraph to set the context for the topics covered in the module. They also have a short set of questions that you should already be able to answer from your previous science courses or from your general knowledge.
- **Double-page spreads** filled with information about each topic together with some questions you should be able to answer when you have worked through the spread. The final question may be more challenging.
- **End-of-module summary pages** to help you link all the topics within each module together.
- **End-of-module examination questions** selected to show you the types of question that may appear in your examination.

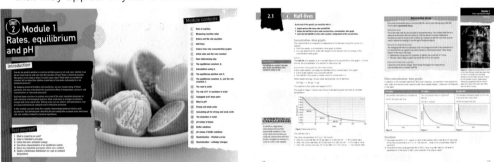

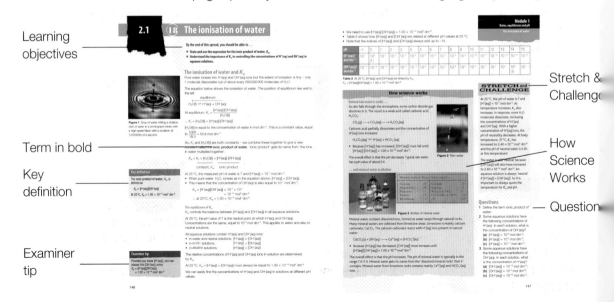

Within each double-page spread you will find other features to highlight important points.

Learning objectives

Term in bold

Key definition

Examiner tip

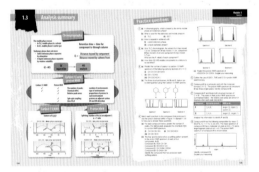

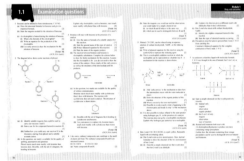

Stretch & Challenge

How Science Works

Question

- **Learning objectives** – these are taken from the Chemistry A2 specification to highlight what you need to know and to understand.
- **Key definitions** – these terms appear in the specification. You must know the definitions and how to use them.
- **Terms in bold** – these draw attention to terms that you are expected to know. These are important terms with specific meanings used by chemists. Each term in bold is listed in the glossary at the end of the book.
- **Examiner tips** – these will help you avoid making common errors in the examination.
- **Worked examples** – these show you how calculations should be set out.
- **How Science Works** – this book has been written in a style that reflects the way that scientists work. Certain sections have been highlighted as good examples of How Science Works.
- **Stretch and Challenge** – these boxes will help you to develop the skills you need to tackle Stretch and Challenge questions on your final examination. You do not have to learn extra information, but you will need to make links across work you've met at AS and A2 and to be able to apply your existing knowledge in unfamiliar contexts.

The examination

It is useful to know some of the language used by examiners. Often this means just looking closely at the wording used in each question on the paper. When you first read a question, do so carefully. Once you have read something incorrectly, it is very difficult to get the incorrect wording out of your head.

- Look at the number of **marks allocated** to a part question – ensure you write enough points down to gain these marks. Do not repeat yourself. Different marks are for different ideas. The number of marks is a guide to the depth of treatment required for the answer.
- Look for words in **bold**. These are meant to draw your attention.
- **Diagrams, tables and equations** often communicate an idea better than trying to explain everything in sentences. Diagrams can usually be drawn freehand to save time.
- Write legibly. You cannot be given any marks for something that is illegible.

Look for the **action word**. Make sure you know what each word means, and answer what it asks. The meanings of some action words are listed below.

- *Define:* only a formal statement of a definition is required.
- *Explain:* a supporting argument is required. The depth of the answer should be judged from the mark allocated.
- *State:* a concise answer is expected, with little or no supporting argument.
- *List:* give a number of points with no elaboration. If you are asked for *two* points then only give two!
- *Describe:* state in words, using diagrams where appropriate, the main points of the topic.
- *Discuss:* give a detailed account of the points involved in the topic.
- *Deduce/Predict:* make a logical connection between pieces of information given.
- *Outline:* restrict the answer to essential detail only.
- *Suggest:* you are expected to apply your knowledge and understanding to a 'novel' situation that you may not have covered in the specification.
- *Calculate:* a numerical answer is required. Working should be shown.
- *Determine:* the quantity cannot be obtained directly. A sequence of calculations may be required.
- *Sketch:* a diagram is required. A graph need only be qualitatively correct, but important points on the graph may require numerical values.

Checking your work all the time is essential. Check each line in your working. Check that an answer is realistic; check that its units are possible; check that it answers the question. If you wait until the end of the examination you may not have time and you will have forgotten the detail of a numerical question. If you do have time at the end of the examination, read through your descriptive answers to ensure that what you wrote is what you intended to write.

Module 1
Rings, acids and amines

Introduction

This module builds on the *Chains, Energy and Resources* unit studied in AS chemistry and introduces a new class of organic molecules known as the arenes. The simplest arene, benzene, was isolated by Michael Faraday in 1825. It is now used to make products as diverse as pesticides, detergents, lubricants, rubber and explosives.

The varied properties and reactions of organic molecules result from the functional groups that form part of their structures. A detailed knowledge of the different ways that molecules can react is essential when producing pharmaceuticals, dyes, polymers and agricultural chemicals.

In this module, you will learn about the reactions of a number of key functional groups. You will study the chemistry of carbonyl compounds, carboxylic acids, esters and amines as well as investigating how the properties of these molecules can be harnessed to improve our quality of life.

Test yourself

1 What does *delocalised* mean?
2 What is meant by the term *electrophile*?
3 Name the aldehyde which contains four carbon atoms.
4 Why are amines referred to as bases?
5 What is the difference between a saturated and an unsaturated fat?

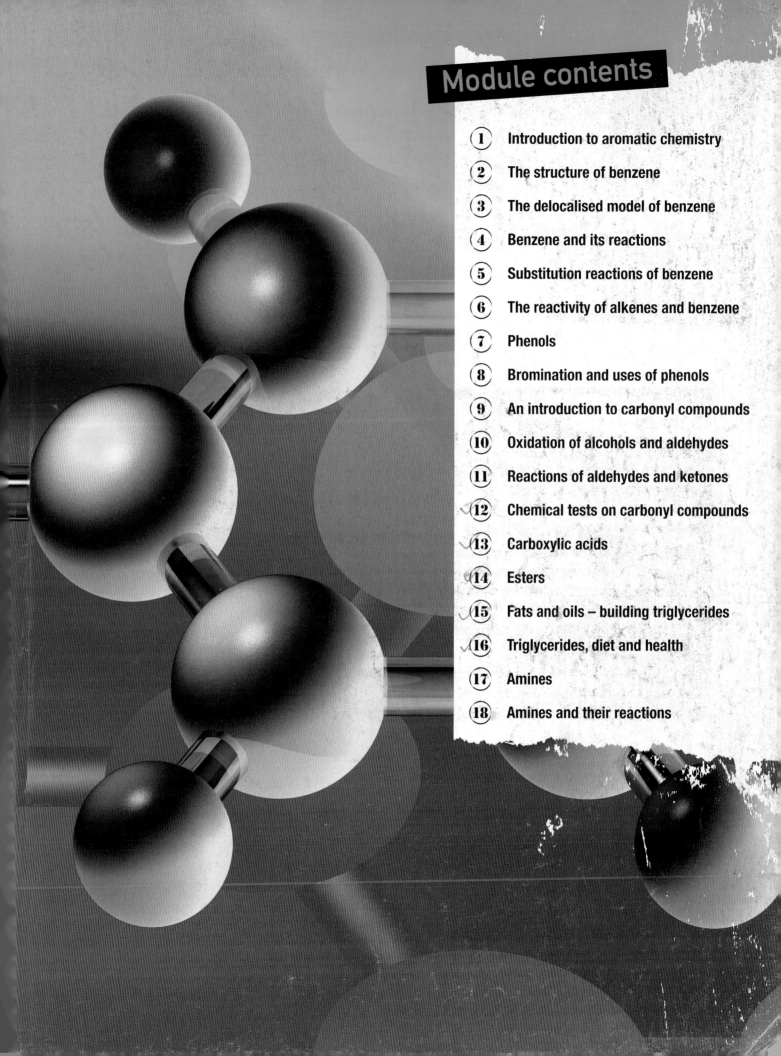

Module contents

By the end of this spread, you should be able to ...

✴ Explain the terms *arene* and *aromatic*.
✴ Describe and explain the models used to describe the structure of benzene.

Arenes

Figure 1 Benzene ring

Arenes are aromatic hydrocarbons containing one or more benzene rings. A benzene molecule is a ring of six carbon atoms, each of which is also bonded to one hydrogen atom. Benzene has the molecular formula C_6H_6. The molecule is commonly represented as a hexagon surrounding a circle, as shown in Figure 1. Any compound with the benzene ring is classified as an aromatic compound.

The term *aromatic* is derived from the Latin 'aroma', meaning fragrance. Early scientists noted that organic materials produced from resins or plant material give off pleasant smells. Chemists analysed many of these sweet-smelling materials and found that they contained benzene rings. However, as scientific knowledge and analytical techniques have developed, chemists have found that the smell generated by a compound has little, if anything, to do with the presence of a benzene ring. In fact aspirin, used as a pain-killer, has no odour and yet its structure, shown in Figure 2, contains a benzene ring.

Arenes occur naturally in materials such as crude oil and coal, and can be produced in refineries from crude oil.

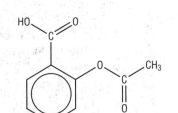

Figure 2 Structure of aspirin – it contains a benzene ring but aspirin is not sweet-smelling; in fact, it has no odour

Naming compounds based on benzene

When one of the hydrogen atoms on a benzene ring is replaced by an atom or group of atoms, a benzene derivative is formed. Groups commonly attached to a benzene ring include –Cl (chloro), –Br (bromo), –NO_2 (nitro) and alkyl groups such as –CH_3 (methyl) and –C_2H_5 (ethyl).

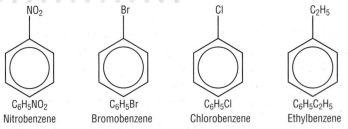

NO_2	Br	Cl	C_2H_5
$C_6H_5NO_2$	C_6H_5Br	C_6H_5Cl	$C_6H_5C_2H_5$
Nitrobenzene	Bromobenzene	Chlorobenzene	Ethylbenzene

Figure 4 Structures of some common benzene derivatives

Figure 3 Aspirin pills spilled from a bottle. Aspirin is a common pain reliever. This analgesic drug has an anti-inflammatory action, and is useful in treating headache, menstrual pain and muscle discomfort. Used in small doses aspirin prevents thrombosis (abnormal blood clots) in patients at risk of having a stroke or heart attack

Sometimes there is more than one group on the ring. It is important that the name reflects the position of each group. Carbon atoms in the benzene ring are numbered to give the smallest possible numbers. Three structural isomers of C_7H_7Br are shown in Figure 5. The name is based on methylbenzene, with the methyl group in position 1.

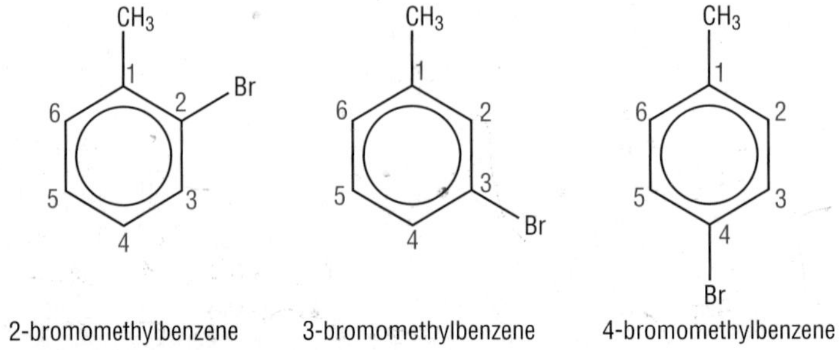

2-bromomethylbenzene   3-bromomethylbenzene   4-bromomethylbenzene

Figure 5 Three isomers of C_7H_7Br

Benzene

Benzene is a starting material for the synthesis of many other aromatic materials. These chemicals include ethylbenzene, phenol and styrene. Smaller quantities of benzene can be used to make a wide range of materials such as detergents, explosives, pharmaceuticals and dyestuffs.

Figure 6 Benzene is classified as a carcinogen – a chemical known or believed to cause cancer in humans. Continued exposure to benzene vapour can result in anaemia or leukaemia

Benzene is a colourless liquid with a sweet odour. It is highly flammable and is formed in both natural processes and human activities. Natural sources of benzene include volcanoes and forest fires. Benzene is a component of crude oil, petrol and cigarette smoke.

Module 1
Rings, acids and amines
Introduction to aromatic chemistry

Figure 7 Compact discs and explosives all find their origin in benzene

Benzene

In 1825, Michael Faraday isolated benzene as a by-product in the manufacture of gas for street lighting from the distillation of whale oil. Faraday also determined that benzene had an empirical formula of CH.

In 1834, Eilhard Mitscherlich found that benzene had a relative formula mass of 78 and a molecular formula of C_6H_6. Other scientists working in different parts of the world produced benzene by various methods. In 1845, Charles Mansfield introduced the first industrial production of benzene, from coal tar.

For many years, scientists speculated over the structure of benzene. Many suggested that benzene has a structure containing several double bonds. Figure 8 shows one such structure.

Many chemical experiments were carried out on benzene. Surprisingly, benzene was rather unreactive for a molecule that was considered to be highly unsaturated. Scientists discovered that benzene did not take part in many of the reactions of alkenes. This evidence suggested that the structure in Figure 8 was unlikely to be correct.

Figure 8 Suggested linear structure for benzene, with several double bonds

In 1865, a German chemist, Friedrich August Kekulé von Stradonitz, published a paper suggesting that the structure of benzene was based on a six-membered ring of carbon atoms joined by alternate single and double bonds. Figure 9 shows this structure.

This proposed structure was based on evidence gained from carrying out many experiments. In 1890, Kekulé claimed that he had discovered the ring shape of benzene while daydreaming about a snake seizing its own tail.

Displayed   Skeletal

Figure 9 Kekulé structure of benzene

Questions

1. Define the term *aromatic*.
2. Give two natural sources of aromatic compounds.
3. Benzene is flammable and carcinogenic. What do you understand by the term *carcinogenic*?
4. Name two compounds that can be made easily from benzene.
5. Benzene has a molecular formula of C_6H_6 and an empirical formula of CH. Define the terms *empirical formula* and *molecular formula*.

5

By the end of this spread, you should be able to ...

✳ **Review the evidence for a delocalised model of benzene.**

Problems with the Kekulé structure

Kekulé's proposed structure of benzene, introduced in spread 1.1.1, fitted with the molecular formula of C_6H_6. However, not all chemists accepted the Kekulé structure because it failed to explain the chemical and physical properties of benzene fully. This spread looks at how scientific evidence prompted changes to Kekulé's proposed structure.

Benzene's low reactivity

Kekulé's model failed to explain benzene's low chemical reactivity:
- if C=C double bonds were present then benzene would react in a similar way to alkenes
- each C=C double bond would be expected to react with bromine water, decolourising it.

However, this does not happen. Nor does benzene take part in other electrophilic addition reactions expected from the C=C bond in alkenes.

Kekulé's equilibrium model of benzene

To account for this lack of reactivity, Kekulé refined his structure of benzene. He suggested that benzene had two forms, differing only by the positions of the double bonds. He suggested that these two forms of benzene were in such a rapid equilibrium that an approaching bromine molecule could not be attracted to a double bond before the structure changed. Hence, bromine could not react with the double bonds.

Figure 1 Friedrich August Kekulé von Stradonitz discovered the ring structure of the molecule benzene, another major advance in organic chemistry

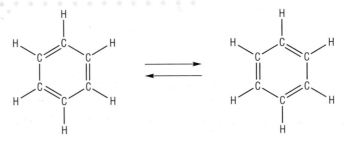

Figure 2 Kekulé's equilibrium model

Kathleen Lonsdale

In 1922, Kathleen Lonsdale graduated from the University of London and was recruited by Sir William Henry Bragg to join his research group in London. His work was in the determination of crystal structures, and Lonsdale became an expert in the new field of X-ray crystallography. This technique involved the analysis of crystal structures using X-rays. Lonsdale's greatest discovery came five years later while analysing some crystals of hexachlorobenzene at the University of Leeds. Here she determined the structure of the benzene ring using a process called X-ray diffraction. She found that all C–C bonds in the ring were the same length, and all the internal C–C–C bond angles were the same: 120°. This discovery had an enormous impact on organic chemistry.

Figure 3 Kathleen Lonsdale is credited with the determination of the structure of benzene

The carbon–carbon bond lengths in benzene

The Kekulé structure of alternating single and double bonds was represented as a symmetrical molecule. But C–C single bonds and C=C double bonds have different lengths.

Many years later, X-ray studies revealed that all six of the carbon–carbon bond lengths in benzene are the same length: 0.139 nm. This is between the C–C bond length of 0.153 nm and the C=C bond length of 0.134 nm. This was important evidence that suggested that the Kekulé structure of benzene was incorrect.

Hydrogenation of benzene

The Kekulé structure of benzene contained three C=C double bonds. The chemical name for this proposed structure is cyclohexa-1,3,5-triene.
- When an alkene reacts with hydrogen the energy change is called the enthalpy change of hydrogenation. When cyclohexene, with one C=C bond, reacts with hydrogen, the enthalpy change of hydrogenation is -120 kJ mol^{-1}.
- Kekulé's benzene, with three C=C bonds, would be expected to have an enthalpy change of hydrogenation of -360 kJ mol^{-1}, three times that of cyclohexene.

These hydrogenation reactions are shown in Figure 4.

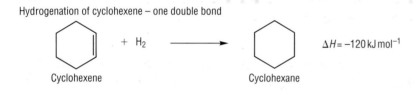

Hydrogenation of cyclohexene – one double bond

Cyclohexene $+$ H$_2$ $\longrightarrow$ Cyclohexane $\Delta H = -120$ kJ mol^{-1}

Hydrogenation of Kekulé's benzene: cyclohexa-1,3,5-triene – three double bonds

Cyclohexa-1,3,5-triene $+$ 3H$_2$ $\longrightarrow$ Cyclohexane Expected $\Delta H = -360$ kJ mol^{-1}

Figure 4 Hydrogenation enthalpies for cyclohexene and Kekulé's benzene

When benzene is hydrogenated, the experimental value for the enthalpy change is -208 kJ mol^{-1}. This enthalpy change is 152 kJ mol^{-1} less than the expected enthalpy change of hydrogenation of cyclohexa-1,3,5-triene. This is shown in Figure 5.

The most important conclusion is that the actual structure of benzene has much less energy than the proposed Kekulé structure.
- The real structure of benzene is 152 kJ mol^{-1} more stable than the Kekulé structure.
- This energy is known as the delocalisation energy, or resonance energy, of benzene.

This evidence suggests that the real structure of benzene is more stable than a structure containing C=C bonds. It helps to explain why benzene is less reactive than alkenes.

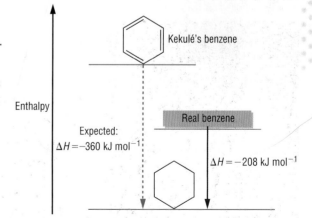

Figure 5 Enthalpy changes for the hydrogenation of cyclohexene and benzene

Questions

1 What would you expect to see if cyclohexene reacted with bromine? Write an equation for this reaction.
2 What type of reaction takes place between alkenes and bromine?
3 What technique was used to show that all the bond lengths in benzene are the same?

7

By the end of this spread, you should be able to ...

❋ Compare the Kekulé and delocalised models of benzene in terms of p-orbital overlap forming π-bonds.

The delocalised model of benzene

The weaknesses identified in the Kekulé model of benzene led to the development of the **delocalised** model for the structure of benzene that is discussed in this spread.

The delocalised model has the following features.

- Benzene is a cyclic hydrocarbon with six carbon atoms and six hydrogen atoms.
- The six carbon atoms are arranged in a planar hexagonal ring. Each carbon atom is bonded to two other carbon atoms and one hydrogen atom.
- The shape around each carbon atom is trigonal planar with a bond angle of 120°.
- Each carbon atom has four outer shell electrons. Three of these electrons bond to two other carbon atoms and one hydrogen atom. The three bonds in this plane are called *sigma* (σ) bonds. This leaves a fourth outer shell electron in a 2p orbital above and below the plane of the carbon atoms.
- The electron in a p-orbital of a carbon atom overlaps with the electrons in the p-orbitals of the carbon atoms on either side. This results in a ring of electron density above and below the plane of the carbon atoms.
- This overlap produces a system of π-bonds (pi bonds) which spread over all six carbon atoms. The p electrons are no longer held between just two carbon atoms. The p electrons are now spread over the whole ring and are said to be delocalised. See Figure 2.

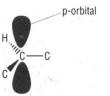

Figure 1 Bonding around a carbon atom in a section of a benzene ring. Each carbon atom provides three of its four outer shell electrons to bond to three other atoms. This leaves one electron in a p-orbital

Examiner tip

Remember these key bullet points because the structure and bonding in benzene is often asked in exams.

Examiner tip

Get the terminology right! The p-orbitals on the carbon atoms overlap sideways to produce a π-cloud of electron density above and below the plane of the carbon atoms containing six electrons. The electron density is spread out.

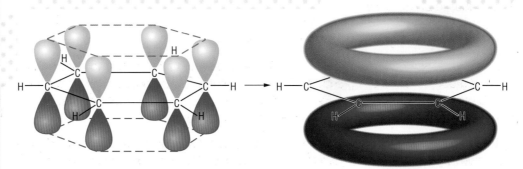

Figure 2 The delocalised structure of benzene forms when the p-orbitals overlap sideways forming a π-electron cloud above and below the plane of the carbon atoms. Each carbon atom contributes one electron to the cloud, giving a delocalised system of six electrons in total

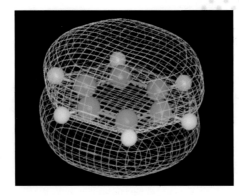

Figure 3 Computer graphic representation of the structure of benzene, C_6H_6, illustrating the delocalised electrons which give great stability to the molecule. These are called π-electrons and appear as the yellow and blue cages above and below the flat plane of the molecule

Figure 4 Benzene is often represented in this way in equations and mechanisms

- The delocalised structure of benzene is the origin of the representation of benzene as a circle inside a hexagon, as shown in Figure 4.

STRETCH and CHALLENGE

In spread 1.1.2 it was seen that benzene has 152 kJ mol^{-1} less energy than predicted. This missing energy is known as resonance energy and is the result of the overlap of p-orbitals – see Figure 2.

In 1931, Linus Pauling proposed the resonance theory. Resonance theory was important because it led to an acceptance by scientists that the structure of benzene was a resonance hybrid of the two Kekulé structures (see spread 1.1.2). You can read more about Pauling's discoveries on the website for this book (see page ii).

Modern physical analysis confirms the resonance hybrid nature of benzene, shown in Figure 5.

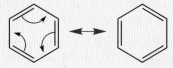

Figure 5 Resonance hybrids of benzene (left) and the common representation of benzene (right)

The delocalised model of benzene and chemical reactivity

In spread 1.1.2 it was shown that benzene is more stable than expected. This increased stability means that benzene struggles to take part in addition reactions, the typical reactions of alkenes with localised π-bonds (see Figure 6).

Under normal conditions, benzene does not:
- decolourise bromine water
- react with strong acids such as HCl
- react with the halogens chlorine, bromine or iodine.

In an addition reaction, electrons from the delocalised system would need to bond to the atom or group of atoms being added. This would result in the product being less stable than benzene, and the reaction would not then be energetically favourable. Addition reactions would disrupt the delocalisation of the ring structure.

Instead of **addition**, benzene and its derivatives typically take part in **substitution** reactions – one of benzene's hydrogen atoms is replaced by another atom or group of atoms. The organic product formed retains the delocalisation, and hence the stability, of the benzene ring. This makes substitution the most common type of reaction in the chemistry of benzene. Substitution reactions of benzene will be discussed in spread 1.1.4.

Questions

1 State the shape and bond angles around a carbon atom in a benzene ring.
2 How is a delocalised ring of electrons formed in benzene?
3 What do you understand by the term *delocalised*?
4 Why does benzene react by substitution reactions rather than addition reactions?

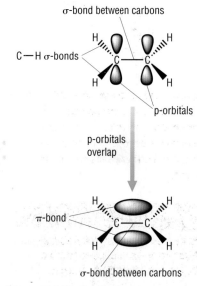

Figure 6 A localised π-bond

Key definition

An **addition reaction** is one in which a reactant is added to an unsaturated molecule to make a saturated molecule.

A **substitution reaction** is one in which an atom or group of atoms is replaced with a different atom or group of atoms

By the end of this spread, you should be able to ...

* Describe the electrophilic substitution of arenes with concentrated nitric acid in the presence of concentrated sulfuric acid.
* Describe the electrophilic substitution of arenes with halogens in the presence of a suitable halogen carrier.

Reactivity

In benzene, the region of high electron density above and below the plane of the carbon atoms attracts electrophiles. To preserve the stability of the benzene ring, benzene takes part in substitution reactions rather than addition reactions (see spread 1.1.3).

In its typical reactions, the benzene ring reacts with **electrophiles** and takes part in electrophilic substitution reactions.

Electrophilic substitution by nitration

Benzene is not used for experiments in schools and colleges because it is a suspected carcinogen. During your course you may carry out nitration using a safer aromatic compound, such as methyl benzoate. For simplicity, the reactions of *benzene* are studied at A level – you can apply these principles to other aromatic compounds.

Nitration of benzene

In nitration, one of the hydrogen atoms on the benzene ring is replaced by a nitro (–NO_2) group.
Benzene reacts with a nitrating mixture of concentrated nitric acid and concentrated sulfuric acid at a temperature of 50 °C. Concentrated sulfuric acid acts as a catalyst.

The equation for the preparation of nitrobenzene is:

$$C_6H_6 + HNO_3 \longrightarrow C_6H_5NO_2 + H_2O$$

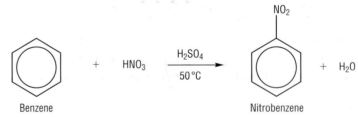

Benzene   Nitrobenzene

Figure 1 Equation for the preparation of nitrobenzene

The nitrating mixture is usually prepared in a round-bottomed or pear-shaped flask. Concentrated nitric acid and concentrated sulfuric acid are first mixed together carefully while cooling the mixture in a beaker of cold water. Benzene is then added carefully to the nitrating mixture, keeping the temperature below 50 °C.

Once all of the benzene has been added, a reflux condenser is fitted to the flask and the mixture is heated at 50 °C in a water bath. If the reaction mixture gets hotter, then more than one nitro group may be substituted onto the benzene ring.

Nitrobenzene is a pale yellow liquid and is an important starting material in the preparation of dyes, pesticides and pharmaceuticals (such as paracetamol).

Key definition

An **electrophile** is an atom (or group of atoms) that is attracted to an electron-rich centre, where it accepts a pair of electrons to form a new covalent bond.

Examiner tip

It is often easier to show the equation by drawing the benzene ring. To attach a group to the benzene ring, a hydrogen atom has been replaced. The ring is now only C_6H_5-, not C_6H_6. In exams, marks are lost if nitrobenzene is represented as $C_6H_6NO_2$ instead of $C_6H_5NO_2$.

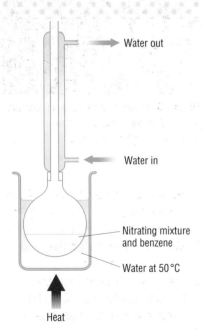

Figure 2 Preparing nitrobenzene by heating benzene, concentrated nitric and concentrated sulfuric acid in a water bath

Figure 3 Nitrobenzene is needed to make dyes and pharmaceuticals

Module 1
Rings, acids and amines
Benzene and its reactions

Nitration of methylbenzene

Methylbenzene (also called toluene) can also be nitrated with a mixture of concentrated nitric and sulfuric acids. The reaction is faster than the nitration of benzene, and can lead to the formation of 2,4,6-trinitromethylbenzene (trinitrotoluene, TNT). This is an explosive.

Figure 4 2,4,6-trinitromethylbenzene (trinitrotoluene or TNT) is a powerful explosive

Halogenation of benzene

Although benzene doesn't react with halogens on their own, it does react in the presence of a special type of catalyst called a halogen carrier. Examples of common halogen carriers include $FeCl_3$, $FeBr_3$, $AlCl_3$ and $AlBr_3$. Iron metal can also be used because it will react with any halogen present to form the required iron(III) halide.

As with nitration, this is an electrophilic substitution reaction in which one of the hydrogen atoms on the benzene ring is replaced by a halogen atom.

The reaction with chlorine

Chlorine reacts with benzene at room temperature and pressure in the presence of $AlCl_3$, $FeCl_3$ or Fe (as the halogen carrier) to produce chlorobenzene, as shown in Figure 5: Chlorobenzene is used as a solvent and in the production of pesticides.

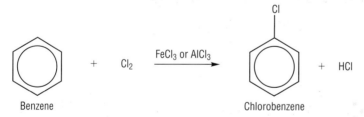

Figure 5 Benzene reacts with chlorine to produce chlorobenzene

The reaction with bromine

Benzene reacts with bromine in the same way that it reacts with chlorine. This time, the halogen carrier catalyst needed is $AlBr_3$ $FeBr_3$ or Fe. The organic product is bromobenzene, a compound used in the preparation of pharmaceuticals. The equation for this reaction is shown in Figure 7.

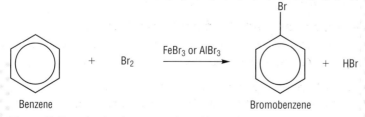

Figure 7 Equation for the preparation of bromobenzene

Figure 6 Spraying insecticide on carrots – chlorobenzene is used in the manufacture of insecticides

Questions

1 Benzene can be nitrated to form nitrobenzene. State the reagents needed for this reaction.

2 Write an equation for the nitration of benzene.

3 State the name of a suitable halogen carrier that could be used to chlorinate benzene.

4 Benzene reacts with electrophiles – define the term *electrophile*.

By the end of this spread, you should be able to ...

✳ Outline the mechanism of electrophilic substitution in arenes.
✳ Outline the mechanism for the mononitration and monohalogenation of benzene.

General mechanism of electrophilic substitution

Electrophilic substitution reactions occur in benzene chemistry because of the delocalised ring of electrons above and below the plane of the carbon atoms.

- The electron-dense ring attracts an electrophile – the electrophile accepts a pair of the π electrons from the delocalised ring to form a covalent bond.
- An intermediate forms that contains both the electrophile and the hydrogen atom that is being substituted. The delocalised π-electron cloud has been disrupted and the intermediate is less stable than benzene.
- The unstable intermediate rapidly loses the hydrogen as an H^+ ion. The delocalised ring of electrons reforms and stability is restored.

The general **reaction mechanism** of electrophilic substitution on benzene is shown in Figure 1. The electrophile is represented by the symbol A^+. Note the careful and exact positioning of each **curly arrow**. (See also Figure 2.)

Key definitions

In a **substitution** reaction, an atom or group of atoms is replaced with a different atom or group of atoms.

Electrophilic substitution is a type of substitution reaction in which an electrophile is attracted to an electron-rich centre or atom, where it accepts a pair of electrons to form a new covalent bond.

A **reaction mechanism** is a series of steps that, together, make up the overall reaction.

A **curly arrow** is a symbol used in reaction mechanisms to show the movement of an electron pair in the breaking or formation of covalent bonds.

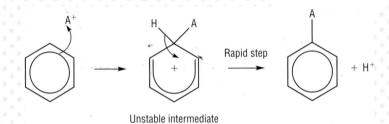

Figure 1 Electrophilic substitution in benzene

Examiner tip

When an exam question asks for a description of a mechanism, an annotated diagram will score most, if not all, of the marks.

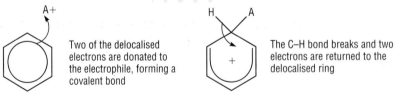

Two of the delocalised electrons are donated to the electrophile, forming a covalent bond

The C–H bond breaks and two electrons are returned to the delocalised ring

Figure 2 Curly arrows are used to represent the movement of electron pairs

Nitration

Nitration of the benzene ring was introduced in spread 1.1.4.

- A mixture of concentrated nitric acid and concentrated sulfuric is used.
- The sulfuric acid is needed to generate an electrophile from the nitric acid.
- The electrophile is the nitryl cation, or nitronium ion, and has the formula NO_2^+.

The equation for the formation of NO_2^+ is

$$HNO_3 + H_2SO_4 \longrightarrow NO_2^+ + HSO_4^- + H_2O$$

The NO_2^+ electrophile then reacts with benzene. The mechanism is shown in Figure 3.

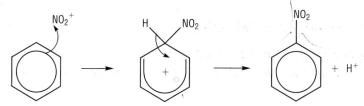

Figure 3 Mechanism for the nitration of benzene

Finally, the H^+ reacts with the HSO_4^- from the first step to reform H_2SO_4. So the sulfuric acid is acting as a catalyst.

$$H^+ + HSO_4^- \longrightarrow H_2SO_4$$

Halogenation

Halogenation of the benzene ring was introduced in spread 1.1.4. This example studies bromination, but the same principle can be applied to substitution of any halogen.

Unlike alkenes, benzene is too stable to react with bromine, Br_2, on its own (see spreads 1.1.2 and 1.1.3). However, benzene does react with bromine in the presence of a halogen carrier, such as $FeBr_3$. The halogen carrier generates the bromonium ion, Br^+, which is a more powerful electrophile than Br_2.

The equation for the formation of Br^+ is

$$Br_2 + FeBr_3 \longrightarrow Br^+ + FeBr_4^-$$

The Br^+ electrophile now reacts with benzene as shown in Figure 4.

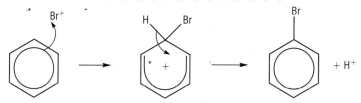

Figure 4 Mechanism for the bromination of benzene

Finally, the H^+ reacts with $FeBr_4^-$ from the first step to form the other product of the overall reaction, HBr, and to reform $FeBr_3$. So the iron(III) bromide is acting as a catalyst.

$$H^+ + FeBr_4^- \longrightarrow FeBr_3 + HBr$$

Different halogen carriers

For bromination, the halogen carrier is $AlBr_3$, $FeBr_3$ or Fe.

For chlorination, $AlCl_3$, $FeCl_3$ or Fe would be used.

You can use Fe for both chlorination and halogenation because Fe reacts with halogens to form the iron(III) halide.

Questions

1 Write an equation for the reaction between chlorine and benzene, and suggest a suitable catalyst.
2 What type of reaction takes place when benzene reacts with concentrated nitric acid in the presence of concentrated sulfuric acid?
3 Use curly arrows to show the mechanism for the reaction between benzene and chlorine.

Examiner tip

Take care – carelessly drawn curly arrows cost marks!

Examiner tip

A catalyst is not consumed in an overall reaction, but it must be involved in the reaction mechanism. A catalyst reacts in one step and is regenerated in a further step.

Examiner tip

Compare the electrophilic substitution mechanisms for nitration and bromination – try to understand what is going on. The generation and reforming of the catalyst is the only real difference between the nitration and bromination mechanisms.

Understand the chemistry and you not only have less to learn, but you are far more likely to get the answers right in exams.

The reactivity of alkenes and benzene

By the end of this spread, you should be able to ...

✱ Explain the relative resistance to bromination of benzene compared with alkenes.

Cyclohexene and bromine water

When aqueous bromine is added dropwise to an alkene, such as cyclohexene, the bromine solution changes colour from orange to colourless.

Cyclohexene behaves like a typical alkene, with the bromine molecule adding across the double bond in an electrophilic addition reaction. This is shown in Figure 1.

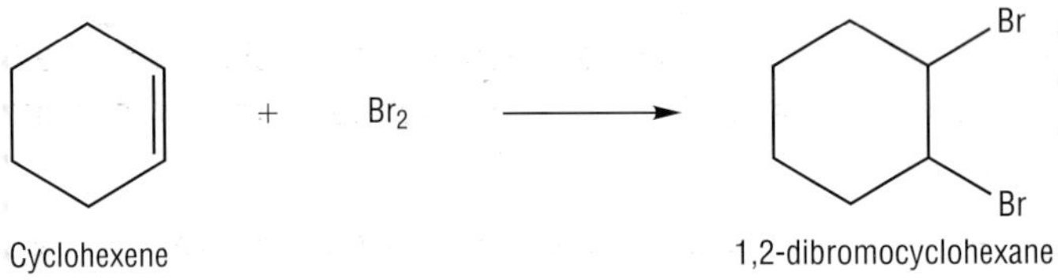

Cyclohexene   1,2-dibromocyclohexane

Figure 1 Reaction of cyclohexene with bromine

Even though Br_2 is a non-polar molecule, it still reacts with the double bond in an alkene.
- The π-bond contains localised electrons, sited above and below the two carbon atoms in the double bond. This produces a region of high electron density.
- When bromine approaches an alkene, the electrons in the π-bond repel the electrons in the Br–Br bond. This induces a dipole in the Br_2 molecule, with one end of the molecule becoming slightly negative and the other end becoming slightly positive. The Br_2 molecule is now polar.

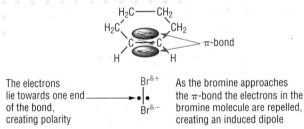

The electrons lie towards one end of the bond, creating polarity

As the bromine approaches the π-bond the electrons in the bromine molecule are repelled, creating an induced dipole

Figure 2 A double bond in cyclohexene induces a dipole in a bromine molecule

- The π-electron pair from the double bond is attracted to the slightly positive bromine atom, causing the double bond to break. A new bond forms between one of the carbon atoms and a bromine atom, forming a positively charged carbocation. The bond between the two Br atoms breaks by heterolytic fission, forming a bromide ion, Br^-.
- Finally the Br^- ion is attracted towards the intermediate carbocation, forming a covalent bond. We now have the final addition product, 1,2-dibromocyclohexane.

The mechanism is shown in Figure 3.

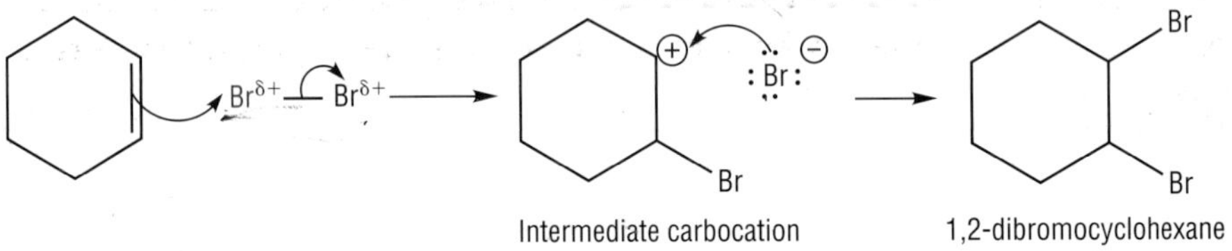

Intermediate carbocation   1,2-dibromocyclohexane

Figure 3 Electrophilic addition in cyclohexene

Benzene and bromine

When bromine is added to benzene, no reaction takes place and the bromine mixture remains orange. This is shown in Figure 4. However, if some iron filings or iron(III) bromide is added, the bromine is decolourised and white fumes of hydrogen bromide gas are observed.

These observations show that benzene is unable to react with bromine without the help of a halogen carrier catalyst (Fe or $FeBr_3$).

- Benzene has delocalised π-electrons spread over all six carbon atoms in the ring structure. Alkenes have π-electrons localised above and below the two carbon atoms in the double bond.
- Benzene has a lower π-electron density than alkenes.
- When a non-polar molecule, such as bromine, approaches the benzene ring there is insufficient π-electron density above and below any two carbon atoms to cause the necessary polarisation of the bromine molecule. This makes benzene resistant to reactions with non-polar halogens.
- A halogen carrier is needed to generate the more powerful electrophile Br^+. The greater charge on the Br^+ ion is able to attract the π-electrons from benzene so that reaction can take place.

In general, molecules in which electrons are evenly distributed over their whole structure tend to be more stable and less reactive than molecules with localised electrons. Delocalised structures require more energy to react. In the case of benzene, energy is needed to disrupt the delocalised π-cloud of electrons.

Focus on PCBs

Polychlorinated biphenyls (PCBs) are a family of compounds made up of the elements chlorine, carbon and hydrogen. A molecule of biphenyl is composed of two benzene rings – see Figure 5. PCBs were first synthesised in 1881 and were found to be fire resistant, relatively unreactive and poor conductors of electricity.

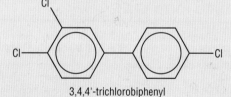

3,4,4'-trichlorobiphenyl

Figure 5 Structure of a PCB

PCBs have a wide range of uses such as coolants and insulating fluids for transformers and capacitors, plasticisers in paint and cement, and in carbonless copy paper.

Unfortunately some PCBs have found their way into the environment. The Swedish chemist Sören Jensen was studying the accumulation of toxins such as mercury in Baltic fish. He found high levels of chemicals that were later identified as PCBs. A particular problem with PCBs is their fat solubility, allowing them to be stored in the body.

PCBs were banned from manufacture in the late 1970s but we still find low levels of PCBs in the environment and in our food, especially in animal fat and fish. Thus, everyone is exposed to very low levels of PCBs. Low-level exposure to PCBs does not appear to affect human health, but the presence of these chemicals is an obvious cause for concern.

Module 1
Rings, acids and amines
The reactivity of alkenes and benzene

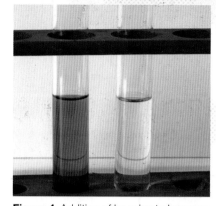

Figure 4 Addition of bromine to benzene (left) and to cyclohexene (right) shows the different reactivities

Examiner tip

The key point here is that the π-electrons in benzene are delocalised, or spread, throughout the molecule. The stability of the benzene ring is caused by the delocalisation of electrons. The halogen cannot be polarised sufficiently and reaction is resisted.

Figure 6 In the Upper Mississippi region of the USA, tree swallows, which feed on aquatic insects that hatch from river water and sediments, have been studied to determine how PCB contamination affects wildlife

Questions

1 What do you understand by the term *electrophile*?
2 If bromine is added to cyclohexene and to benzene:
   (a) What would you see?
   (b) Write an equation for any reaction that takes place.

By the end of this spread, you should be able to . . .

* Describe the reactions of phenol with aqueous alkalis and with sodium to form salts.
* Discuss the role of phenol as an early antiseptic.

Phenols

The phenols are a class of organic compounds in which a hydroxyl group, –OH, is attached directly to a benzene ring. If an aromatic compound has an –OH group attached to a side chain on the benzene ring, it is described as an aromatic alcohol. This is shown in Figure 1.

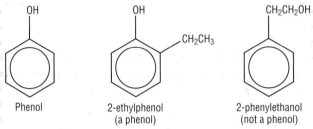

Phenol

2-ethylphenol
(a phenol)

2-phenylethanol
(not a phenol)

Figure 1 Phenols must have an –OH group attached directly to a benzene ring

Phenol

Phenol is the simplest of the phenols, with one of the hydrogen atoms on a benzene ring replaced by an –OH group. Phenol is a solid at room temperature and pressure. It is slightly soluble in water because the –OH group forms hydrogen bonds with water molecules. However, the presence of a benzene ring makes phenols less soluble in water than alcohols.

Figure 2 Phenol is a pink crystalline solid at room temperature

Reactions of phenol to form salts

Reaction with sodium hydroxide

When dissolved in water, phenol forms a weak acidic solution by losing H^+ from the –OH group:

$$C_6H_5OH + aq \rightleftharpoons C_6H_5O^- + H^+$$

Phenol is neutralised by aqueous sodium hydroxide to form the salt sodium phenoxide, $C_6H_5O^-Na^+$, and water:

$$C_6H_5OH + NaOH \longrightarrow C_6H_5O^-Na^+ + H_2O$$

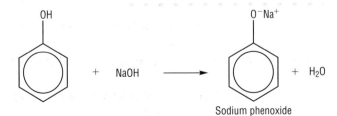

Figure 3 Phenol forms a phenoxide on reaction with sodium hydroxide

Examiner tip

Notice that the product, sodium phenoxide, is ionic. Do not draw a bond between the sodium and the oxygen because that would indicate covalent bonding. Sodium phenoxide is an ionic salt because H^+ from the –OH group has been replaced by a metal ion to form $C_6H_5O^-Na^+$.

Reaction with sodium

When a reactive metal such as sodium is added to phenol, the metal effervesces producing hydrogen gas. The organic product sodium phenoxide, $C_6H_5O^-Na^+$, is a salt.

$$2C_6H_5OH + 2Na \longrightarrow 2C_6H_5O^-Na^+ + H_2$$

Figure 4 Phenol reacts with sodium producing hydrogen gas

Figure 5 British surgeon Joseph Lister, inventor of antiseptic surgery

---

Focus on phenol

In the 1800s, going into hospital for surgery was a very risky business. Without effective anaesthetics, patients were often given alcohol to numb the pain. In addition, there were no effective antiseptics and surgeons did not realise the importance of clean equipment, clothing or the environment in which they were working. A patient who agreed to surgery was more likely to die than to survive!

Enter Joseph Lister.

In 1870, Lister was the first person to treat wounds with dressings coated in carbolic acid, or phenol as we now know it. He insisted that surgeons washed their hands and sterilised their instruments with a solution of 5% carbolic acid before carrying out operations. Lister developed a carbolic acid spray which was used to drench the whole operating theatre and hospital wards, killing bacteria before any wounds could be infected.

Lister increased the success rate for amputations, the most common operation of the time, from 60% to over 90%. However, carbolic acid sprays were found to be caustic to the skin and body tissues, and some surgeons became so ill that they had to give up using the spray. Lister carried out further experiments and found that boracic (boric) acid was a better and safer antiseptic.

Two postage stamps were released in September 1965 to commemorate Lister's contribution to the field of antiseptics.

Rumour suggests that the anti-bacterial mouthwash Listerine was named in honour of the father of antiseptics!

Figure 6 Listerine antiseptic

Figure 7 Postage stamps that commemorate Lister's contribution to antiseptics

Questions

1. Write an equation for the reaction between phenol and potassium hydroxide. Name the organic product formed in this reaction.
2. Some phenol was melted and a small piece of potassium added to it. What would you observe?
3. Which of the structures in Figure 8 are phenols and which are alcohols?

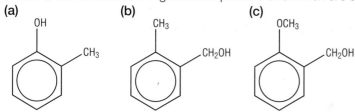

Figure 8

By the end of this spread, you should be able to ...

* Describe the reaction of phenol with bromine to form 2,4,6-tribromophenol.
* Explain the relative ease of bromination of phenol compared with benzene.
* State the uses of phenols in plastics, antiseptics, disinfectants and paints.

Reaction with bromine

Like benzene, phenol undergoes electrophilic substitution with bromine. But unlike benzene, the reaction takes place at room temperature without the need for a halogen carrier catalyst.

When bromine water is added to an aqueous solution of phenol, the orange bromine colour disappears and a white precipitate of 2,4,6-tribromophenol is formed.

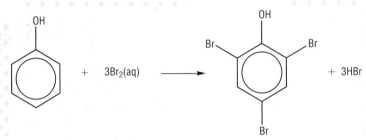

Figure 1 Bromine reacts with aqueous phenol to produce 2,4,6-tribromophenol

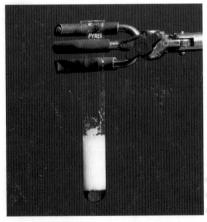

Figure 2 Bromination of phenol. The white precipitate is 2,4,6-tribromophenol

The relative ease of bromination of phenol

Bromine reacts much more readily with phenol than with benzene.
* The increased reactivity occurs because a lone pair of electrons (occupying a p-orbital) on the oxygen atom in the phenol group is drawn into the benzene ring.
* This creates a higher electron density in the ring structure and the ring is activated.
* The increased electron density polarises bromine molecules, which are then attracted more strongly towards the ring structure than in benzene.

The increased electron density of phenol's ring of electrons increases the reactivity towards all electrophiles, not just with bromine.

Non-bonding electrons on the oxygen atom in the phenol group are drawn into the benzene ring

The lone pair becomes part of the delocalisation increasing the electron density in the ring

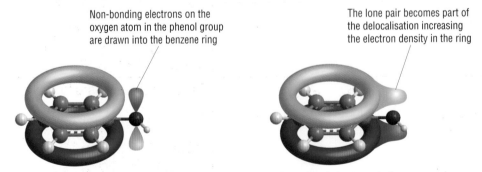

Figure 3 Overlap of a lone pair on the oxygen with the delocalised electrons in benzene

Uses of phenols

In spread 1.1.7 we saw that phenol was used as an antiseptic for many years. However, chemists have found many other uses for phenol. The use of phenol in the synthesis of dyes and for the manufacture of pharmaceuticals, such as aspirin, has improved our quality of life. As far back as 1842, chemists successfully nitrated phenol, synthesising a compound called picric acid, which is an explosive.

Phenol can also be reacted with the aldehyde methanal to form resins, which are the basis of some plastic materials, such as bakelite and melamine. These resins are also used as binding materials in the manufacture of plywood and MDF.

Other uses of phenols

Phenol	Use
alkyl phenols 2-ethylphenol	Found in surfactants and detergents
chlorophenols 2,4-dichlorophenol	Found in antiseptics and disinfectants such as TCP and Dettol
salicylic acid	Preparation of aspirin and other pharmaceuticals
bisphenol	Production of epoxy resins for paints

Questions

1 What would you expect to see when bromine is added to aqueous phenol?
2 Why does bromine react more readily with phenol than with benzene?
3 Give three uses of phenol.

By the end of this spread, you should be able to ...

✱ **Recognise and name aldehydes and ketones.**

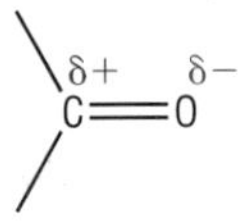

Aldehydes and ketones

Aldehydes and ketones both contain the carbonyl **functional group**, C=O. The carbonyl group is shown in Figure 1.

The double bond in the carbonyl and alkene functional groups comprises a σ-bond and a π-bond. The σ-bond is formed by overlap of orbitals between the C and O atoms. The π-bond is formed by the overlap of the p-orbitals on the carbon and oxygen atoms in the carbonyl bond.

However, the carbonyl bond does not react in the same way as the double bond in an alkene because the oxygen atom is more **electronegative** than the carbon atom. This creates a dipole in the C=O bond because the bonded electrons are attracted towards the oxygen – see Figure 2.

Figure 1 The carbonyl functional group

Key definitions

The **functional group** is the part of an organic molecule responsible for its chemical reactions.

Electronegativity is a measure of the attraction of a bonded atom for the pair of electrons in a covalent bond.

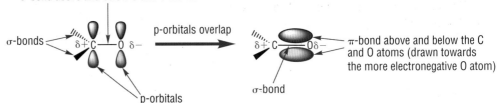

Figure 2 Formation of the C=O group π-bond

Naming aldehydes and ketones

Aldehydes and ketones are named using the same set of rules that you learned in AS chemistry.

Naming aldehydes

- In aldehydes, the carbon atom of the carbonyl group is joined to at least one hydrogen atom.
- Aldehydes are named by adding the suffix '-al' to the stem name of the parent chain.
- The parent chain is the longest unbranched carbon chain present in the molecule.

Key definitions

The **stem** is the longest carbon chain present in an organic molecule.

A **suffix** is the part of the name added *after* the stem.

Worked example

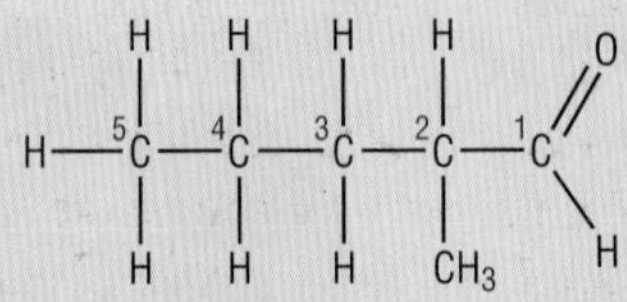

Figure 3
2-methylpentanal

- There are five carbons in the longest unbranched chain, so the stem is *pent-*.
- There is a methyl side chain on carbon-2, hence *2-methyl*.
- There is a suffix – the alkane chain ending is shortened to *-an-*.
- The compound is an aldehyde, so the suffix is *-al*.
- The name is *2-methylpentanal*.

Examiner tip

Notice that the position of the aldehyde is never recorded in the name because it is always on carbon-1.

Naming ketones

The aldehyde carbon atom is always on the end of the chain and is counted as carbon-1. In aldehydes, naming always starts at the carbonyl carbon atom.

In ketones, the carbon atom of the carbonyl group is joined to two other carbon atoms or carbon chains.

Ketones are named by adding the suffix '-one' to the stem name of the parent chain. The parent chain is the longest chain containing the C=O group. The C=O carbon atom is numbered to show its position on the carbon chain so that the smallest possible number is used.

Examiner tip

In a ketone, the carbonyl group can be at any position on the chain except at the end, so it is important to include its position in the name. If the carbonyl group is at the end of the chain, the compound is an aldehyde.

Worked example

H—^{1}C—^{2}C—^{3}C—^{4}C—^{5}C—H (with H, CH$_3$, H, H substituents and O on carbon-3)

Figure 4 The structure of 2-methylpentan-3-one

- There are five carbons in the longest unbranched chain, so the stem is *pent-*.
- The carbonyl is on carbon-3 and there is a methyl group on carbon-2, hence *2-methyl* and *-3-* for the ketone
- There is a suffix – the alkane chain ending is shortened to *-an-*.
- The compound is a ketone, so the suffix is *-one*.
- The name is *2-methylpentan-3-one*.

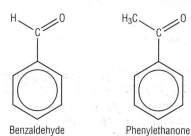

Benzaldehyde   Phenylethanone

Figure 5 Structures of benzaldehyde and phenylethanone

Figure 6 Mature almond fruits hanging on an almond tree. Almond nuts are the kernels of the fruit, which splits open when ripe. Almonds are grown for food and for their oil

Aromatic aldehydes and ketones

Aromatic aldehydes and ketones are compounds containing both a benzene ring *and* a carbonyl group. The simplest aromatic aldehyde is benzaldehyde, and the simplest aromatic ketone is phenylethanone. These can be seen in Figure 5.

At room temperature, benzaldehyde is a colourless liquid with a pleasant almond-like odour. It is used to make almond essence, an ingredient in Bakewell tarts. Benzaldehyde gives the flavour to the marzipan found under the icing on traditional Christmas cakes. Phenylethanone is used to create fragrances that resemble cherry, honeysuckle, jasmine and strawberry.

The essential oil of cinnamon bark contains about 80 to 90% cinnamaldehyde. For hundreds of years, cinnamon has been used in medicines to treat coughing, hoarseness and sore throats.

Questions

1 Show the displayed formula of the following carbonyl compounds:
   (a) hexan-3-one;   (b) butanal;   (c) 2-methylpentan-3-one.
2 What is the difference between the structures of an aldehyde and a ketone?
3 The aldehydes and ketones contain the carbonyl functional group. What do you understand by the term *functional group*?
4 Name the following carbonyl compounds:
   (a)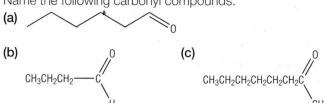
   (b) CH$_3$CH$_2$CH$_2$—C(=O)H   (c) CH$_3$CH$_2$CH$_2$CH$_2$CH$_2$CH$_2$C(=O)CH$_3$

Figure 7 Structure of cinnamaldehyde. Cinnamon is the dried bark of the tree *Cinnamomum zeylanicum*. It is widely used as a spice in cookery

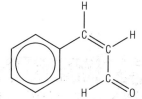

By the end of this spread, you should be able to ...

* Describe the oxidation of primary alcohols to form aldehydes and carboxylic acids.
* Describe the oxidation of secondary alcohols to form ketones.
* Describe the oxidation of aldehydes to form carboxylic acids.

Key definition

A **redox reaction** is one in which both reduction and oxidation take place.

Oxidation of alcohols

You will recall from your AS chemistry studies that primary and secondary alcohols can be oxidised using an oxidising agent.

- A suitable oxidising agent is a solution containing acidified dichromate ions, $H^+/Cr_2O_7^{2-}$.
- The oxidising mixture can be made from potassium dichromate, $K_2Cr_2O_7$, and sulfuric acid, H_2SO_4.

During the reaction, the acidified potassium dichromate changes from orange to green.

Remember that tertiary alcohols are *not* oxidised by acidified dichromate ions.

Primary alcohols

Primary alcohols can be oxidised to aldehydes and to carboxylic acids (see *AS Chemistry* spread 2.2.3).

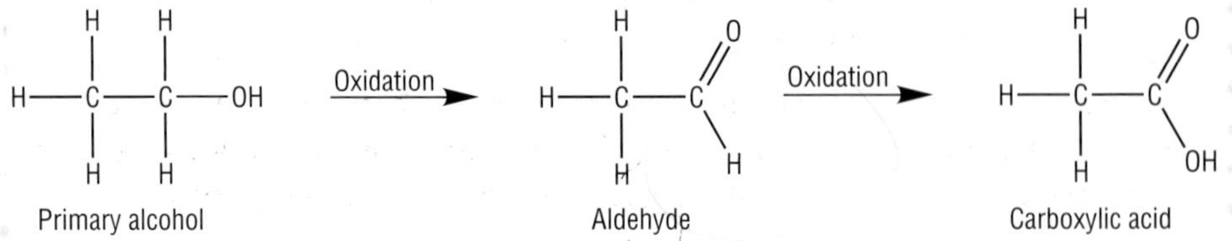

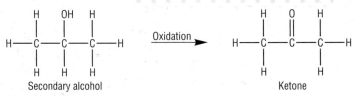

Figure 1 Ethanol oxidised to ethanal, and finally to ethanoic acid

The equation for the oxidation of ethanol to the aldehyde ethanal is shown below. Note that the oxidising agent is shown as [O] – this simplifies the equation.

$$CH_3CH_2OH + [O] \longrightarrow CH_3CHO + H_2O$$

When preparing aldehydes in the laboratory, you will need to distil the aldehyde from the reaction mixture as it is formed. This prevents the aldehyde from being oxidised further to a carboxylic acid.

When making the carboxylic acid, the reaction mixture is usually heated under **reflux** before distilling the product off.

$$CH_3CH_2OH + 2[O] \longrightarrow CH_3COOH + H_2O$$

Key definition

Reflux is the continual boiling and condensing of a reaction mixture to ensure that the reaction takes place without the contents of the flask boiling dry.

Secondary alcohols

Secondary alcohols can be oxidised to ketones (see *AS Chemistry* spread 2.2.3). Unlike aldehydes, ketones are not oxidised further.

Figure 2 Propan-2-ol can be oxidised to propanone

The equation for the oxidation of propan-2-ol to the ketone propanone is:

$$CH_3CHOHCH_3 + [O] \longrightarrow CH_3COCH_3 + H_2O$$

Module 1
Rings, acids and amines
Oxidation of alcohols and aldehydes

Oxidising aldehydes

You learnt in AS chemistry that aldehydes can be oxidised further to carboxylic acids. Acidified potassium dichromate(VI) is a suitable oxidising agent, and the reaction is normally carried out under reflux. The orange potassium dichromate(VI) changes colour, becoming green.

Aldehyde + [O] →(Oxidised) Carboxylic acid

Signifies use of an oxidising agent

Figure 3 Oxidation of an aldehyde to a carboxylic acid

STRETCH and CHALLENGE

Acidified potassium dichromate(VI) is an oxidising agent that oxidises primary alcohols, secondary alcohols and aldehydes. During oxidation, dichromate(VI) ions are reduced and the colour changes from orange to green.

You can write a fully balanced equation for the reaction by combining the oxidation half-equation for the organic compound with the reduction half equation for the dichromate(IV). The example below shows oxidation of the aldehyde ethanal.

$CH_3CHO(l) + H_2O(l) \longrightarrow CH_3COOH(aq) + 2H^+(aq) + 2e^-$ *oxidation*
$Cr_2O_7^{2-}(aq) + 14H^+(aq) + 6e^- \longrightarrow 2Cr^{3+}(aq) + 7H_2O(l)$ *reduction*

The two equations can be combined to give an overall equation. To do this, the first equation is multiplied by 3 to make the number of electrons in each equation the same, and then the two equations are added together. The overall equation becomes

$3CH_3CHO(l) + Cr_2O_7^{2-}(aq) + 8H^+(aq) \longrightarrow 3CH_3COOH(aq) + 2Cr^{3+}(aq) + 4H_2O(l)$

You will learn more about writing equations for redox reactions in spread 2.2.9.

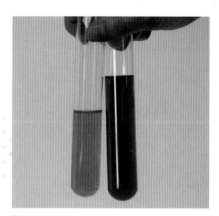

Figure 4 Potassium dichromate(VI) is reduced in its reactions with aldehydes and changes colour from orange to green

'Waiter, my wine is sour'

When a bottle of wine is opened, oxygen enters the bottle and an oxidation process begins. Over time, the taste and colour of the wine change, and often a brownish tint can be seen when the glass containing the wine is held against a white background.

This occurs because compounds in the wine called tannins are oxidised when exposed to air. This oxidation produces small quantities of hydrogen peroxide, which then oxidises ethanol to ethanal. Ethanal produces the oxidised odour and masks the natural fragrance of a wine. After a bottle of wine has been opened, ethanal can form within as little as two hours. White wines and older reds are particularly susceptible to oxidation.

"This one's rubbish as well."

Figure 5 Wine tasting

Figure 6 Wine tasters smell the wine before tasting in order to sample the bouquet

Questions

1 (a) Draw the displayed formulae of a primary alcohol, a secondary alcohol and a tertiary alcohol that would be isomeric with pentan-1-ol.
  (b) What products, if any, would be formed if the three alcohols in part **(a)** were oxidised? Name each organic product.
  (c) Name a suitable oxidising agent and the reaction conditions for the oxidation.
2 What precautions would you need to take when oxidising a solution of propan-1-ol if you were hoping to make propanal?
3 In an oxidation equation, what does the symbol [O] represent?

By the end of this spread, you should be able to . . .

* Describe the reduction of carbonyl compounds to form alcohols.
* Outline the mechanism for nucleophilic addition reactions of aldehydes and ketones with hydrides.

Reducing aldehydes and ketones

Carbonyl compounds are reduced to alcohols using a suitable reducing agent such as sodium tetrahydridoborate(III), $NaBH_4$ (sometimes called sodium borohydride). Water is often used as the solvent. The reaction is usually carried out by warming the carbonyl compound with the reducing agent.

Reduction of aldehydes

Aldehydes are reduced to primary alcohols by $NaBH_4$. The reduction of propanal is shown in Figure 1. The reducing agent is shown as [H] – this simplifies the equation.

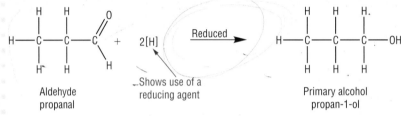

Figure 1 Reduction of an aldehyde produces a primary alcohol

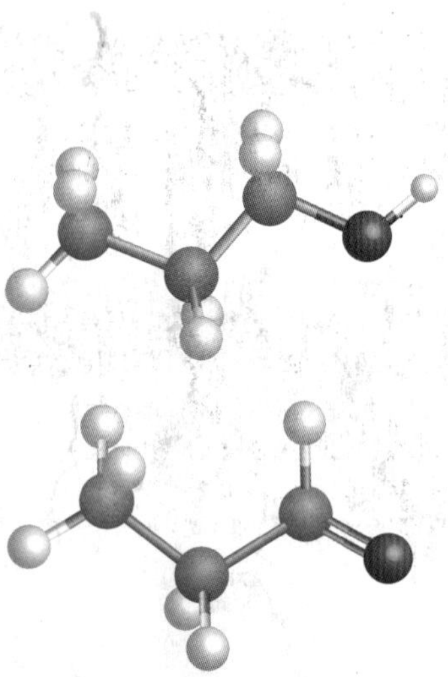

Figure 2 The primary alcohol propan-1-ol and the aldehyde propanal – note the difference in the functional groups at the right-hand end of each molecule

Reduction of aldehydes

Ketones are reduced to secondary alcohols by $NaBH_4$. The reduction of pentan-2-one is shown in Figure 3.

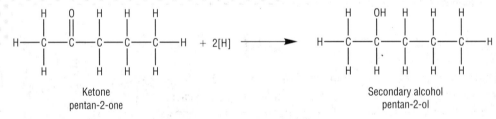

Figure 3 Reduction of a ketone produces a secondary alcohol

Insect repellants

Aldehydes such as citronellal (see Figure 4) give citronella oil a distinctive lemon scent. Citronellal has insect-repellent properties and is effective at repelling mosquitoes.

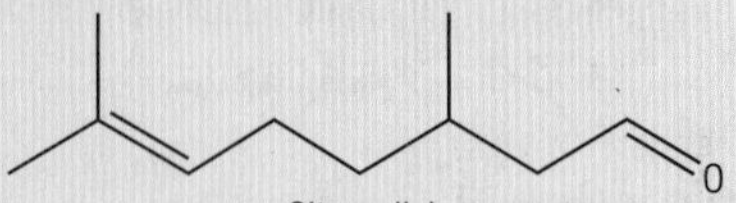

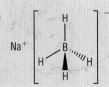

Citronellal

Figure 4 Citronella candles are used to ward off mosquitoes

Discovery of a mild reducing agent

In 1940, Hermann Irving Schlesinger discovered sodium tetrahydridoborate(III), $NaBH_4$, while studying the chemistry of boron at the University of Chicago. He found that $NaBH_4$ was less reactive than other reducing agents available at the time. $NaBH_4$ would reduce only the carbonyl group present in aldehydes and ketones to form alcohols. This meant that $NaBH_4$ could be used to selectively reduce the carbonyl group only in a molecule containing more than one functional group.

The structure of $NaBH_4$ is shown in Figure 5.

Figure 5 $NaBH_4$ readily generates hydride ions – a hydride ion is a negatively charged hydrogen ion with a lone pair of electrons, $:H^-$

Module 1
Rings, acids and amines
Reactions of aldehydes and ketones

Nucleophilic addition reactions

Aldehydes and ketones react with $NaBH_4$ in a **nucleophilic addition** reaction. Essentially the BH_4^- ion acts as a source of hydride ions, $:H^-$. The hydride ion acts as the **nucleophile**.

- In the first stage of the reaction, the electron-deficient carbon atom in the polar $^{\delta+}C=O^{\delta-}$ double bond is attacked by the hydride ion, $:H^-$, which acts as a nucleophile.
- The lone pair of electrons from the $:H^-$ ion forms a bond with the carbon atom.
- At the same time, the π-bond in the C=O bond breaks to produce a negatively charged intermediate.
- The intermediate donates an electron pair to a hydrogen atom of an H_2O molecule, forming a dative covalent bond and a hydroxide ion, OH^-.
- The organic addition product is an alcohol.

The mechanism for the reduction of propanal is shown in Figure 6.

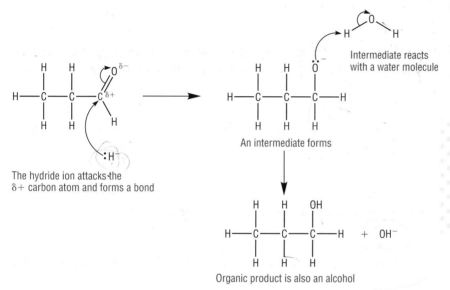

The hydride ion attacks the $\delta+$ carbon atom and forms a bond

An intermediate forms

Intermediate reacts with a water molecule

Organic product is also an alcohol

Figure 6 Reduction of an aldehyde by nucleophilic addition

Chanel No 5

The use of aldehydes in perfumes was first developed by Ernest Beaux. Working for Coco Chanel, he created the perfume Chanel No 5. Many synthetically produced aldehydes are now used in making perfumes. For example, hexyl cinnamaldehyde has a floral, jasmine fragrance and decanal provides a fragrance of violets, roses and oranges. The structures of these aldehydes are shown in Figure 7.

Hexyl cinnamaldehyde

Decanal

Figure 7 Chanel No 5 was the first perfume to include aldehydes

Questions

1. Write equations for the reduction of the following carbonyl compounds:
   (a) propanone;
   (b) pentanal;
   (c) 2-methylhexan-3-one.
2. Define the term *nucleophile*.
3. Show the mechanism for the reaction between propanone and $NaBH_4$.
4. Pent-2-enal reacts with sodium tetrahydridoborate by nucleophilic addition. Draw the displayed formula of the product formed in the reaction.

⑫ Chemical tests on carbonyl compounds

By the end of this spread, you should be able to ...

✳ Describe the use of 2,4-dinitrophenylhydrazine to detect a carbonyl group and to identify a carbonyl compound.

✳ Describe the use of Tollens' reagent to detect the presence of an aldehyde group.

Detecting the presence of a carbonyl group

- Aldehydes and ketones can be detected using the reagent 2,4-dinitrophenylhydrazine, frequently abbreviated to 2,4-DNP or 2,4-DNPH.
- A solution of 2,4-DNP in a mixture of methanol and sulfuric acid is known as Brady's reagent.
- When Brady's reagent is added to an aldehyde or a ketone, a yellow or orange precipitate is formed.
- This precipitate, called a 2,4-dinitrophenylhydrazone derivative, confirms the presence of the carbonyl functional group, C=O, in an organic compound.

The test is positive only for aldehydes and ketones. There is *no* precipitation with a carboxylic acid or with an ester – although both of these classes of compounds contain a C=O bond. The reason is beyond the scope of this course – you might like to research this on the Internet.

The equation for the reaction of propanal with 2,4-dinitrophenylhydrazine is shown in Figure 2.

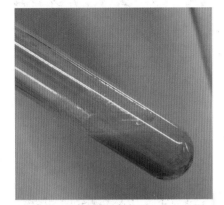

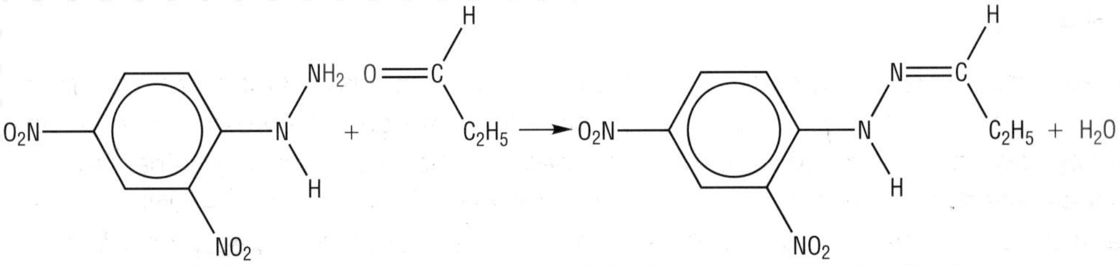

Figure 1 A 2,4-dinitrophenylhydrazone derivative is made by mixing 2,4-DNP with an aldehyde or a ketone

Examiner tip

In exams, you will not be asked to write an equation for a reaction involving 2,4-DNP or to draw the structure of the product.

2,4-dinitrophenylhydrazine   Propanal   2,4-dinitrophenylhydrazone derivative of propanal

Figure 2 Reaction of propanal with 2,4-dinitrophenylhydrazine

Aldehyde or ketone?

Once a compound has been identified as a carbonyl using 2,4-dinitrophenylhydrazine, the compound can be further classified as an aldehyde or a ketone.

- Tollens' reagent is a weak oxidising agent used to distinguish between aldehydes and ketones.
- Aldehydes are easily oxidised to carboxylic acids by Tollens' reagent. Ketones are *not* oxidised by Tollens' reagent.

Tollens' reagent can be made in the laboratory as follows:

- aqueous sodium hydroxide is added to aqueous silver nitrate until a brown precipitate of silver oxide is formed
- dilute aqueous ammonia is then added until the precipitate just dissolves.

This colourless solution is Tollens' reagent, sometimes called 'ammoniacal silver nitrate'.

Table 1 shows the result of adding Tollens' reagent to aldehydes and to ketones.

Carbonyl compound	Observation with Tollens' reagent
Aldehyde	Silver-grey solid or 'silver mirror' is formed
Ketone	No reaction

Table 1 Observations with Tollens' reagent

In Tollens' reagent, the oxidising species is the aqueous silver(I) ion, $Ag^+(aq)$.

The aldehyde is oxidised to a carboxylic acid as shown in Figure 4.

Aldehyde

Carboxylic acid

Figure 4 Oxidation of an aldehyde using Tollens' reagent

Module 1
Rings, acids and amines
Chemical tests on carbonyl compounds

Figure 3 An aldehyde after treatment with Tollens' reagent

The silver ions are reduced to silver metal:

$$Ag^+(aq) + e^- \longrightarrow Ag(s)$$

Identifying a carbonyl compound

Once a carbonyl compound has been identified as an aldehyde or a ketone, it is possible to positively identify the carbonyl compound by carrying out further experiments on the 2,4-dinitrophenylhydrazone derivative.

- The yellow/orange solid 2,4-dinitrophenylhydrazone derivative is slightly impure.
- The impure product is filtered and recrystallised to produce a purified sample of yellow or orange crystals of the 2,4-DNP derivative – this is filtered and allowed to dry.
- The melting point of this purified derivative is measured and recorded.
- The melting point is then compared to a database or data table to identify the original aldehyde or ketone.

Table 2 shows the boiling points of three carbonyl compounds and the melting points of their 2,4-DNP derivatives. Note that it very difficult to distinguish between heptan-2-one and cyclohexanone from their boiling points because they are so close together, and they are affected by changes in atmospheric pressure. However, their 2,4-DNP derivatives have melting points that are many degrees apart, allowing easy identification.

Compound	Boiling point/°C	Melting point of 2,4-DNP derivative/°C
Heptan-2-one	151	90
Cyclohexanone	156	162
Octan-2-one	173	58

Table 2 Boiling points of some ketones and the melting points of their 2,4-DNP derivatives

> **Examiner tip**
>
> Make certain that you learn all the steps. This question is frequently asked in exams. An acceptable answer is to purify the orange solid by recrystallisation, measure its melting point and compare the melting point with data table values.

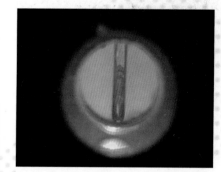

Figure 5 Magnified view, through a lens, of a sample being analysed for its melting point in a melting point apparatus. The crystals of the sample (orange in sample tube) are being slowly electrically heated until they melt

Questions

1 The carbonyl compounds CH_3COCH_3 and CH_3CH_2CHO are structural isomers.
   (a) Name these compounds.
   (b) State the reagents you would use and the observations you would make using a chemical test in which the two compounds give similar results.
   (c) State the reagents you would use and the observations you would make using a chemical test in which the two compounds give different results.
2 Explain how you would use the solid obtained from the reaction of an aldehyde with 2,4-dinitrophenylhydrazine to prove that the original aldehyde was butanal.
3 The reaction of 2,4-DNP with ethanal gives water as one of its products. Suggest the type of reaction taking place.

By the end of this spread, you should be able to ...

* Explain the water solubility of carboxylic acids.
* Describe the reactions of carboxylic acids with metals, carbonates and bases.

Introduction to carboxylic acids

Carboxylic acids are found widely in nature. The sting of ants contains methanoic (formic) acid, HCOOH; vinegar contains ethanoic (acetic) acid, CH_3COOH; rancid butter produces the unmistakeable smell of butanoic acid, $CH_3CH_2CH_2COOH$. Citrus fruits contain citric acid, rhubarb leaves contain oxalic acid and apples contain malic acid. The smell of goats and other farmyard animals is caused by hexanoic acid, $CH_3CH_2CH_2CH_2CH_2COOH$.

Figure 1 Carboxylic acids are widely distributed throughout nature

Naming carboxylic acids

The functional group in a carboxylic acid is the carboxyl group – COOH, shown in Figure 2.

Carboxylic acids are named by removing the final -e from the alkane stem and adding the suffix '-oic acid'.

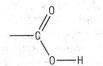

Figure 2 Structure of the carboxyl group

Figure 3 Methanoic acid, HCOOH, makes up more than half of an ant's body mass!

Worked example

The first two members of the carboxylic acid homologous series are methanoic acid and ethanoic acid. They contains one and two carbon atoms, respectively, and their names are derived from methane and ethane. Their structures are shown in Figure 4.

The carboxyl carbon atom in the COOH functional group is always numbered carbon-1. When naming branched-chained carboxylic acids, you start counting from the functional group end of the molecule. In Figure 5, you can see the structure of 4-methylhexanoic acid – note how it is named.

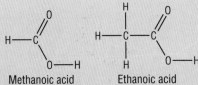

Figure 4 Structures of methanoic and ethanoic acids

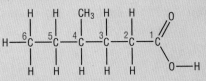

Figure 5 Structure of 4-methylhexanoic acid

Solubility of carboxylic acids

Carboxylic acids with between one and four carbon atoms are very soluble in water. The highly polar C=O and O–H bonds allow carboxylic acid molecules to form hydrogen bonds with water molecules (Figure 6). As the number of carbon atoms in the carboxylic acids increases, the solubility decreases. This is the result of the longer non-polar hydrocarbon chain in the molecule, which does not interact with water molecules.

Acid reactions of carboxylic acids

- Carboxylic acids are weak acids when compared with nitric, sulfuric and hydrochloric acids.
- Carboxylic acids take part in the typical acid reactions – reacting with metals, carbonates and bases to form salts.
- Salts formed from carboxylic acids are known as carboxylates. The names of the carboxylates formed from the first four carboxylic acids are given in Table 1.

In acid reactions of carboxylic acids, the product is a carboxylate and contains the carboxylate ion. The carboxylate ion is ionic and is usually shown as in Figure 7.

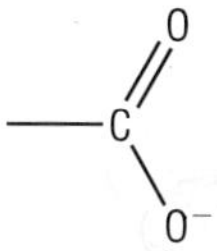

Figure 7 Carboxylate ion

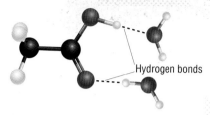

Figure 6 Hydrogen bonds, strong dipole–dipole interactions, form between water and the polar groups on a carboxylic acid

Carboxylic acid	Name of salt formed
methanoic acid	methanoate
ethanoic acid	ethanoate
propanoic acid	propanoate
butanoic acid	butanoate

Table 1 Carboxylates formed from the first four carboxylic acids

Reacting with metals

A carboxylic acid reacts with reactive metals to form a salt and hydrogen gas – during the reaction there will be effervescence. In the following example, ethanoic acid reacts with sodium metal to form the salt sodium ethanoate.

$$CH_3COOH + Na \longrightarrow CH_3COO^-Na^+ + \frac{1}{2}H_2$$
ethanoic acid sodium ethanoate

Reacting with bases

Carboxylic acids react with aqueous bases, such as metal hydroxides, to form a salt and water. In the equation below, propanoic acid is reacting with aqueous potassium hydroxide:

$$CH_3CH_2COOH(l) + KOH(aq) \longrightarrow CH_3CH_2COO^-K^+(aq) + H_2O(l)$$
propanoic acid potassium propanoate

Reacting with carbonates

An aqueous solution of a carboxylic acid reacts with carbonates to form a salt, carbon dioxide and water. In the equation below, methanoic acid is reacting with sodium carbonate:

$$2HCOOH(aq) + Na_2CO_3(s) \longrightarrow 2HCOO^-Na^+(aq) + CO_2(g) + H_2O(l)$$
methanoic acid sodium methanoate

> **Examiner tip**
>
> The product formed in the reaction with sodium is a salt – the acidic hydrogen in the COOH group of the carboxylic acid has been replaced by a metal. In this reaction, we call the salt a carboxylate. Because the acid is ethanoic acid, the product is an ethanoate.

Questions

1 Name each of the following carboxylic acids.

 (a)   (b) 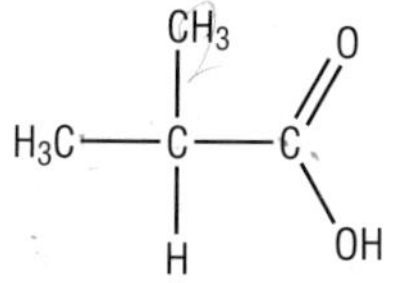   (c)

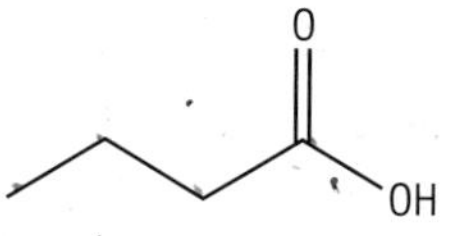

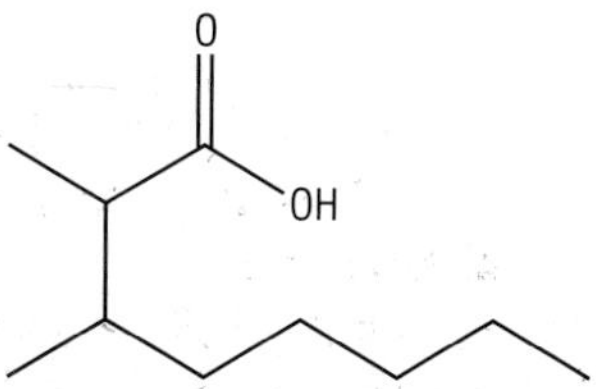

2 Write fully balanced equations for the reaction of butanoic acid with:
 (a) sodium metal; (b) sodium hydroxide; (c) magnesium carbonate.

By the end of this spread, you should be able to ...

* Describe the esterification of carboxylic acids with alcohols in the presence of an acid catalyst, and also of acid anhydrides with alcohols.
* Describe the hydrolysis of esters.
* State the uses of esters in perfumes and flavourings.

Making esters

Esters from carboxylic acids

You will recall from your AS work (*AS Chemistry* spread 2.2.4) that an ester can be prepared by reacting a carboxylic acid with an alcohol in the presence of an acid catalyst.

* Concentrated sulfuric acid is often used as the acid catalyst.
* This reaction is known as **esterification**.

Figure 1 shows the formation of ethyl propanoate from propanoic acid and ethanol.

$$CH_3CH_2COOH + CH_3CH_2OH \longrightarrow CH_3CH_2COOCH_2CH_3 + H_2O$$

Propanoic acid   Ethanol   Ethyl propanoate

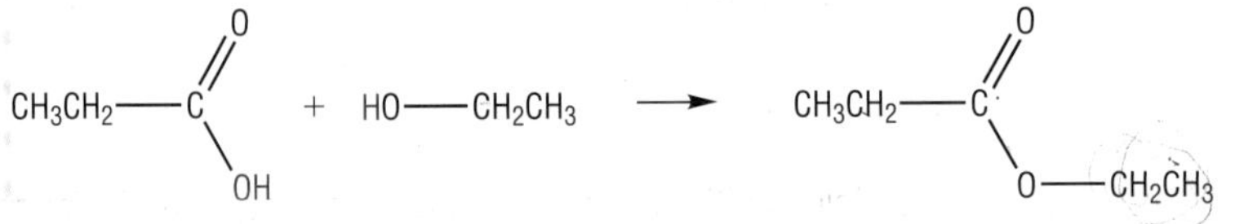

Figure 1 Formation of the ester ethyl propanoate from a carboxylic acid

Esters from acid anhydrides

Esters can also be prepared by gently heating an acid anhydride (see Figure 2) with an alcohol. This method gives a much better yield of the ester than preparations from carboxylic acids.

The esterification reaction between ethanoic anhydride and methanol is shown in Figure 3. Note that the other product is a carboxylic acid.

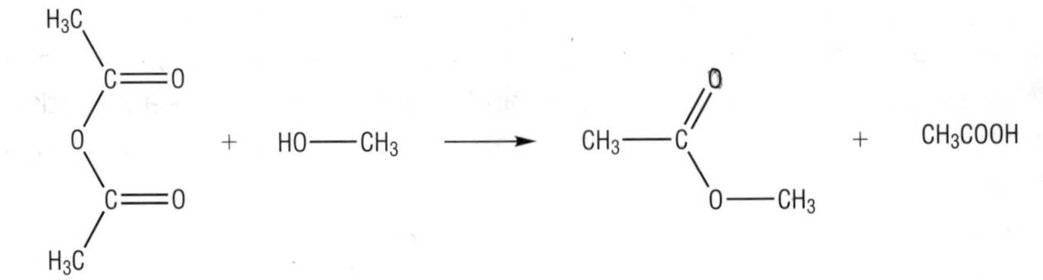

Ethanoic anhydride   Methanol   Methyl ethanoate   Ethanoic acid

Figure 3 Ethanoic anhydride reacts with methanol to make the ester methyl ethanoate

Ester hydrolysis

Hydrolysis is the chemical breakdown of a compound by reaction with water. Hydrolysis of esters is essentially the reverse of esterification. Hydrolysis of esters takes place with aqueous acid or aqueous alkali.

Key definition

Esterification is the reaction of an alcohol with a carboxylic acid to produce an ester and water.

Examiner tip

When naming an ester, the alcohol provides the alkyl part of the name and the carboxylic acid provides the alkanoate part of the name – for example,

ethanol + propanoic acid →
ethyl propanoate + water

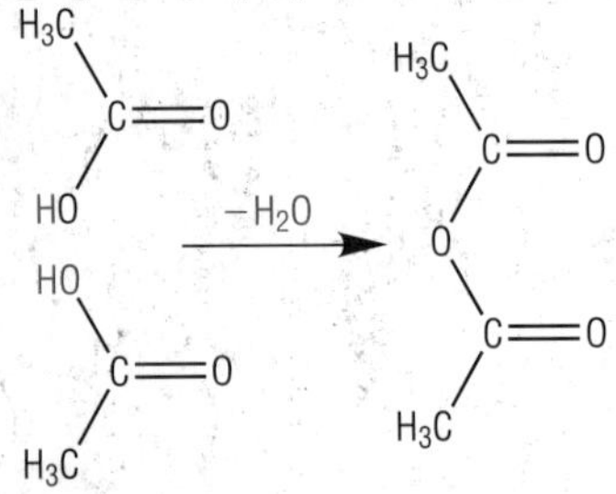

Figure 2 An acid anhydride is formed by removal of a molecule of water from two carboxylic acid molecules. The diagram shows how ethanoic anhydride, $(CH_3CO)_2O$, is related to ethanoic acid, CH_3COOH ('anhydride' just means 'without water')

Key definition

Hydrolysis is a reaction with water or hydroxide ions that breaks a chemical compound into two compounds.

Acid hydrolysis of esters

In acid hydrolysis, the ester is heated under reflux with dilute sulfuric acid or dilute hydrochloric acid. The ester is broken down by water, with the acid acting as a catalyst. The equation for the acid hydrolysis of the ester propyl ethanoate is shown in Figure 4.

Figure 4 Acid hydrolysis of propyl ethanoate

Alkaline hydrolysis of esters

Esters can also be hydrolysed in aqueous alkaline conditions. Aqueous sodium (or potassium) hydroxide is refluxed with the ester. This reaction is non-reversible and leads to the formation of the sodium salt of the carboxylic acid. The reaction is sometimes called 'saponification' and is the basis of soap-making (see spread 2.1.19).

The alkaline hydrolysis reaction of ethyl propanoate is shown in Figure 5.

Figure 5 Alkaline hydrolysis of ethyl propanoate

Esters as perfumes and flavourings

Esters have many uses, both industrially and in catering. Esters are responsible for the flavour of many foods and the pleasant smell of flowers. Esters therefore can be used as perfumes and as flavourings.

Many esters are found in essential oils, obtained by steam distillation of organic plant matter. 'Oil of wintergreen' is the essential oil obtained from the wintergreen plant. The oil can be massaged into painful muscles and joints to offer relief as 'deep heat'. It can also be used before exercising to help warm muscles up, or afterwards to relieve the aches of over-exercising. The smell is distinctive. American Indians used the dried leaves of the plant to make 'wintergreen tea'. Years later, scientists analysed the essential oil found in this plant and discovered the ester methyl salicylate. The structure of methyl salicylate is shown in Figure 6 alongside its source, the wintergreen plant.

Benzyl ethanoate, $CH_3COOCH_2C_6H_5$, is found naturally in many flowers and is the main constituent of the essential oils from the jasmine flowers. It is used widely for its aroma in perfumery and cosmetics, and to give apple and pear flavours in drinks and foods. Benzyl ethanoate is found in perfumes, shampoo, fabric softener, soap, hairspray and deodorants.

Questions

1 Methyl propanoate can be prepared either from a carboxylic acid or from an acid anhydride.
   Write balanced equations for these preparations of methyl propanoate.
2 Butyl hexanoate can be hydrolysed with aqueous acid or with aqueous alkali.
   (a) Draw the structure of butyl hexanoate.
   (b) Write a fully balanced equation for the acid hydrolysis and for the alkaline hydrolysis of the ester.
3 State two commercial uses of esters.

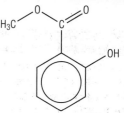

Figure 6 Wintergreen is an evergreen plant found throughout North America – the leaves are hard and brittle and, when broken, give the unmistakable wintergreen aroma and taste of the ester methyl salicylate

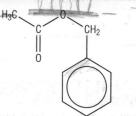

Figure 7 Benzyl ethanoate is responsible for food flavourings that resemble apples and pears

By the end of this spread, you should be able to . . .

* Describe a triglyceride as a triester of glycerol (propane-1,2,3-triol) and fatty acids.
* Compare the structures of saturated fats, unsaturated fats and fatty acids.

Fats and oils

Fats have many important functions in the body.

Fat protects your organs, provides insulation and also acts as a long-term energy store. We all need to eat some fat as part of a balanced diet because fats are important for many functions of the human body.

• Animal and vegetable fats and oils are esters of long-chain carboxylic acids called fatty acids.
• Fats and oils are very similar substances, differing only in their melting points. A melting point above room temperature gives a solid fat, e.g. butter, whereas a melting point below room temperature gives a liquid oil, e.g. olive oil.

Triglycerides – the building blocks

Triglycerides occur naturally in animal and vegetable fats. They are triesters of an alcohol called propane-1,2,3-triol (commonly known as glycerol) and three fatty acid molecules.

Glycerol is an alcohol with three –OH groups (see Figure 2). Notice the 'triol' in its systematic name.

Fatty acids are long-chain carboxylic acids. The hydrocarbon chains can be either saturated or unsaturated. Most naturally occurring fatty acids contain an even number of carbon atoms because of how they are synthesised in nature.

Figure 3 shows the displayed formula of hexadecanoic acid (palmitic acid), $C_{15}H_{31}COOH$, which is a *saturated* fatty acid. This means that there are no double bonds in the hydrocarbon chain.

Figure 1 Assorted fats and oils including olive oil, sunflower oil, butter, goose fat, duck fat, butter, lard and margarine; fats and oils are an essential part of the diet

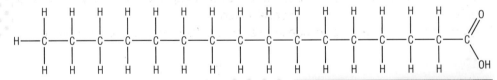

Figure 2 Structure of propane-1,2, 3-triol, commonly known as glycerol

Figure 3 Palmitic acid is believed to have been discovered in palm oil by Edmond Frémy in 1840 – the acid can be obtained from butter, cheese, milk and meat as well as from the fruit of the oil palm

Figure 4 shows the skeletal formula of octadec-9-enoic acid (oleic acid), $C_{17}H_{33}COOH$, which is an *unsaturated* fatty acid. Note the double bond in the hydrocarbon chain.

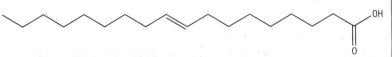

Figure 4 Oleic acid is a mono-unsaturated fatty acid, with one double bond in its hydrocarbon chain. Oleic acid can be obtained from various animal and vegetable sources – its main source is olive oil

Module 1
Rings, acids and amines
Fats and oils – building triglycerides

Unsaturated fats with one double bond in the hydrocarbon chain are called mono-unsaturated fats. The double bond is commonly found between carbon-9 and carbon-10. Remember that the carboxyl carbon is carbon-1, so you must count from the COOH end.

A shorthand notation for fatty acids consists of two numbers separated by a colon:
- the first number indicates the number of carbon atoms
- the second number indicates the number of double bonds
- the number in brackets indicates the position(s) of the double bond(s).

Worked example

Hexadecanoic acid is represented as $16:0$ (see Figure 3)

Octadec-9-enoic acid is represented as $18:1$ (9) (see Figure 4)

Octadeca-9,12-dienoic acid is represented as $18:2$ (9,12) (see Figure 5)

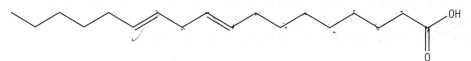

Figure 5 Structure of octadeca-9,12-dienoic acid – this has more than one double bond in the hydrocarbon chain, and is called a polyunsaturated fatty acid

Forming triglycerides

Triglycerides are triesters with three fatty acids joined to each of the alcohol groups on glycerol through an ester linkage.
- A molecule of a simple triglyceride is derived from three molecules of the same fatty acid.
- Natural triglycerides are mostly mixed triglycerides derived from two or three different fatty acids.
- Triglycerides can be saturated or unsaturated, depending on whether they are derived from fatty acids that are saturated or unsaturated.
- The formation of a simple triglyceride from saturated fatty acids is shown in Figure 7.

Figure 6 Octadeca-9,12-dienoic acid is also known as linoleic acid; it is a component of evening primrose oil, which is made from pressing the seeds of the evening primrose and is used in moisturising creams and lip balms

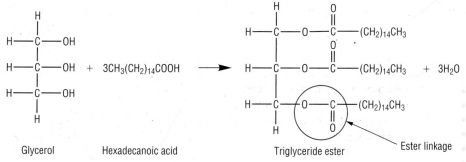

Glycerol   Hexadecanoic acid   Triglyceride ester   Ester linkage

Figure 7 Formation of a triglyceride

Examiner tip

This reaction is a simple esterification reaction. This type of reaction is dealt with in spread 1.1.14.

Questions

1 Capric acid ($10:0$) is a saturated fatty acid.
   (a) Draw the displayed formula of the acid.
   (b) Name the functional group present in the acid.
   (c) Draw the structure of the triglyceride formed between glycerol and capric acid – the fatty acid carbon chain can be shown as $-C_9H_{19}$

2 Explain why some triglycerides are fats while others are oils.

3 What is the difference between a simple triglyceride and a mixed triglyceride?

4 What is the shorthand representation of octadeca-9,12,15-trienoic acid?

By the end of this spread, you should be able to . . .

* Compare the structures of *cis* and *trans* isomers of unsaturated fatty acids.
* Compare the link between *trans* fatty acids, the possible increase in 'bad' cholesterol and the resultant increased risk of coronary heart disease and strokes.
* Describe and explain the increased use of esters of fatty acids as biodiesel fuels.

Isomerism in unsaturated fatty acids

In *AS Chemistry* (spread 2.1.7), you studied *E/Z* isomerism in alkenes. This type of stereoisomerism occurs because of the restricted rotation around the double bond. *Cis–trans* isomerism is a special type of *E/Z* isomerism in which each carbon of the C=C double bond carries one atom or group that is the same.

Fats and oils are derived from saturated or unsaturated fatty acids. Foods high in saturated fats have been known for many years to increase the risk of heart disease by raising blood cholesterol levels.

Many people believe that foods containing unsaturated fats are better for us than foods high in saturated fats. Unsaturated fats, however, can exist in two isomeric forms – *cis* and *trans*. We hear very little about *cis*-fats because they present little danger to our health, but it is now thought that *trans*-fats increase the risk of coronary heart disease.

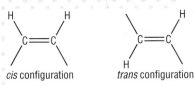

cis configuration   *trans* configuration

Figure 2 *Cis* and *trans* configurations around a double bond

Figure 1 Common sources of saturated fats in the diet include meat products, butter, lard, pastry, cakes and biscuits – they are found in many popular fast foods

In nature, unsaturated fatty acids normally exist in the *cis* form.
* In the *cis* form, the unsaturated fatty acid molecules cannot pack closely together, and *cis* fatty acids exist as liquids at room temperature.
* The *trans* form of an unsaturated fatty acid is quite linear in its structure and the fatty acid chains can pack together closely. Consequently a *trans*-unsaturated fatty acid will always have a higher melting point than its *cis* isomer.

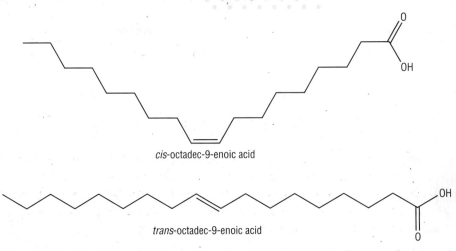

cis-octadec-9-enoic acid

trans-octadec-9-enoic acid

Figure 3 Structures of *cis*- and *trans*-octadec-9-enoic acid. Molecules of the *trans* isomer can pack closer together than those of the *cis* isomer – therefore the *trans* fatty acid is more solid at room temperature

Module 1
Rings, acids and amines
Triglycerides, diet and health

The food industry uses many unsaturated fats and oils in its products. However, often some of the double bonds are removed by partial hydrogenation to make the fats and oils more solid, as in margarine manufacture (see *AS Chemistry* spread 2.1.17). A side effect of hydrogenation is that the remaining double bonds may be structurally changed, with many being converted from the *cis* form to the less healthy *trans* form.

Trans fats and cholesterol

Lipoproteins are particles that carry lipids, such as cholesterol and fats, through the blood.

- **High-density lipoproteins (HDLs)** are believed to transport cholesterol out of the blood, and eventually out of the body. HDLs are commonly referred to as 'good' lipoproteins.
- **Low-density lipoproteins (LDLs)** carry about 65% of the cholesterol in blood. LDLs can deposit lipids onto artery walls, building up fatty deposits that restrict blood flow.

Trans fatty acids are thought to behave like saturated fats in the body. They raise LDL levels and increase the risk of heart disease. They also tend to lower HDL levels, so they are potentially even more damaging.

Fatty acids as biodiesel

Soaring oil prices, the non-renewable nature of fossil fuels and the desire to protect our environment have led scientists to develop alternative renewable sources of fuel. Biodiesel is one such fuel and is seen as a potential solution to our ever-increasing demand for fuel.

Biodiesel is easily produced using waste cooking oil. However, natural sources of suitable oils also occur in crops such as palm or soya bean – in the UK, rapeseed has the greatest potential for biodiesel production.

Making biodiesel

The chemical structure of biodiesel is an ethyl or methyl ester of a fatty acid. Biodiesel is made by a process known as transesterification. Triglycerides in fats and oils are reacted with methanol or ethanol in the presence of a sodium (or potassium) hydroxide catalyst. Glycerol is also formed and is sold to the pharmaceutical or cosmetics industries, improving the atom economy of the process. An equation for the process is shown in Figure 6.

Figure 6 Transesterification for the production of biodiesel (R = fatty acid carbon chain)

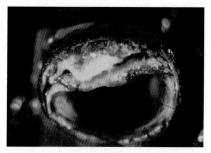

Figure 4 An artery that has been partially blocked by atherosclerosis, the build-up of fatty plaques of atheroma on its inner wall. One of the main components of atheroma is the lipid cholesterol. Blockage of a coronary artery, one of the arteries that supply oxygen to the heart muscle, can cause chest pain or a heart attack

Figure 5 Rapeseed, a raw material in the production of biodiesel

Biodiesel: friend or foe?

Biodiesel has been claimed to be *carbon neutral* because when the 'fuel crop' grows it absorbs the same amount of carbon dioxide as is released when the fuel is burned. However, the use of biodiesel is currently under scrutiny – it has been discovered that some poor countries have been using large areas of land, traditionally used for local food crops, for producing biodiesel for sale to richer countries.

The carbon neutrality of biodiesel is also being questioned. At present, every claimed benefit of biodiesel is being combated by a counter-claim.

Questions

1. Explain the difference between saturated fats and unsaturated fats.
2. Draw the skeletal formulae of *cis*- and *trans*-hex-3-enoic acid.
3. Name three foods high in saturated fats.
4. What do you understand by the term *carbon neutral*?

By the end of this spread, you should be able to . . .

* Explain the basicity of amines in terms of proton acceptance by the nitrogen lone pair.
* Describe the reactions of amines with acids to form salts.

Something fishy about amines?

Amines are all derivatives of ammonia. In amines, the hydrogen atoms from the ammonia molecule have been replaced one at a time by hydrocarbon chains shown by R in Figure 1 below.

This give three types of amine: primary, secondary and tertiary.

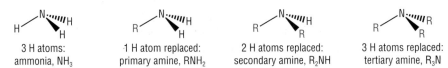

3 H atoms: ammonia, NH_3

1 H atom replaced: primary amine, RNH_2

2 H atoms replaced: secondary amine, R_2NH

3 H atoms replaced: tertiary amine, R_3N

Figure 1 Relationship of ammonia with primary, secondary and tertiary amines

Amines occur commonly in nature, many being well known for their physiological effects and actions.

* Amphetamine is a complex amine used to treat daytime drowsiness and chronic fatigue syndrome.
* Phenylephrine is an amine commonly used as a decongestant. Phenylephrine is found in many over-the-counter cold and flu preparations, commonly in combination with paracetamol.
* Adrenaline is the 'fight or flight' amine helping the body to deal with sudden stress.

The structures of these three amines are shown in Figure 3.

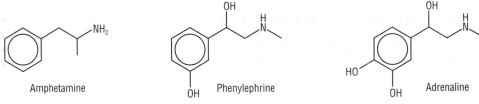

Amphetamine   Phenylephrine   Adrenaline

Figure 3 Amphetamine, phenylephrine and adrenaline are all amines

* Amines are also known for their unpleasant odours – often being described as having a fishy smell.

Naming amines

A primary amine is one with only one carbon chain attached to the nitrogen.

$CH_3CH_2CH_2CH_2NH_2$ is a primary amine.

There are four carbons in the alkyl chain, so it is given the name *butyl*.

The ending -amine is added as a suffix to give the name *butylamine*.

A secondary amine is one in which the nitrogen atom is attached to two carbon chains.

The secondary amine shown in Figure 5 has two alkyl groups connected to the amine nitrogen atom.

The longest alkyl group contains three carbons: *-propyl*.

The suffix is added: *-amine*.

The shorter alkyl chain contains one carbon connected to the N atom, so *N-methyl* is added as the prefix.

The compound is *N-methylpropylamine*.

A tertiary amine is one in which the nitrogen atom is attached to three carbon chains.

The structure of *N,N-diethylpropylamine*, a tertiary amine, is shown in Figure 6.

Figure 2 Lemsip cold and flu remedy – this lemon-flavoured soluble powder contains the analgesic (painkiller) paracetamol, the decongestant phenylephrine and vitamin C

Figure 4 Fish drying generates a smell similar to the common amines

Examiner tip

A primary amine is one in which the nitrogen atom is attached to one carbon chains.

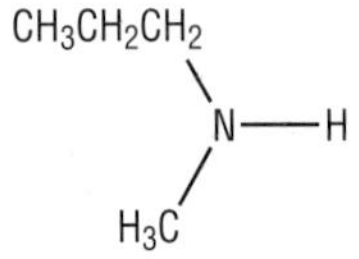

Figure 5 *N*-methyl propylamine

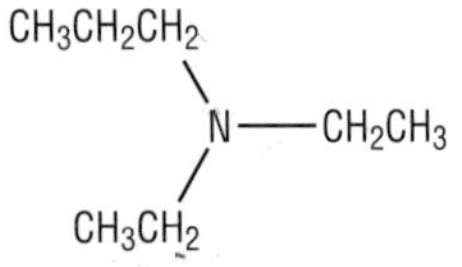

Figure 6 *N,N*-diethylpropylamine

Basicity in amines

Amines are weak bases. (Remember that ammonia is also a weak base.)

A base is defined as a proton acceptor. Ammonia and some common amines are shown in Figure 7. Each of these:

- has a lone pair of electrons on the nitrogen atom
- can accept a proton, H^+.

When a base accepts a proton, a dative covalent bond forms between the lone pair of the nitrogen atom and the proton. This is shown for methylamine in Figure 8.

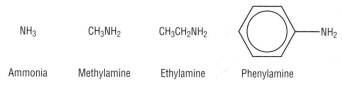

NH3
CH3NH2
CH3CH2NH2
NH2

Ammonia Methylamine Ethylamine Phenylamine

Figure 7 Ammonia and some common amines

H
H3C
N
H
+
H+
H
H3C
N
H
H
+

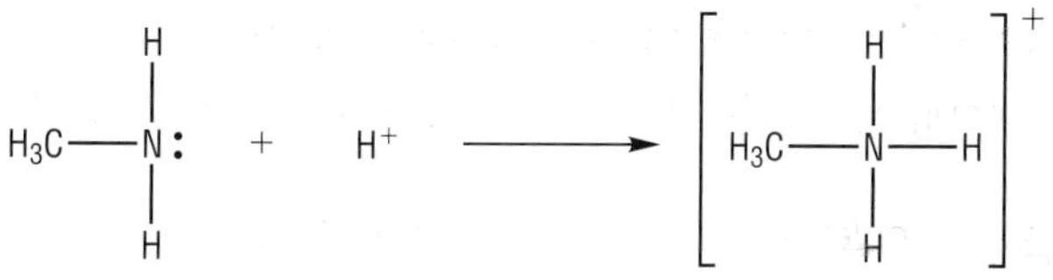

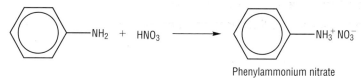

Methylamine Proton Methylammonium ion

Figure 8 Reaction of methylamine as a base

Base reactions of amines

Amines are basic in character, reacting with acids to make salts. In general, for amines:

base + acid ⟶ salt

Ethylamine reacts with hydrochloric acid to form an alkylammonium salt:

$$CH_3CH_2NH_2(aq) + HCl(aq) \longrightarrow CH_3CH_2NH_3^+Cl^-(aq)$$

ethylamine ethylammonium chloride

The reaction of phenylamine and nitric acid is shown in Figure 9.

NH2
+
HNO3
NH3+ NO3−

Phenylammonium nitrate

Figure 9 Reaction of phenylamine and nitric acid

> **Examiner tip**
>
> Remember from AS chemistry that ammonia reacts with protons to form an ammonium ion:
>
> $NH_3 + H^+ \rightarrow NH_4^+$
>
> Note the similarity of ammonia to amines as bases.

> **Examiner tip**
>
> Alkylammonium salts from sulfuric acid are named 'sulfate', 'nitrate' from nitric acid and 'chloride' from hydrochloric acid.

Questions

1 An amine was analysed and was found to contain 2.90 g of carbon, 0.84 g of hydrogen and 1.70 g of nitrogen.

(a) Calculate the empirical formula of the amine.

(b) The relative molecular mass of the amine is 45.0. Deduce its molecular formula.

(c) The amine is a secondary amine. Suggest a possible structure for the amine.

2 Some reactions of propylamine are shown in Figure 10.

$$\mathbf{A} \xleftarrow{HNO_3} CH_3CH_2CH_2NH_2 \xrightarrow{HCl} \mathbf{B}$$

Figure 10 Reactions of propylamine

(a) Draw the structures of compounds A and B.

(b) For the reaction between $CH_3CH_2CH_2NH_2$ and hydrochloric acid:

(i) name the product; **(ii)** write an equation.

By the end of this spread, you should be able to ...

* Describe the preparation of aliphatic amines by substitution of halogenoalkanes.
* Describe the preparation of aromatic amines by reduction of nitroarenes.
* Describe the synthesis of an azo dye by diazotisation and coupling.
* State the use of these reactions in the formation of dyestuffs.

Preparation of primary aliphatic amines

Aliphatic amines can be prepared by warming halogenoalkanes gently with an excess of ammonia, using ethanol as the solvent.

The formation of propylamine by the reaction of 1-chloropropane with ammonia is shown below. This reaction is nucleophilic substitution.

$$CH_3CH_2CH_2Cl + NH_3 \longrightarrow CH_3CH_2CH_2NH_2 + HCl$$
1-chloropropane   propylamine

Further ammonia then reacts with the hydrogen chloride formed:

$$NH_3 + HCl \longrightarrow NH_4^+Cl^-$$

As a nucleophile, ammonia has a lone pair of electrons and attacks the $\delta+$ carbon atom in the polar carbon–halogen bond. However, the product, propylamine, also has a lone pair of electrons. This can attack another molecule of 1-chloropropane, causing further substitution:

$$CH_3CH_2CH_2Cl + CH_3CH_2CH_2NH_2 \longrightarrow (CH_3CH_2CH_2)_2NH + HCl$$

To prepare a primary amine, an excess of ammonia is used to minimise this further substitution.

Preparing aromatic amines

Nitrobenzene and other nitroarenes can be reduced using a mixture of tin and concentrated hydrochloric acid, heated under reflux, followed by neutralisation of the excess hydrochloric acid. Aromatic amines are formed as products in these reactions.

The mixture of tin and hydrochloric acid acts as the reducing agent, shown in equations as [H] for simplicity:

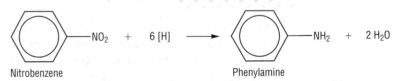

Nitrobenzene   Phenylamine

Figure 1 Reduction of nitrobenzene to make phenylamine (aniline)

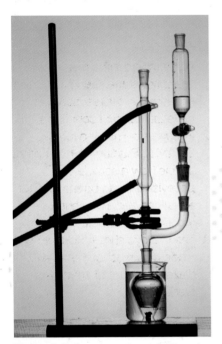

Figure 2 Reflux apparatus used to make phenylamine. Concentrated hydrochloric acid, in the dropping funnel on the right-hand side, is added to the flask containing tin and nitrobenzene

Synthesis of dyes from phenylamine

There are two steps in the industrial preparation of dyestuffs:
* diazotisation
* coupling reactions.

Diazotisation

Diazotisation is the first stage in the process, responsible for the formation of the diazonium ion. When a mixture of phenylamine and nitrous acid is kept below 10 °C, a diazonium salt is formed.

Module 1
Rings, acids and amines
Amines and their reactions

Nitrous acid, HNO_2, is generated in the reaction mixture (in situ) by reacting together sodium nitrite, $NaNO_2$, and excess hydrochloric acid, HCl:

$$NaNO_2 + HCl \longrightarrow HNO_2 + NaCl$$

The cold nitrous acid then reacts with an aromatic amine to form a diazonium salt. In the example shown in Figure 3, phenylamine is used as the aromatic amine.

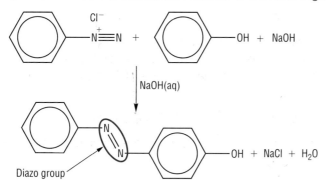

Figure 3 Diazotisation reaction of phenylamine

Coupling

A coupling reaction occurs when the diazonium salt, benzenediazonium chloride, is reacted with a phenol (or other aromatic compound, such as an amine) under alkaline conditions. In this reaction, two benzene rings are linked together through an azo functional group, –N=N–.

The reaction of phenol with benzenediazonium chloride is shown in Figure 4.

Figure 4 Reaction between benzenediazonium chloride and phenol

The product is a brightly coloured compound that is used as an azo dye.

Case study

Sir William Henry Perkin, an English chemist, is credited with the synthesis of the first 'aniline dye' called mauveine. Aniline is the common name for phenylamine, $C_6H_5NH_2$, and the dyes made from aniline are commonly called aniline dyes.

Perkin reacted aniline with potassium dichromate, a strong oxidising agent, and the resulting black solid made a beautiful purple solution in alcohol. Variations on the original synthesis produced dozens of dyes from aniline – aniline reds, aniline violets, greens, yellows, browns and blues. Substituting other benzene-containing compounds for phenylamine produced many other artificial colours.

Figure 5 Plaque commemorating the work of William Perkin in Stepney, East London. Nowadays, scientists use Perkin's original ideas to synthesise new colours for dyes in the textile industry

Sir William Henry PERKIN, F.R.S.
discovered the first aniline dyestuff,
March 1856,
while working in his home laboratory
on this site and went on to
found science-based industry.
1838–1907
STEPNEY HISTORICAL TRUST

Questions

1 Butylamine can be made from 1-bromobutane. Give a balanced equation for this reaction and state the reaction conditions.

2 Phenylamine has the formula $C_6H_5NH_2$. Starting from benzene, outline how a sample of phenylamine can be made. Give equations and all essential reagents and conditions.

3 The azo dye shown in Figure 6 is made by a reaction between a diazonium ion and another arene.

Figure 6

(a) Give a balanced equation to show the formation of the diazonium ion from phenylamine.

(b) What conditions would be needed for this reaction to occur?

(c) Identify the compound with which the diazonium ion joins to form the azo dye.

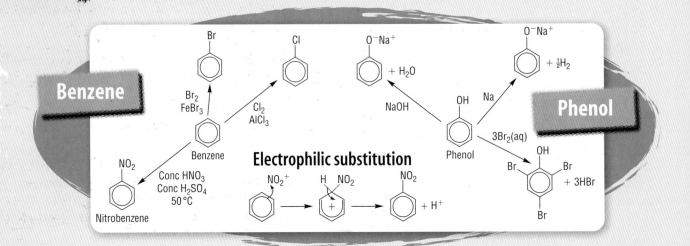

Electrophilic substitution

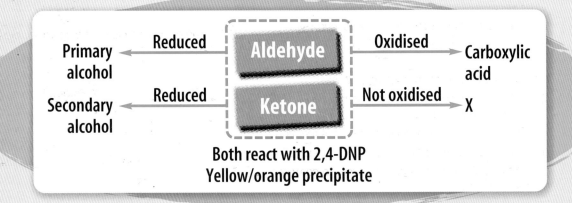

Both react with 2,4-DNP
Yellow/orange precipitate

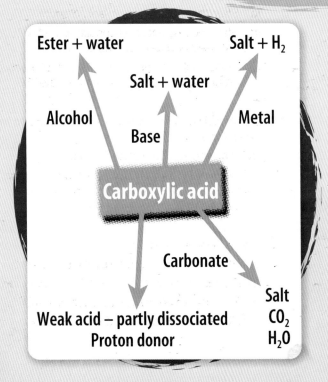

Weak acid – partly dissociated
Proton donor

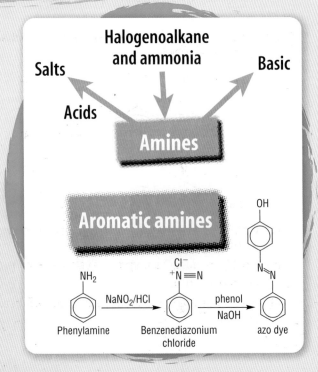

Practice questions

1 Describe the nature of the bonding and structure of benzene.

2 The nitration of benzene is an important industrial reaction.

(a) Name **two** types of commercially important materials whose manufacture involves the nitration of benzene.

(b) State the conditions required for the nitration of benzene using nitric acid.

(c) Write a balanced equation for the nitration of benzene.

(d) (i) Outline the mechanism for the nitration of benzene.

(ii) Explain what a curly arrow means in this type of mechanism.

(iii) Give the name of the NO_2^+ ion and state its function in the mechanism shown above.

3 Benzene and cyclohexene react differently with bromine.

(a) (i) Write an equation for the reaction between bromine and cyclohexene.

(ii) Name the product formed in the reaction.

(iii) What would you observe if you were to carry out the reaction?

(iv) What type of reaction is taking place?

(b) (i) Write an equation for the reaction between bromine and benzene.

(ii) Name a suitable halogen carrier for this reaction.

(iii) Name the organic product formed in the reaction.

(iv) What would you observe if you were to carry out the reaction?

(v) What type of reaction is taking place?

4 Some common reactions of phenol are shown in the reaction scheme below.

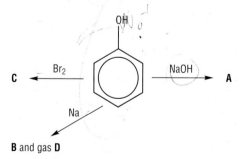

B and gas D

(a) Write an equation for the reaction between phenol and sodium hydroxide to form A.

(b) Identify product C and give its systematic name.

(c) Identify B and D.

(d) Phenol has many important uses. State **two** uses for phenol.

5 Alcohols can be oxidised with acidified potassium dichromate.

(a) Show the structure of a primary, a secondary and a tertiary alcohol that are structural isomers of butan-1-ol.

(b) Each of the alcohols drawn in part (a) is heated under reflux with an excess of acidified potassium dichromate. For each alcohol:

(i) state what you would see;

(ii) draw the skeletal formula of the organic product formed.

6 Methanal takes part in a nucleophilic addition reaction with sodium tetrahydridoborate.

(a) Draw the displayed formula of the organic product formed in this reaction.

(b) Use curly arrows to show the mechanism for this reaction.

7 Some of the chemistry of propanoic acid is shown in the diagram below.

C ←(KOH)— H—C—C—C(OH)=O ... —(Reaction A)→ H—C—C—C(O—CH₃)=O

↓ Na₂CO₃

B

(a) State the name of the organic product **C**.

(b) State the reagents and conditions and write a fully balanced equation for reaction A.

(c) What would you observe in the reaction between propanoic acid and sodium carbonate?

8 Lauric acid (12:0) is a saturated fatty acid that can form triglycerides.

(a) Draw the structure of lauric acid.

(b) Write an equation for the formation of a triglyceride from glycerol and lauric acid. The carbon chain in lauric acid can be shown as $-C_{11}H_{23}$.

9 Amines behave as bases.

(a) What do you understand by the term *base*?

(b) Propylamine was treated with hydrochloric acid.

(i) Write an equation for the reaction that takes place.

(ii) Name the salt formed in this reaction.

1 Benzene can be nitrated to form nitrobenzene, $C_6H_5NO_2$.
  (a) Draw the structural formula for benzene and give its empirical formula. [2]
  (b) State the reagents needed for the nitration of benzene. [2]
  (c) An electrophile is formed during the nitration of benzene.
    (i) What is the formula of this electrophile? [1]
    (ii) Write an equation for the production of the electrophile. [1]
    (iii) Use curly arrows to show the mechanism for the nitration of benzene. [4]
                        [Total: 10]
                      (OCR 2814 Jan02)

2 The diagram below shows some reactions of phenol.

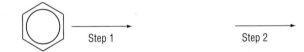

  (a) **(i)** Identify suitable reagents that could be used to carry out reactions I and II. [2]
    (ii) State a use for the compound formed in reaction III. [1]
    (iii) Outline how you could carry out reaction III in the laboratory starting from phenol and a suitable aromatic amine. [5]
  (b) In this question, one mark is available for the quality of spelling, punctuation and grammar.
    Phenol reacts much more readily with bromine than benzene does.

Describe, with the aid of a diagram, the bonding in benzene.
Explain why electrophiles, such as bromine, react much more readily with phenol than with benzene. [8]
                        [Total: 16]
                      (OCR 2814 Jun07)

3 Bromine will react with benzene in the presence of a catalyst.
  (a) **(i)** Give the name or formula of a suitable catalyst for this reaction. [1]
    (ii) State the general name of this type of catalyst. [1]
    (iii) Write a balanced equation for this reaction. [2]
    (iv) State the name of the organic product. [1]
  (b) The reaction between bromine and benzene is electrophilic substitution. Complete the scheme below to show the likely mechanism for this reaction. Assume that the electrophile is Br^+ (you do not need to show the action of the catalyst.) Show clearly all the curly arrows as well as the structures of the intermediate and the products.

    Intermediate
                        [4]

  (c) In this question, two marks are available for the quality of written communication.
    Bromine reacts much more readily with cyclohexene than it does with benzene. The reaction with cyclohexene does not need a catalyst. The structure of cyclohexene is shown below.

Cyclohexene

    (i) Describe with the aid of diagrams the π-bonding in cyclohexene and benzene. [2]
    (ii) Use your answer to explain why bromine reacts much more readily with cyclohexene than it does with benzene. [7]
                        [Total: 18]
                      (OCR 2814 Jun04)

4 Like esters, carbonyl compounds can contribute to the smell of plants and food. The carbonyl compounds **D** and **E** are structural isomers.
  D CH_3COCH_3 **E** CH_3CH_2CHO
  (a) Name compounds **D** and **E**. [2]

(b) State the reagents you would use and the observations you would make for a simple chemical test

  (i) in which **D** and **E** behave in the same way. [2]

  (ii) which can be used to distinguish between **D** and **E**. [3]

  [Total: 7]

  (OCR 2814 Jan06)

5 Ethanal, CH_3CHO, can be reduced using an aqueous solution of sodium borohydride, $NaBH_4$, as the reducing agent.

  (a) Write a balanced equation for this reaction using the symbol [H] to represent the reducing agent. [2]

  (b) This is a nucleophilic addition reaction in which the nucleophile can be represented as a hydride ion, H^-. A mechanism for the reaction is shown below.

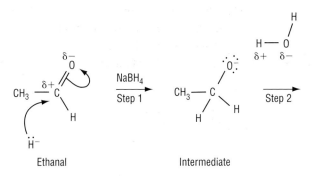

Ethanal   Intermediate

  (i) Add 'curly arrows' to the mechanism to show how the intermediate reacts with the water molecule in step 2. [2]

  (ii) Draw the structure of the organic product of this reaction. [1]

  (iii) What is meant by the term nucleophile? [1]

  (iv) Describe in words exactly what is happening to the electron pairs and bonds in step 1 of the mechanism above. [3]

  (c) It is also possible to reduce ethanal to the same product using hydrogen gas, H_2, in the presence of a catalyst. This reaction does not go by a nucleophilic mechanism. Explain why hydrogen gas cannot act as a nucleophile. [1]

  [Total: 10]

  (OCR 2814 Jan04)

6 But-2-enal, $CH_3CH=CHCHO$, is a pale yellow, flammable liquid with an irritating odour.

  (a) But-2-enal exists as two stereoisomers. Draw skeletal formulae to show the structures of the two stereoisomers of but-2-enal. [2]

  (b) (i) Describe a simple chemical test that would show that but-2-enal is an aldehyde. [2]

  (ii) Explain why this test gives a different result with aldehydes than it does with ketones. [2]

  (c) But-2-enal also reacts with sodium borohydride, $NaBH_4$.

  (i) Identify the organic compound formed in this reaction. [1]

  (ii) State the type of chemical reaction occurring. [1]

  (d) Precautions must be taken to prevent but-2-enal, C_4H_6O, from catching fire.

  Construct a balanced equation for the complete combustion of but-2-enal, C_4H_6O. [1]

  [Total: 9]

  (OCR 2814 Jun07)

7 An unknown colourless liquid with molecular formula C_4H_8O was thought to be one of butanal, but-3-en-1-ol, or butanone.

Butanal

But-3-en-1-ol

Butanone

  (a) State a simple chemical test that would positively identify:

  (i) butanal only

  reagent

  observation

  organic product

  (ii) but-3-en-1-ol only [3]

  reagent

  observation

  type of reaction [3]

  (b) Butanal and butanone both react with 2,4-dinitrophenylhydrazine to produce mixtures containing orange precipitates.

  Outline how the mixtures containing these orange precipitates can be used to distinguish between butanal and butanone. [3]

  [Total: 9]

  (OCR 2814 Jan07)

8 Compounds with the formula C_4H_9OH are alcohols.

   (a) Draw formulae to show the four structural isomers of alcohols with the molecular formula $C_4H_{10}O$.   **[4]**

   (b) One of the isomers in **(a)**, compound **D**, reacts with $K_2Cr_2O_7$ in the presence of H_2SO_4, to give **E**. When **E** is heated with ethanol in the presence of concentrated H_2SO_4, compound **F** is formed.

      (i) In this question, one mark is available for the quality of written communication.

      State the reaction, if any, of **each** of your alcohols in (a) with acidified $K_2Cr_2O_7$. Use this information and the reactions above to identify **D** and **E**. Give your reasoning.   **[5]**

     (ii) Write the equation for the formation of **F** from **E**.   **[1]**

   (c) Compound **F** and compound **G** (shown below) are both esters. Draw the structure of the product of the reaction of **G** with hot, aqueous NaOH.

                     [2]

[Total: 12]

(OCR 2814 Jan03)

9 (a) Esters are well known as compounds providing the flavour in many fruits and the scents of some flowers. The ester $CH_3(CH_2)_2COOCH_3$ contributes to the aroma of apples.

      (i) Name the ester $CH_3(CH_2)_2COOCH_3$.   **[1]**

     (ii) State the reagents and conditions for the hydrolysis of this ester in the laboratory.   **[3]**

   (b) Leaf alcohol, **B**, is a stereoisomer that can form when insects eat leaves.

      (i) Draw the skeletal formula of **B**.   **[1]**

     (ii) Draw the other *E/Z* isomer of **B**.   **[1]**

   (iii) Draw a structure for the ester expected when **B** reacts with ethanoic acid in the presence of an acid catalyst. Show **all** the bonds in the ester group.   **[2]**

[Total: 8]

(OCR 2814 Jan02)

10 An ester, **D**, is used as a solvent for paints and varnishes.

   (a) Ester **D** can be manufactured by heating an alcohol under reflux with ethanoic acid and a catalyst.

      (i) State a suitable catalyst for this reaction.   **[1]**

     (ii) Explain why the reaction is carried out under reflux.   **[1]**

   (b) Ester **D** has the structural formula $CH_3COOCH(CH_3)CH_2CH_3$.

      (i) Draw the displayed formula of ester **D**.   **[2]**

     (ii) State the name of the alcohol used to make ester **D**.   **[1]**

   (c) Apart from being a good solvent, suggest another use for ester **D**.   **[1]**

[Total: 6]

(OCR 2814 Jun04)

11 Amines are commonly occurring compounds. Ethylamine, $C_2H_5NH_2$, is a primary amine responsible for the smell of decaying fish.

   (a) Explain the meaning of the term *primary amine*.   **[1]**

   (b) Ethylamine and phenylamine are bases. Write an equation to show ethylamine acting as a base.   **[2]**

   (c) Aromatic amines such as phenylamine are intermediates in the synthesis of many other compounds such as **A** below.

      (i) Complete the scheme by writing the reagents for stages I, II and III.

                     [5]

     (ii) Write the equations for stages I and III.   **[2]**

   (iii) State a general use for compounds such as **A**.   **[1]**

[Total: 11]

(OCR 2814 Jan02)

12 E110 is a yellow colouring agent that is commonly added to a variety of foods. E110 contains an azo dye made from an amine and a phenol.

(a) Describe how you would prepare a sample of an azo dye in the laboratory from an amine, a phenol and any other necessary reagents. Include in your answer essential reagents and conditions for each stage and name any functional groups formed during the process.
[7]

(b) The structure of E110 is shown below.

(i) On the structure above, draw a circle around the functional group that identifies this molecule as an azo dye.
[1]

(ii) Deduce how many carbon and hydrogen atoms are in a molecule of E110.
[2]

(c) The solubility of E110 in water can be improved by converting the phenolic –OH group into a charged –O⁻ group. Suggest a suitable reagent that will convert the –OH group in E110 into an –O⁻ group.
[1]

(d) Draw the structures of a phenol and an amine that could be used to make E110 by the method in part (a). Assume that the $SO_3^-Na^+$ groups do not change during the process.
[2]

[Total: 13]
(OCR 2814 Jan05)

13 1,4-Diaminobenzene is used in the manufacture of a variety of materials including dyes and polymers.

1,4-diaminobenzene

(a) Explain what is meant by the term 1,4-diamino in the name of this compound.
[2]

(b) 1,4-Diaminobenzene can be manufactured from 1,4-dinitrobenzene.

1,4-dinitrobenzene   1,4-diaminobenzene

(i) What type of reaction is this?
[1]

(ii) State reagents and conditions that could be used to carry out this reaction.
[2]

(iii) Complete and balance the equation below for this reaction.

[2]

(c) 1,4-Diaminobenzene is used to make permanent black dye for hair. 1,4-Diaminobenzene can irritate the skin because it is basic. Therefore, it is sometimes neutralised with excess hydrochloric acid to give the salt.

(i) Explain how the amino acid groups in a primary amine such as 1,4-diaminobenzene allow the molecule to act as a base.
[2]

(ii) Draw the structure of the salt formed in this reaction.
[2]

[Total: 11]
(OCR 2814 Jun05

14 Scientists have recently investigated an ester called ethyl oleate, the structure of which is shown below. This ester may affect the release of neurotransmitters in the brain. It is possible that the body could make this ester from the ethanol present in alcoholic drinks and the natural fatty acid oleic acid.

(a) Write the structural formula of oleic acid.
[1]

(b) (i) Write the molecular formula of oleic acid, ethanol and ethyl oleate.

(ii) Construct a balanced equation for the formation of ethyl oleate from ethanol and oleic acid.
[4]

(c) Suggest how oleic acid could be obtained from the glyceryl ester shown below.

[1]

[Total: 6]
(OCR 4821 Jun99)

Module 2
Polymers and synthesis

Introduction

If your passion is jumping out of a plane, smashing a shuttlecock, riding a bike, skiing down a piste or dressing in the latest fashion then nylon will play an important part in your life. Nylon, one of a number of synthetic polymers developed by research chemists, is used in clothing fabrics, for making parachutes, in ski wear and in the strings used in badminton racquets.

Other polymers occur naturally. Proteins, for example, play a key role in many of the biological processes that keep us healthy. Some proteins act as biological catalysts while others protect us from disease or allow our muscles to move.

Organic chemists work at the cutting edge of new developments and spend many hours trying to create new materials by modifying the structure of existing molecules. This branch of chemistry is known as organic synthesis.

In this module, you will study the production and use of condensation polymers and learn about some of the techniques that chemists use to synthesise new materials.

Test yourself

1 What functional groups are found in amino acids?
2 Define the term *polymer*.
3 Define *biodegradable*.
4 State a use for polyesters.

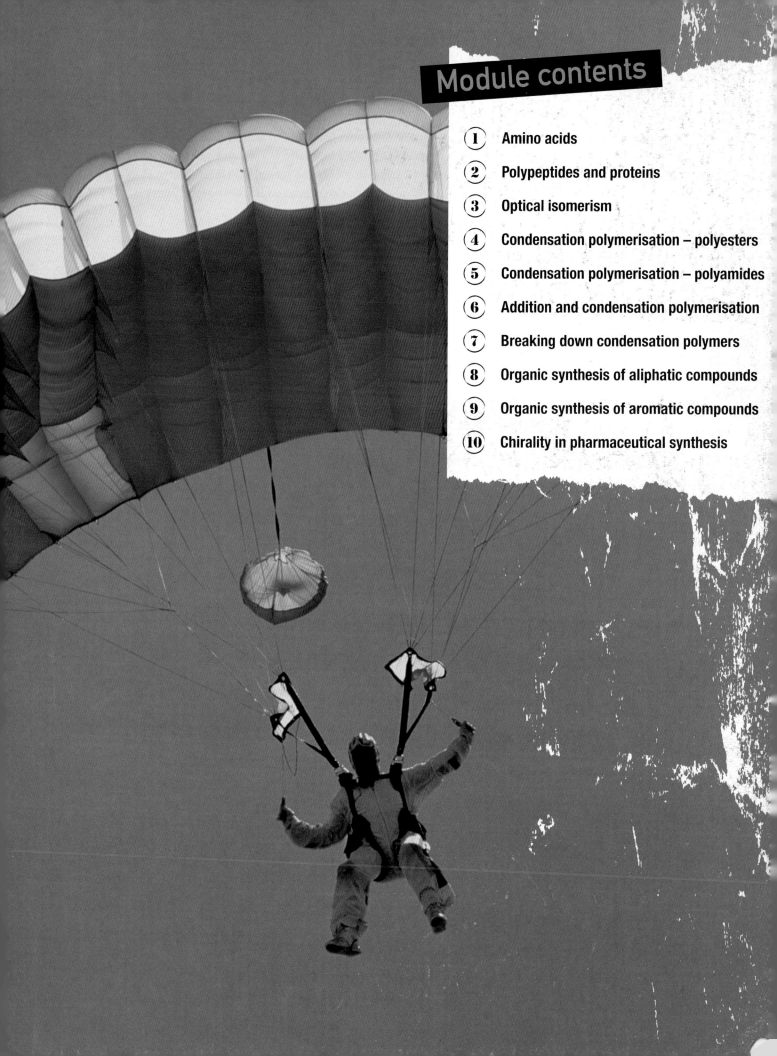

Module contents

By the end of this spread, you should be able to ...

* **State the general formula for an α-amino acid as RCH(NH₂)COOH.**
* **State that an amino acid exists as a zwitterion at a pH value called the isoelectric point.**
* **State that different R groups in α-amino acids may result in different isoelectric points.**
* **Describe the acid–base properties of α-amino acids at different pH values.**

Amino acids

Amino acids are the building blocks for two important groups of biological molecules – **peptides** and proteins. Proteins act as enzymes, hormones and antibodies. Proteins (polypeptides) transport substances such as oxygen, vitamins and minerals to cells throughout the body. Structural proteins, such as collagen and keratin, are responsible for the formation of bones, teeth, hair and the outer layer of skin, and they help to maintain the structure of blood vessels and other tissues.

Figure 1 Coloured scanning electron micrograph of hair shafts growing from the surface of human skin. The shafts of hair (red) are anchored in individual hair follicles (not seen) below the surface of the skin. Hair is made up of a fibrous protein called keratin

> **Key definition**
>
> A **peptide** is a compound made of amino acids linked by peptide bonds (see spread 1.2.2).

The body has 20 different amino acids that can be assembled into proteins. These amino acids are α-amino acids having both a basic amine group and an acidic carboxyl group connected to the same carbon atom.

Amino acids are soluble in both acids and bases. The general formula of an α-amino acid is shown in Figure 2 – this can be written as RCH(NH₂)COOH.

The R– group is usually a carbon-containing group that may also include an –OH, –SH, –COOH or –NH₂ group. The exception is glycine, the simplest amino acid, which has H– as the R– group.

Examples of some α-amino acids are shown in Figure 3. The different R– groups are coloured in red.

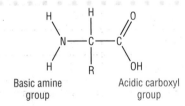

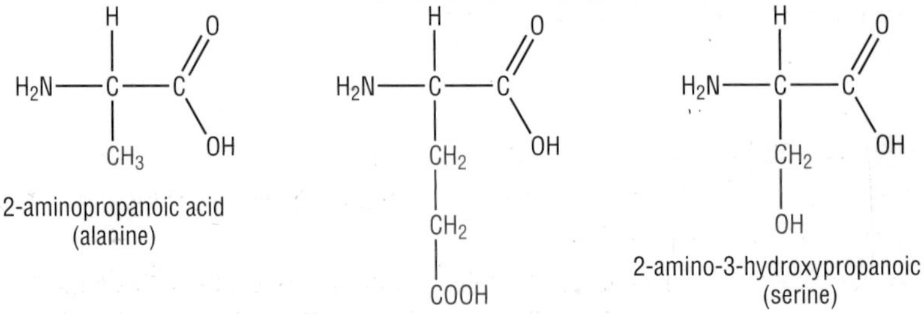

Figure 2 General formula of an α-amino acid

Figure 3 Structures of some amino acids

2-aminopropanoic acid
(alanine)

2-aminopentanedioic acid
(glutamic acid)

2-amino-3-hydroxypropanoic acid
(serine)

Zwitterions and the isoelectric point

The acidic carboxyl group and the basic amine group in an amino acid can interact with each other to form an internal salt, known as a **zwitterion**.

- A proton is transferred from the acid carboxyl group to the basic amine group.
- A zwitterion has no overall charge because the positive and negative charges cancel each other out.

The zwitterion formed from glycine is shown in Figure 4.

The **isoelectric point** is the pH at which there is no net electrical charge. The zwitterion of an amino acid exists at this pH value. You might expect that all amino acids have their isoelectric points at the neutral pH of 7. However, each amino acid has its own special isoelectric point. Many amino acids have their isoelectric point close to pH 6, but some isoelectric points have much higher or lower pH values.

The isoelectric points of some common amino acids are shown in Table 1.

Acid and base properties of amino acids

Amino acids are amphoteric, which means that they can react with both acids and bases.

At a pH that is more acidic than the isoelectric point:

- the amino acid behaves as a base and accepts a proton from the acid
- the amino acid forms a positively charged ion.

At a pH that is more alkaline than the isoelectric point:

- the amino acid behaves as an acid and donates a proton to the base
- the amino acid forms a negatively charged ion.

Figure 5 shows how an amino acid reacts in solutions with pH values that are well above and well below its isoelectric point.

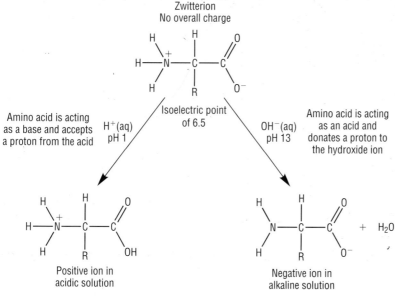

Figure 5 Acid–base reactions of an amino acid

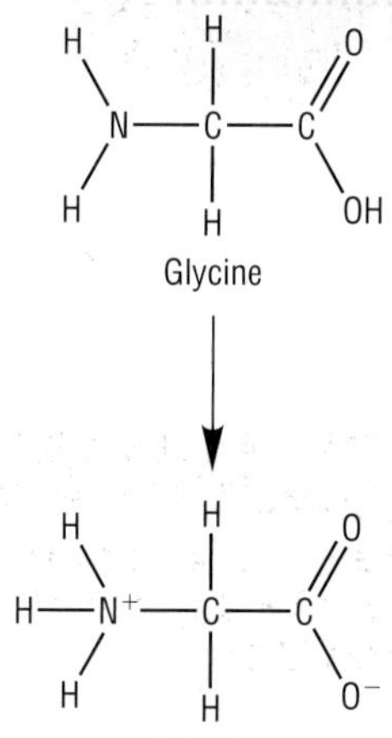

Figure 4 Formation of a zwitterion from glycine

Key definitions

A **zwitterion** is a dipolar ionic form of an amino acid that is formed by the donation of a hydrogen ion from the carboxyl group to the amino group. Because both charges are present, there is no overall charge.

The **isoelectric point** is the pH value at which the amino acid exists as a zwitterion.

Amino acid	Isoelectric point
Glycine	5.97
Alanine	6.01
Leucine	5.98
Serine	5.68
Proline	6.48

Table 1 Isoelectric points of some amino acids

Examiner tip

Questions based on the acid–base nature of the amino acids are quite common in exams. You must read the questions carefully – you may be asked for an equation in which the amino acid is reacting as an acid; in others you may be asked for an equation of an amino acid reacting with a base. The equation is the same for both questions!

Questions

1 Define the terms **(a)** isoelectric point; **(b)** zwitterion.

2 **(a)** Draw the structure of the zwitterion of 2-aminopropanoic acid (alamine) that is formed at its isoelectric point.

(b) Write an equation for the reaction between this zwitterion and NaOH(aq).

3 Glutamic acid has the formula $HOOCCH_2CH_2CH(NH_2)COOH$ with an isoelectric point of pH 3.22. Write the formula of the ionic species of glutamic acid present at:

(a) pH 3.22; **(b)** pH 1; **(c)** pH 13.

By the end of this spread, you should be able to ...

* Explain the formation of a peptide (amide) linkage between α-amino acids to form polypeptides and proteins.
* Describe the acid and alkaline hydrolysis of proteins and peptides to form α-amino acids or carboxylates.

Amino acids and condensation reactions

Amino acids join together by peptide linkages to form **peptides**. When two amino acids join together, a dipeptide is formed with the elimination of a water molecule – this is a **condensation reaction**. The 'peptide linkage' or 'peptide bond' is the name used for the amide bond in proteins and polypeptides. In Figure 1, the amino acids glycine and alanine are joined together to form a dipeptide.

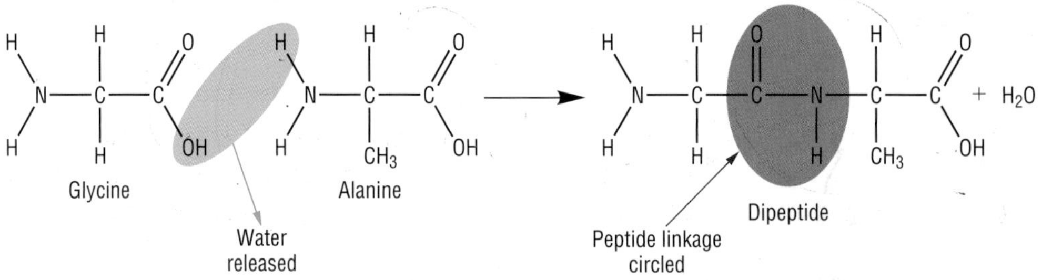

Key definitions

A **peptide** is a compound containing amino acids linked by peptide bonds. The number of amino acids is often indicated by the prefix di-, tri- or tetra-:

dipeptide = 2 amino acids
tripeptide = 3 amino acids
tetrapeptide = 4 amino acids

A **condensation reaction** is one in which two small molecules react together to form a larger molecule with the elimination of a small molecule such as water.

Figure 1 Formation of a dipeptide between glycine and alanine

It is possible to form a different dipeptide from the same two amino acids. For the example shown in Figure 1, instead of the carboxyl group of glycine reacting with the amino group of alanine, the carboxyl group of alanine could react with the amino group of glycine (Figure 2). The dipeptide formed is a structural isomer of that shown in Figure 1.

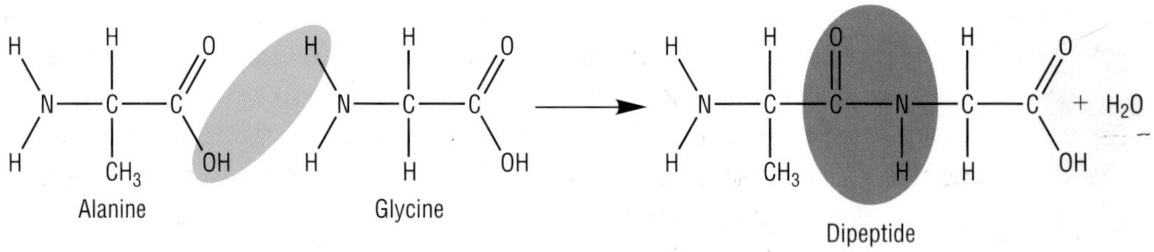

Figure 2 Alternative reaction of alanine and glycine

Forming polypeptides and proteins

A protein or polypeptide is a long chain of amino acids joined together by peptide linkages. The chain is made by condensation reactions of amino acids (see Figure 1 for the formation of a dipeptide by condensation). For each additional amino acid added to the chain, a molecule of water is formed. A polypeptide contains many linked amino acids. Proteins are simply long polypeptides, generally with more than 50 amino acids. A section of a polypeptide is shown in Figure 3.

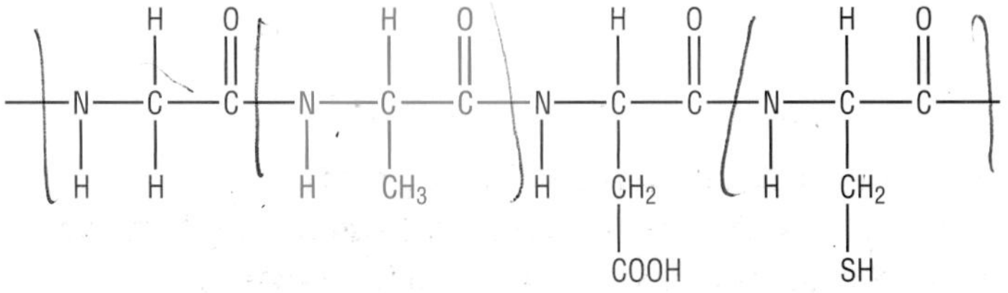

Examiner tip

When you are asked to draw a section of a peptide, remember that one end of the structure contains a C=O group and the other end an N–H group. Polypeptides are drawn with their ends open.

Figure 3 Section of a polypeptide showing four different amino acids

Hydrolysis of polypeptides and proteins

Acid hydrolysis

Polypeptides and proteins can be hydrolysed using aqueous acid. A water molecule is needed to break each peptide link to form a mixture of the amino acids that made up the protein. During acid **hydrolysis**, the amino acids formed are positively charged because of the presence of H^+ ions from the acid. Traditionally the protein or polypeptide is heated under reflux with 6 mol dm^{-3} HCl for 24 hours.

In Figure 4, a dipeptide is hydrolysed to show the general principle of acid hydrolysis. This reaction can be applied to polypeptides.

Figure 4 Acid hydrolysis of a dipeptide

Alkaline hydrolysis

A solution of alkali, in the form of aqueous sodium hydroxide, is used at just above 100 °C. The polypeptide or protein is broken down into amino acids in the form of their sodium salts. Figure 5 shows a section of a protein chain undergoing alkaline hydrolysis.

Figure 5 Alkaline hydrolysis of a polypeptide chain

Module 2
Polymers and synthesis
Polypeptides and proteins

Figure 6 Meat can be made more tender by soaking it in an acid solution such as vinegar or lemon juice before cooking – the acid starts to hydrolyse the proteins in the muscle

Questions

1. Glycine is aminoethanoic acid.
   (a) Write the equation for the formation of the dipeptide formed from two molecules of glycine. Show the peptide linkage clearly.
   (b) Name the two functional groups that come together to form the peptide linkage.
2. The structures of two amino acids are shown in Figure 7. Draw the displayed formula of the two dipeptides that could be formed when the amino acids link together.

Figure 7 Structures of two amino acids

3. Figure 8 shows part of a polypeptide chain.

— NHCH$_2$CONHCHCONHCHCHCONHCHCHCONHCHCHCO —

CH$_2$OH CH$_2$ CH$_2$SH (CH$_2$)$_4$

C$_6$H$_5$ $^+$NH$_3$

Figure 8 Structure of a polypeptide

   (a) How many amino acids are linked in this fragment?
   (b) The polypeptide could be split up into small fragments by hydrolysis. Give the reagents and conditions needed to carry out this process.
   (c) Give the structure of the right-hand amino acid after hydrolysis at pH 1.

③ Optical isomerism

By the end of this spread, you should be able to ...

* Describe *optical isomers* as non-superimposable mirror images about an organic chiral centre.
* Identify chiral centres in a molecule of given structural formula.
* Explain that optical isomerism and *E/Z* isomerism are types of stereoisomerism.

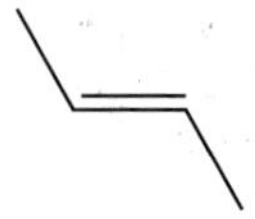

E-but-2-ene

Z-but-2-ene

Figure 1 *E* and *Z* isomers of but-2-ene

Key definitions

Stereoisomers are species with the same structural formula but with a different arrangement of the atoms in space.

A **chiral carbon** is a carbon atom attached to four different atoms or groups of atoms.

Optical isomers (or **enantiomers**) are stereoisomers that are non-superimposable mirror images of each other.

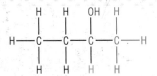

Figure 2 The butan-2-ol molecule has a chiral carbon, because the red carbon atom is bonded to four different groups. It is joined to –H, –CH$_3$, –OH and –C$_2$H$_5$

Stereoisomerism

In **stereoisomerism**, the atoms making up the isomers are joined up in the same order but have different arrangements in space. You met stereoisomerism in *AS Chemistry* spread 2.1.7. Alkenes can show *E/Z* isomerism provided that each of the carbon atoms in the double bond is attached to two different groups.

There is a second type of stereoisomerism, known as optical isomerism – so called because the isomers have different effects on plane-polarised light.

Optical isomerism

* Optical isomerism arises in organic molecules that contain a carbon atom attached to four different atoms or groups of atoms.
* A carbon atom with four different groups attached is said to be **chiral**.
* If a molecule has a chiral centre in its structure, then two mirror-image arrangements are possible in space. Neither arrangement can be placed over the other, just as a right hand cannot be placed over a left hand. These arrangements are *non-superimposable* mirror images and are called **optical isomers** (or **enantiomers**).
* Optical isomers are chemically identical. However, optical isomers rotate plane-polarised light in different directions.
* There are only ever two optical isomers formed for each chiral carbon. If there are two chiral carbons, then there will be two pairs of optical isomers.

Identifying chiral carbon atoms

You must be able to identify chiral carbon atoms in organic molecules. Many α-amino acids (all except glycine, H$_2$NCH$_2$COOH) have optical isomers with the chiral carbon attached to a carboxyl group, an amine group, an R– group and a hydrogen atom.

Worked example

Which of the following molecules have chiral centres?
(a) CH$_3$CH$_2$CH$_2$CH$_2$CH$_2$OH;
(b) CH$_3$CH$_2$CH(NH$_2$)CH$_3$;
(c) CH$_3$CH(OH)Cl

The easiest way to decide if the molecules have optical isomers is to draw their displayed formulae.

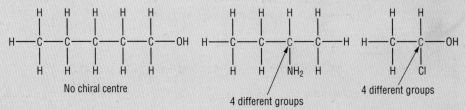

No chiral centre   4 different groups   4 different groups

Figure 3

Both **(b)** and **(c)** have optical isomers.

Representing optical isomers

It is important to be able to draw a pair of optical isomers showing the three-dimensional arrangement around the central chiral carbon atom. Once one isomer has been drawn, the other isomer can be drawn as its mirror image. A line is often drawn between the two isomers to show that they are mirror images of each other. Figure 4 shows the optical isomers of $CH_3CH_2CH(NH_2)CH_3$.

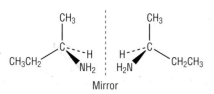

Figure 4 Optical isomers of $CH_3CH_2CH(NH_2)CH_3$

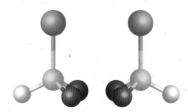

Figure 5 Simplified three-dimensional representation of two optical isomers – note that they are mirror images of each other

Properties of optical isomers

Optical isomers rotate plane-polarised light differently. One of the optical isomers rotates the light clockwise and the other rotates the light anti-clockwise. A mixture containing equal amounts of each isomer is known as a *racemic* mixture. A racemic mixture has no effect on plane-polarised light because the rotations cancel each other.

Optical activity in amino acids

Naturally occurring α-amino acids are optically active – the one exception is glycine, which has two hydrogen atoms on the central carbon atom. Many sugars also display optical activity. Optical activity is important in biological systems where frequently only one of the optical isomers is biologically active. In plants and animals, only one of the optical isomers is synthesised naturally, and only one will interact with an enzyme due to the stereospecific nature of enzymes.

In Lewis Carroll's *Alice Through The Looking Glass*, Alice innocently says, 'Perhaps looking-glass milk isn't good to drink.' Looking-glass milk probably would not be good to drink – especially if it contained the mirror image of a molecule that was good to the taste!

Questions

1 2-aminopropanoic acid (alanine), $CH_3CH(NH_2)COOH$, has a chiral centre and so can exist as two optical isomers.
   (a) State what is meant by a *chiral centre*.
   (b) Draw three-dimensional diagrams to show the two optical isomers of 2-aminopropanoic acid.
   (c) State the bond angle around the chiral carbon.
2 Which of the following molecules can form optical isomers?
   A $CH_3CH_2CH_2CHClCH_3$
   B $CH_3CH_2CH(OH)CH_2CH_3$
   C $CH_3CH_2CH_2C(OH)_2CH_2CH_2Cl$
3 How many pairs of optical isomers does the molecule in Figure 7 have?

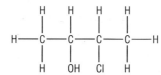

Figure 7 3-chlorobutan-2-ol

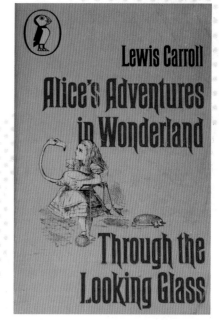

Figure 6 Did Lewis Carroll know about optical activity when he wrote his children's classic?

By the end of this spread, you should be able to . . .

✱ Describe condensation polymerisation forming polyesters, such as Terylene and poly(lactic acid).
✱ State the use of polyesters as fibres in clothing.

Introduction

Condensation polymerisation is the joining of two monomers with the elimination of a small molecule, such as water or HCl.

- Condensation polymerisation requires monomers with two different functional groups.
- A functional group on one monomer unit bonds to a different functional group on another monomer unit.

Polyesters and polyamides are common condensation polymers. These are discussed over the next two spreads.

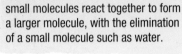

Key definition

Condensation is a reaction in which two small molecules react together to form a larger molecule, with the elimination of a small molecule such as water.

Polyesters

You have learnt that an ester can be made from a carboxylic acid and an alcohol (see spread 1.1.14). In forming the polyester, monomer units are bonded together by ester linkages and water is eliminated.

For a polyester, the monomers have a carboxyl group, –COOH, and a hydroxyl group, –OH. Examples of polyesters include Terylene and poly(lactic acid).

There are two common types of polyester:

1 Polyesters made by reacting two different types of monomer units:
  - one monomer is a dicarboxylic acid with two –COOH groups
  - the other monomer is a diol with two –OH groups.
2 Polyesters made by reacting just one type of monomer unit containing –COOH and –OH groups.

Both types are shown in Figure 2. A and B represent the carbon skeletons separating the functional groups in the structures.

Figure 1 Polyester is used to make shirts and ties and is blended with cotton in many other items of clothing

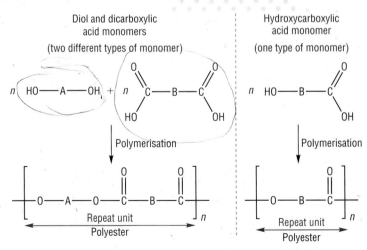

Figure 2 Formation of polyesters from two types of monomer (left) and from one type of monomer (right)

Polyesters from two types of monomer units

Terylene is made from the monomers ethane-1,2-diol and benzene-1,4-dicarboxylic acid. Figure 3 shows the formation of Terylene from its monomers, with the ester group coloured in red.

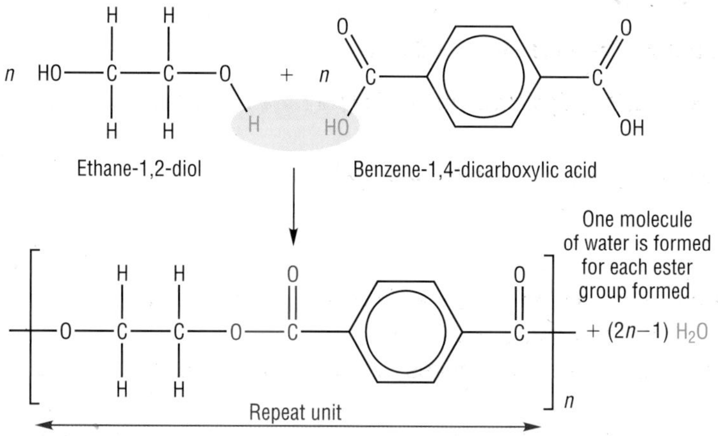

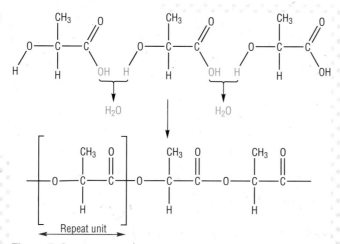

Figure 3 Formation of Terylene, a polyester

During condensation polymerisation:

- a bond forms between a hydroxyl group on the diol and a carboxyl group on a dicarboxylic acid
- an ester linkage forms, and water is eliminated.

Polyesters are used in making carpets, sports clothing and shirts. In clothing and bedding, polyesters are often blended with cotton. Almost all polyesters can be machine-washed and machine-dried. Polyester is a strong fibre which is resistant to stretching, shrinking and chemical attack. Unfortunately it burns easily and should not be exposed to naked flames.

Polyesters from one type of monomer unit

Polyesters can also be formed from hydroxycarboxylic acids, which contain both –COOH and –OH groups.

Poly(lactic acid) (PLA) is made from the monomer lactic acid (2-hydroxypropanoic acid). Figure 5 shows the formation of a section of PLA from its monomer, lactic acid.

During condensation polymerisation:

- a bond forms between a hydroxyl group on one lactic acid molecule and a carboxyl group on a different lactic acid molecule
- an ester linkage forms, and water is eliminated.

Poly(lactic acid) is biodegradable and there has been much recent interest in its use for food and drink cartons. See spread 1.2.7 for more details.

Figure 5 Condensation polymerisation of lactic acid

Key definition

The **repeat unit** is the specific arrangement of atoms that occurs in the structure over and over again. Repeat units are included in brackets, outside which is the letter *n*.

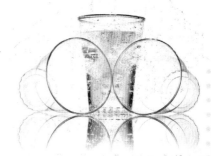

Figure 4 These clear cups for cold drinks are manufactured from a biopolymer, poly(lactic acid), derived from maize

Questions

1 What two functional groups must be present in monomers in order to form a polyester?

2 Draw the repeat unit of the polymer that can be made from ethanedioic acid and butane-1,4-diol.

3 Name two polyesters.

By the end of this spread, you should be able to . . .

* Describe condensation polymerisation to form the polyamides nylon-6,6 and Kevlar.
* State the use of polyamides as fibres in clothing.

Figure 1 The amide bond is present in polyamides and joins the monomer units together

Polyamides

You have learnt that an ester can be made from a carboxylic acid and an alcohol (see spread 1.1.14). In forming the polyester, monomer units are bonded together by ester linkages and water is eliminated (see spread 1.2.4).

As with polyesters, polyamides are condensation polymers that can be made from either two different types of monomer units or just one type of monomer unit. Amide linkages join the monomer units together to make a polyamide, with the elimination of water.

In a polyamide, the monomers have a carboxyl group, –COOH, and an amine group, –NH_2. Examples of polyamides include nylon-6,6, Kevlar and proteins.

There are two types of polyamide:

1 Polyamides made by reacting together two different types of monomer units:
   • one monomer is a dicarboxylic acid with two –COOH groups
   • the other monomer is a diamine with two –NH_2 groups.
2 Polyamides made by reacting just one type of monomer unit containing both –COOH and –NH_2 groups. Amino acids (see spread 1.2.1) are this type of monomer, and polypeptides and proteins are this type of polyamide (see spread 1.2.2).

Both types are shown in Figure 2. A and B represent the carbon skeletons separating the functional groups in the structures.

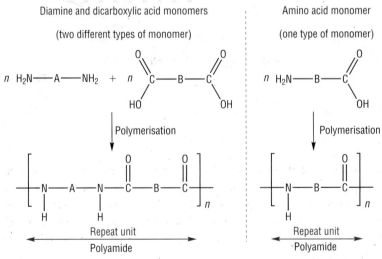

Figure 2 Formation of polyamides from two types of monomer (left) and from one type of monomer (right)

Figure 3 Balloon fabric is made from nylon

Polyamides from two types of monomer units

Polyamides can be formed by reacting together two different types of monomer units:
• one monomer is a dicarboxylic acid with two –COOH groups
• the other monomer is a diamine with two –NH_2 groups.

Nylon-6,6 is made from the monomers 1,6-diaminohexane and hexane-1,4-dioic acid Figure 4 shows the formation of nylon-6,6 from its monomers, with the amide group coloured in blue.

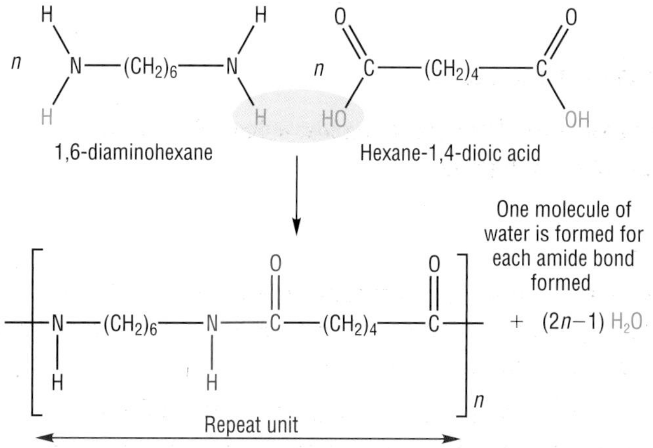

One molecule of water is formed for each amide bond formed

$+ (2n-1) H_2O$

Figure 4 Formation of the polyamide nylon-6,6

During condensation polymerisation:
- a bond forms between an amine group on a diamine and a carboxyl group on a dicarboxylic acid
- an amide linkage forms and water is eliminated.

Polyamides such as nylon-6,6 and Kevlar are used widely in clothing. Kevlar has some remarkable properties, including fire resistance and a higher strength than steel. It is used to make protective clothing – for example for firefighters, bulletproof vests and crash helmets. Kevlar is formed from the polymerisation of benzene-1,4-diamine and benzene-1,4-dioic acid. Figure 6 shows the formation of Kevlar from its monomers.

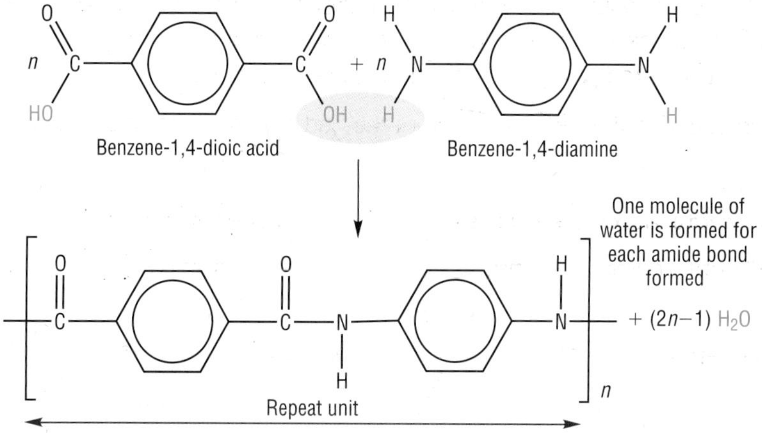

Figure 5 Formula One racing drivers use helmets made of carbon fibre with a layer of Kevlar, which provides greater protection than a traditional helmet for a fraction of the weight

One molecule of water is formed for each amide bond formed

$+ (2n-1) H_2O$

Figure 6 Formation of the polyamide Kevlar

Body armour

Kevlar can be spun into knife-resistant fabric sheets five times stronger than steel, used in the body armour worn by the police. It protects the wearer from shots by spreading the force of the bullet over a larger area of the body. The individual layers of fabric catch the bullet and deform, preventing the bullet from reaching the skin. In addition, a layer of resin-hardened glass fibre between the Kevlar layers prevents knife blades from entering the jacket.

Figure 7 Kevlar in action as a bullet-proof vest

Questions

1 What two functional groups must be present in monomers in order to form a polyamide?
2 Draw the structural formula of the polymer that can be made from propanedioic acid and 1,4-diaminobutane.
3 Name two polyamides.

By the end of this spread, you should be able to . . .

* ✳ Compare condensation polymerisation with addition polymerisation.
* ✳ Suggest the type of polymerisation from a given monomer or pair of monomers, or a given section of a polymer molecule.
* ✳ Identify the monomer(s) required to form a given section of a polymer (and *vice versa*).

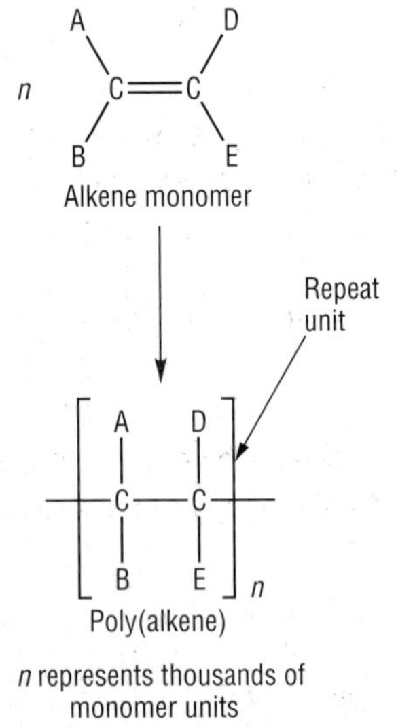

Addition polymers

Addition polymerisation was discussed in detail in *AS Chemistry* (spread 2.1.18). Alkenes undergo addition polymerisation to produce saturated chains containing no double bonds.

* Addition polymers are made from one type of monomer only.
* There is no product other than the polymer.

By using different alkene monomers, many different addition polymers can be formed, each with their own specific properties. A general equation for addition polymerisation is shown in Figure 1.

Condensation polymers

Condensation polymerisation is the joining of monomers with the elimination of a small molecule, such as water or HCl. The monomers must have two functional groups in their structure. The formation of polyesters and polyamides as condensation polymers is discussed in detail in spreads 1.2.4 and 1.2.5.

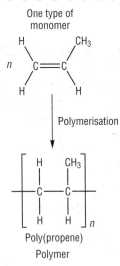

Figure 1 Addition polymerisation of an alkene

Figure 2 Propene monomers forming poly(propene)

Comparison of addition and condensation polymers

Feature	Addition polymer	Condensation polymer	
		Polyester	**Polyamide**
Functional group(s)	C=C	–COOH and –OH	–COOH and –NH$_2$
Type(s) of monomer	1	1 or 2	1 or 2
Product(s)	Poly(alkene)	Polyester + water	Polyamide + water

Identifying types of polymerisation

It should be possible to identify the type of polymerisation involved from either the monomer(s) or by examining a length of the polymer chain.

* If the monomer contains a double bond, then the polymerisation is addition.
* If there are two monomers, each with two functional groups, then the polymerisation is condensation.
* If there is one monomer with two different functional groups, then the polymerisation is condensation.
* If the polymer contains ester or amide linkages, then the polymerisation is condensation.
* If the backbone of the polymer is a continuous chain of carbon atoms, then the polymer is likely to have been made by addition and the monomer will be an alkene.

Worked example 1 – identifying the type of polymerisation from monomer(s)

What type of polymerisation is possible from the monomers shown?

(a) The molecule contains a carbon–carbon double bond and so would undergo addition polymerisation.

(b) There are two monomers – one containing two hydroxyl groups and the other two carboxyl groups. The monomers react together by condensation to produce a polyester.

(c) There is only one monomer but it does not contain a C=C bond. It has two functional groups and so could react with another monomer of the same type to give a polyester, a condensation polymer.

Worked example 3 – identifying the type of polymerisation from polymer chains

For each of the following polymer chains, suggest the type of polymerisation taking place and identify the monomer(s) used to make the polymer.

(a) can also be written as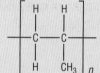

This is a poly(alkene): an addition polymer; replace C–C by C=C to give the monomer.

(b)

This is a polyamide formed by a condensation reaction, so break it at the C–N of the amide bond (in blue) and add H₂O to get the monomers.

dicarboxylic acid   diamine

Worked example 2 – polymer repeat units

Draw the repeat unit for each of the polymers in example 1.

(a) The poly(alkene) is formed by addition of monomers with C=C bonds to form a carbon chain.

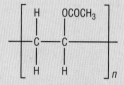

(b) A condensation reaction between hydroxyl and carboxyl groups forms a polyester with many ester linkages.

(c) As in **(b)**, a condensation reaction forms a polyester, but this time all monomers are the same.

Questions

1 Using equations, show the formation of two named addition polymers.
2 Given the following sections of polymer chains, suggest the type of polymer and identify the monomer(s) from which it is obtained.

(a)

(b)

(c)

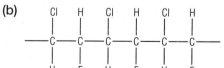
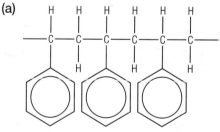

Figure 3 Structures of some polymers

1.2 7 Breaking down condensation polymers

By the end of this spread, you should be able to …

- **Describe the acid and base hydrolysis of polyesters and polyamides.**
- **Outline the role of chemists in the development of degradable polymers.**
- **Explain that condensation polymers may be photodegradable and may be hydrolysed.**

Hydrolysis of polyesters

Esters can be hydrolysed under acidic or basic conditions (see spread 1.1.14). Polyesters are readily hydrolysed by hot aqueous alkalis, such as aqueous sodium hydroxide. Each ester linkage is hydrolysed to:

- the sodium salt of a carboxylic acid, $-COO^-Na^+$
- a hydroxyl group, –OH.

Polyesters can also be hydrolysed with hot aqueous acid, such as aqueous hydrochloric acid, although the reaction is much slower than with an aqueous alkali:

- the monomer units of the polyester are produced.

Figure 1 shows the base and acid hydrolysis of Terylene.

Examiner tip

The conditions for hydrolysis are always aqueous acid or aqueous base in the presence of heat. *Aqueous* implies water and *hydrolysis* means the breaking of a bond with a water molecule. This means that water must be present!

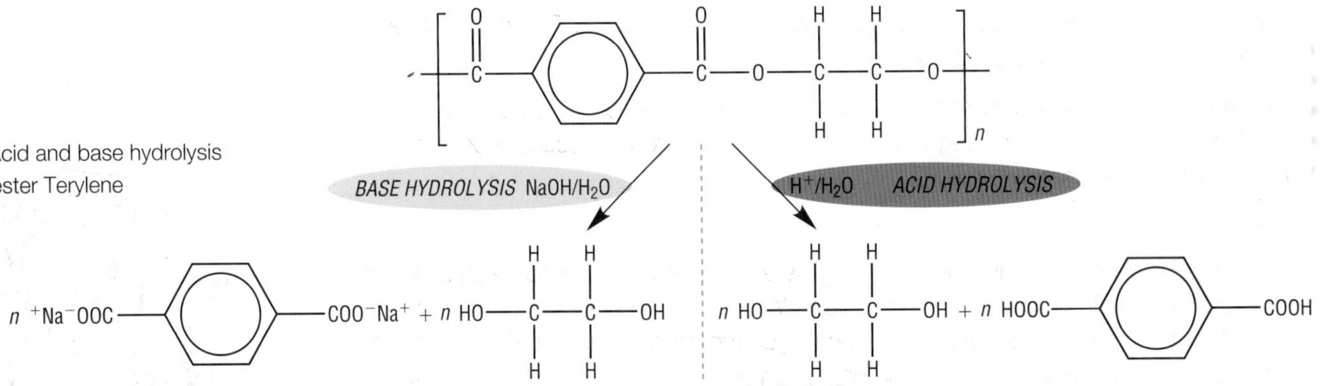

Figure 1 Acid and base hydrolysis of the polyester Terylene

Hydrolysis of polyamides

Like polyesters, polyamides can be hydrolysed by either hot aqueous acid, such as aqueous hydrochloric acid, or hot aqueous alkali, such as aqueous sodium hydroxide.

- In acid conditions, the dicarboxylic acid is produced together with an ammonium salt of the diamine.
- In basic conditions, the sodium salt of the dicarboxylic acid and the diamine are formed.

Figure 2 shows the base and acid hydrolysis of nylon-6,6.

You have to be careful in the laboratory when carrying out practical work – if you spill sodium hydroxide on clothing made out of polyester, your clothes are likely to dissolve!

Key definitions

A **biodegradable polymer** is a polymer that breaks down completely into carbon dioxide and water.

A **degradable polymer** is a polymer that breaks down into smaller fragments when exposed to light, heat or moisture.

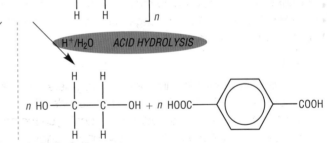

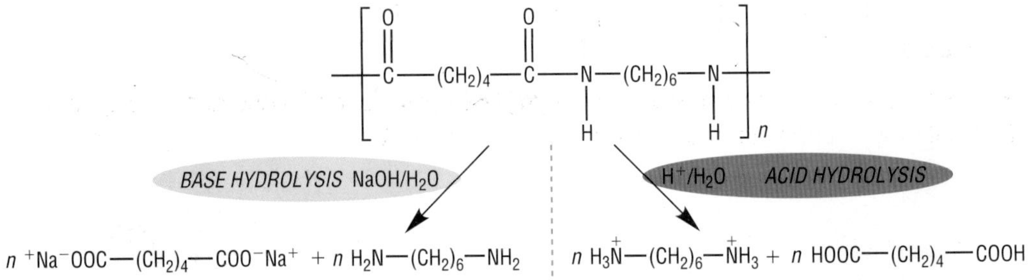

Figure 2 Acid and base hydrolysis of the polyamide nylon-6,6

Degradable polymers

Take a look inside an average shopping trolley. It is not only brimming with the week's necessities, but also overflowing with environmentally damaging polymers. Our fruit is packed into expanded polystyrene trays covered in shrink-wrap poly(vinyl chloride) films. Even when a product is supplied inside a box, manufacturers seem compelled to wrap the box in plastic. And, of course, after we have done the shopping we pack everything into plastic carrier bags. When our rubbish is collected a week later, the wrapping, packaging and carrier bags are taken off to a landfill site for burial.

Customer demand and environmental pressures, coupled with the spiralling cost of crude oil, have led scientists to investigate and develop renewable and sustainable polymers for packaging.

Biodegradable polymers often have chemical bonds that can undergo hydrolysis. Many are based on polyesters, for example:

- poly(lactic acid), made from lactic acid derived from corn starch
- poly(glycolic acid), made from glycolic acid isolated from sugar cane and unripe grapes.

These polyesters have been developed to replace traditional oil-based products. Poly(lactic acid) is by far the most widely used of these 'bioplastics' and is used for compost bags, food packaging and disposable tableware. Poly(glycolic acid) is used primarily in surgery for stitches.

Synthetic biodegradable polyesters also play a role in thermoformed trays for fresh produce and meat, as well as disposable plates, bowls and cups. These plastics are made from cellulose and potato starch.

Supermarkets are now doing their bit to reduce polymer waste. In April 2008, Sainsbury's launched carrier bags made from tapioca starch – these biodegrade in 28 days when composted.

Photodegradable polymers

Photodegradable plastics are synthetic polymers designed to become weak and brittle when exposed to sunlight for prolonged periods. They can be made by blending the polymer with light-sensitive additives that catalyse the breakdown of the polymer in the presence of UV radiation. Alternatively, photodegradable plastics can be manufactured by incorporating carbonyl bonds, C=O, within the backbone of the polymer. The carbonyl bonds in the polymer chain absorb light energy and break, fracturing the polymer chain. Photodegradable polymers are initially converted to waxy compounds when exposed to light, before being converted to carbon dioxide and water in the presence of bacteria.

Figure 3 According to government figures, between 10 and 15 billion carrier bags are used in the UK each year, with the average household working its way through 323 of them. The bags frequently end up in landfill sites or clog up rivers and harm wildlife

Figure 4 Compostable plastic bags are now in use in some British supermarkets

Questions

1 Draw the products of the base hydrolysis of the polymer shown in Figure 5.

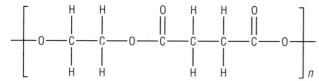

Figure 5

2 What is the difference between a biodegradable polymer and a degradable polymer?

3 What do you understand by the term *photodegradable*?

By the end of this spread, you should be able to ...

* Identify the functional groups in an aliphatic molecule containing several functional groups.
* Predict properties and reactions of aliphatic molecules containing several functional groups.
* Devise multi-stage synthetic routes for preparing aliphatic organic compounds.

What is organic synthesis?

Organic synthesis is a specialised branch of chemistry concerned with producing organic compounds by chemical reactions. Organic synthesis may be simple, requiring only one step, or it may require more complex, multi-step reactions involving many stages.

In industry, a multi-step synthesis of a complex organic molecule starts from a commercially available raw material. Crude oil is the raw material for many organic compounds, but other compounds derived from crude oil are also used as starting materials.

Organic synthesis is often presented as a puzzle. The solution to the puzzle lies in the chemistry you have studied during your course. For example, you might be asked to convert a compound A into a compound D, in a certain number of reaction steps.

Synthetic routes in aliphatic chemistry

The flowchart in Figure 2 links together the aliphatic functional groups studied in A2 chemistry. The flowchart can be used to solve simple synthesis questions. In exams, you will be expected to know the reactions from the A2 part of your course. You may also be supplied with other reactions, including some you will not have seen before. Using both familiar and unfamiliar reactions, you could be asked to suggest a multi-step synthesis of an organic compound from a supplied starting material. This is the puzzle of organic synthesis.

Figure 1 Robert Burns Woodward was regarded as the father of modern organic synthesis. In 1965 he received the Nobel Prize for Chemistry for his contributions to the chemical synthesis of organic substances, including quinine (1944), patulin (1950), cholesterol (1951) and cortisone (1951)

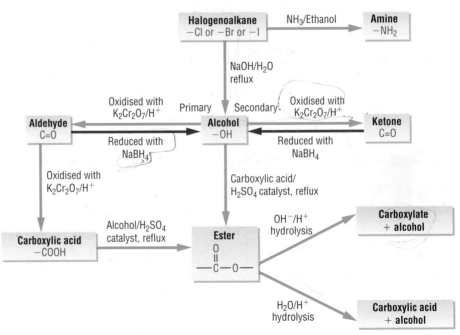

Figure 2 Summary of the reactions of aliphatic functional groups

Module 2
Polymers and synthesis
Organic synthesis of aliphatic
compounds

Stepwise approach to organic synthesis

You will need to follow a number of steps to solve a synthetic problem.

- You will need a good knowledge of all the chemical reactions, reagents and conditions in the specification.
- You should check the carbon skeleton of the starting material and the product – a longer chain usually means that two carbon-containing compounds must be reacted together.
- Examine the starting material and the product to see if you can spot what has changed – it may be that you need to change only the functional group.
- Think about working backwards – starting with the product can often help to solve the problem. This is called retro-synthesis.
- Finally check your answer and every step of the synthesis. Are the equations balanced? Can you account for every atom?

Retro-synthesis

Retro-synthesis is a synthesis that is planned backwards, starting at the product – often called the target molecule. The process involves breaking up the target molecule into simple, available reactant molecules. Retro-synthesis requires good problem-solving skills and an excellent knowledge of chemical reactions.

Worked example

State the reagents and conditions and give suitable equations for the preparation of 3-aminopropan-1-ol from 3-chloropropanal.

First, draw the structures of the starting material and the product.

Now use your knowledge of functional groups to propose a route. This conversion cannot be carried out in one step. However, using your chemical knowledge you should remember that an alcohol can be made by reducing an aldehyde, and that halogenoalkanes can be converted to amines using ammonia.

Write out your proposed answer remembering that the question asks for reagents, conditions and equations.

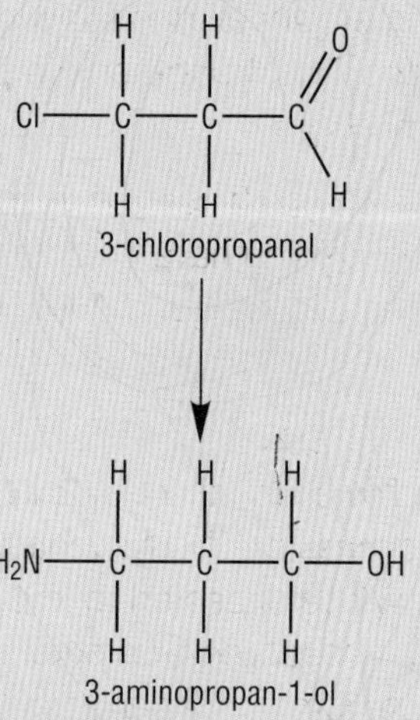

Figure 3 Synthesis of 3-aminopropan-1-ol from 3-chloropropanal

Step 1: Reagents: $NaBH_4/H_2O$ Conditions: heat

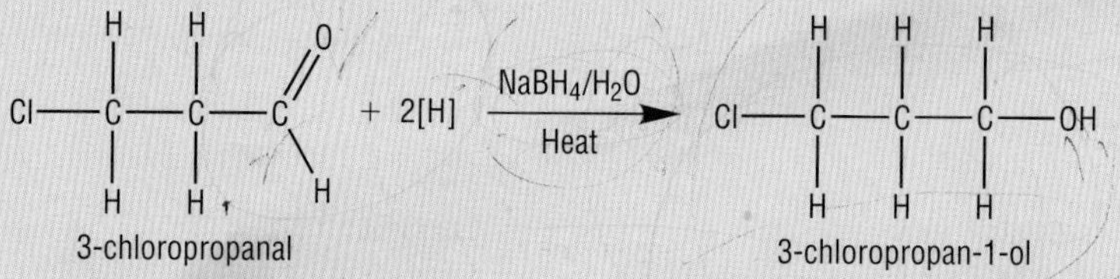

Figure 4 Converting the aldehyde group to an alcohol functional group

Step 2: Reagents: excess ammonia, NH_3 Conditions: ethanol solvent/reflux

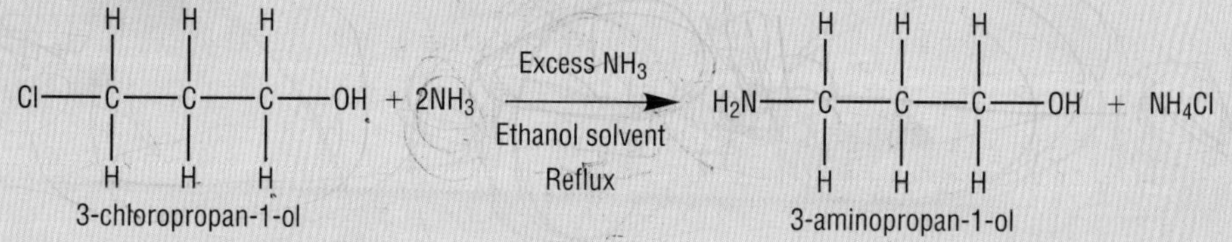

Figure 5 Conversion of the halogenoalkane to the amine

Question

1 Suggest a two-step synthesis for each of the following reactions:
  (a) 2-aminoethanoic acid from 2-chloroethanol;
  (b) ethanoic acid from ethyl methanoate.

By the end of this spread, you should be able to ...

* ✳ Identify the functional groups in an aromatic molecule containing several functional groups.
* ✳ Predict the properties and reactions of an aromatic molecule containing several functional groups.
* ✳ Devise multi-stage synthetic routes for preparing aromatic organic compounds.

Reacting aromatic compounds

In spread 1.2.8 there was a summary of the reactions of the aliphatic functional groups studied in the course. This spread looks at synthetic routes towards aromatic compounds. Some conversions of aromatic compounds are shown in Figure 1.

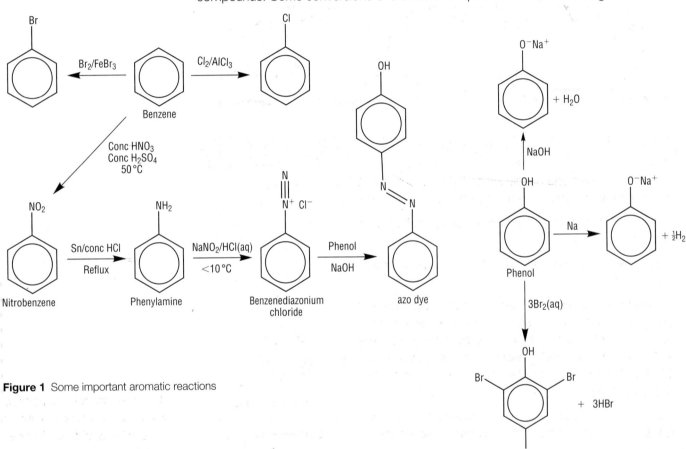

Figure 1 Some important aromatic reactions

Worked example

Devise a synthetic route to prepare a sample of compound B from compound A, as shown in Figure 2.

Compound A contains the aldehyde and halogenoalkane functional groups. Compound B contains an ester and an amine.

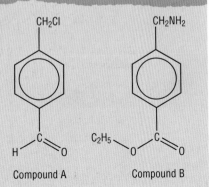

Figure 2 Structures of compounds A and B

Module 2
Polymers and synthesis
Organic synthesis of aromatic compounds

The following chemical knowledge will be needed:
- An ester can be made from an alcohol and a carboxylic acid.
- Aldehydes can be oxidised to carboxylic acids and reduced to primary alcohols.
- Amines can be made from halogenoalkanes using excess ammonia.

Compound B is an ethyl ester – the ethyl group must have come from ethanol. This means that compound B must have been prepared from a carboxylic acid. This is a three-step synthetic route:
- convert the –Cl group to –NH_2
- oxidise the aldehyde to a carboxylic acid
- react the aromatic carboxylic acid with ethanol in an esterification reaction.

The synthetic route, reagents and conditions are shown in Figure 3.

Figure 3 Synthetic route for conversion of A into B

Compound A → (Excess NH_3, Ethanol) → (K_2Cr_2O_7/H^+, Reflux) → (C_2H_5OH, H_2SO_4/Heat) → Compound B

Worked example

State the reagents and conditions and give suitable equations for the preparation of 3-chloronitrobenzene from benzene.

First, draw out the structures of the starting material and product.

Now, use your knowledge of functional groups to propose a synthetic route.

This reaction cannot be carried out in one step. However, using your chemical knowledge you should remember that –NO_2 can replace an –H in benzene by reacting benzene with concentrated HNO_3, helped by sulfuric acid acting as a catalyst. Chlorine can also be substituted onto a benzene ring using a halogen carrier.

Write out your proposed answer, remembering that the question asks for reagents, conditions and equations.

Step 1:
Reagents: concentrated nitric acid
Conditions: concentrated sulfuric acid at 50 °C

Figure 5 Converting benzene into nitrobenzene

Step 2:
Reagents: chlorine
Conditions: $AlCl_3$ as a halogen carrier

Figure 6 Chlorination of an aromatic ring

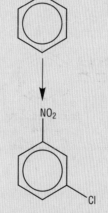

Figure 4 Structures of benzene and 3-chloronitrobenzene

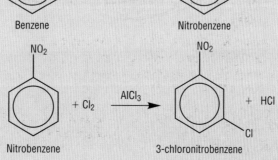

Questions
1 State the reagents, conditions and equations for each step in the synthesis of phenylamine from benzene.
2 Devise a synthetic route for the synthesis of compound X, starting with benzaldehyde, C_6H_5CHO, as the only organic compound.

Figure 7 Compound X, an aromatic ester

By the end of this spread, you should be able to . . .

* Explain that the synthesis of pharmaceuticals often requires the production of a single optical isomer.
* Explain that synthetic molecules often contain a mixture of optical isomers, whereas natural molecules often have only one optical isomer.
* Explain that the synthesis of a pharmaceutical that is a single optical isomer increases costs, reduces side effects and impoves pharmacological activity.
* Describe strategies for the synthesis of a pharmaceutical with a single optical isomer.

The importance of chirality in drug synthesis

In 1954, a German pharmaceutical company developed a new drug called thalidomide. In the late 1950s and the early 1960s, doctors prescribed thalidomide to prevent morning sickness in pregnant women. It was also used in some cold and flu remedies.

The drug was a chiral compound (see spread 1.2.3) and while one of the stereoisomers had the required therapeutic effect, unfortunately its mirror image led to a number of deformities in developing babies. It is estimated that approximately 10 000 babies were affected in Europe. The drug was never marketed in the USA because the Food and Drug Administration demanded further tests before granting a licence. During these tests, the relationship between thalidomide and birth defects surfaced in Germany.

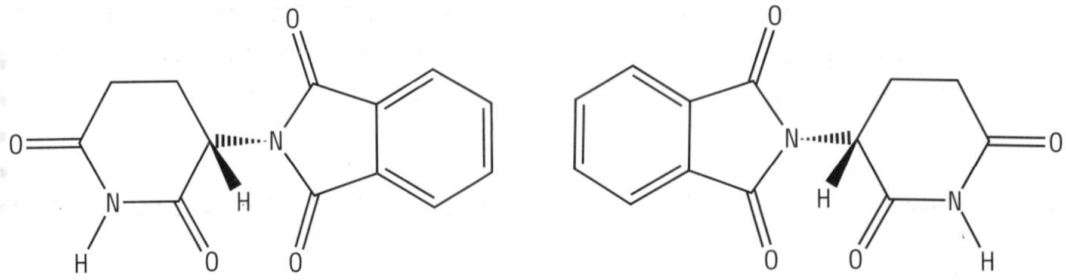

(−)-Thalidomide   (+)-Thalidomide

Figure 1 Stereoisomers of thalidomide – both optical isomers are sedatives, but unfortunately the optical isomer on the right leads to birth defects

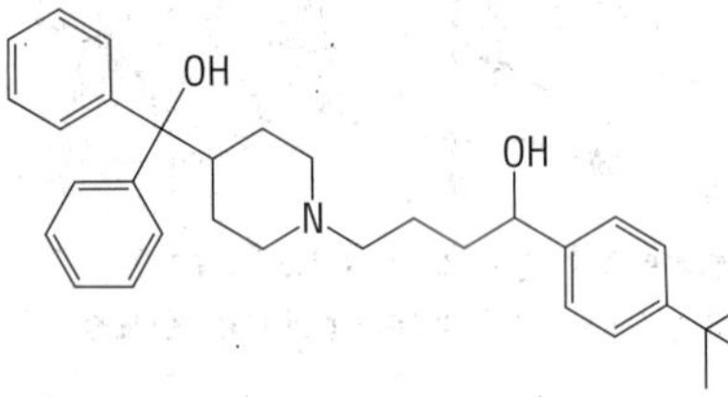

Another drug called Seldane was one of the first antihistamines, used to relieve symptoms associated with hay fever, such as sneezing and nasal congestion. Seldane has a chiral carbon and forms a mixture of optical isomers, with one of these being pharmacologically active. However, after testing and licensing, it was found that the other 'inactive isomer' caused a potentially fatal heart condition in some patients.

Pharmaceutical companies now have greater awareness of the importance of chiral carbon atoms and the potential effects of optical isomers on the human body. Drug-licensing agencies have introduced new testing regimes to check the effect of each optical isomer separately, rather than just testing the drug alone. This testing is costly for the drug companies and has led to research into the development and synthesis of drugs containing just one of the optical isomers.

Synthesising pharmaceuticals

In the body, drugs and medicines act by interacting with materials such as proteins, nucleic acids and with biological systems such as cell membranes. Biological molecules have complex three-dimensional structures that bind to a drug molecule in only one possible way. The three-dimensional structure of the drug determines the **pharmacological activity** of a drug, and whether a pharmaceutical will have the desired therapeutic effect or not.

Figure 2 Seldane (structure above top) was one of the first antihistamine drugs used to combat the effects of hay fever. Antihistamines block the effect of the chemical histamine, which is produced by the body during an allergic reaction to pollen or dust. Can you spot the chiral carbon in the structure of Seldane above?

Key definition

Pharmacological activity is the beneficial or adverse effects of a drug on living matter.

Figure 3 Pharmacological activity depends on whether or not a drug can interact with a receptor site in a biological system. Only one of the optical isomers has the correct stereochemistry to bind to the receptor site

The production of a single isomer with the correct pharmacological activity presents significant advantages to pharmaceutical companies.

- *Risks from undesirable side effects are reduced.* If thalidomide had been used as the 'correct' single optical isomer, morning sickness would have been prevented without the deformities caused by the other optical isomer.
- *Drug doses are reduced.* When a drug containing a mixture of two optical isomers is prescribed, half of the drug is often wasted in the body because only one of the isomers has the desired therapeutic effect. Making a drug containing one of the optical isomers reduces the required dose by half.

When a chiral compound is synthesised in the laboratory, a mixture of optical isomers is usually formed. But when the same material is made naturally in a living system, it is produced as a single optical isomer. In spread 1.2.3 you saw that natural α-amino acids are optically active, existing as single optical isomers in nature.

To prepare a chiral compound in the laboratory, complicated separation techniques are needed to isolate the pharmacologically active isomer. Separation is difficult because optical isomers usually have the same or very similar physical properties – melting points, boiling points and solubilities. Separation techniques include the use of enzymes, electrophoresis and chromatography. These take a significant amount of time and are costly. Frequently the non-active isomer has to be disposed of if it cannot be sold on to other users.

Modern chiral synthesis

A number of methods are currently being used or are being developed to prepare single chiral isomers.

- *Using enzymes as biological catalysts.* Nature is good at making single optical isomers. If a synthetic route can be designed using enzymes, then a single isomer can be produced. Drug manufacturers can use purified enzymes or enzyme-containing microorganisms to biocatalyse any reaction step that produces a chiral carbon.
- *Chiral pool synthesis.* This technique makes use of the pool of naturally occurring chiral molecules within the synthetic route. The chirality of the original molecule can lead to the formation of a product that is a single, pure optical isomer. Common starting materials used in chiral synthesis include natural α-amino acids and sugars.
- *Use of transition element complexes.* In 1968 William Knowles introduced the use of transition metal complexes to produce chiral catalysts. These could then transfer their chirality to synthesise a single isomer product.

Chiral drugs at home

- Ibuprofen is a type of anti-inflammatory drug that targets both bone and muscle pain. Ibuprofen controls pain by blocking messages to the brain and reducing swelling or inflammation. Ibuprofen is commonly used to relieve headache, back pain and period pain, and to treat cold and flu symptoms and arthritis.

The structure of ibuprofen has one chiral carbon, so two optical isomers are possible. One of the optical isomers of ibuprofen relieves pain much more effectively than the other. Fortunately in the body the less active form of ibuprofen is converted to the pharmacologically active isomer, so the whole dose given to a patient is active. This minimises any possible side effects. Ibuprofen is sold as a mixture of both its optical isomers.

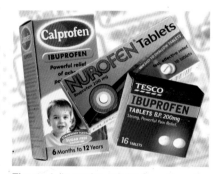

Figure 4 Ibuprofen can be found in many over-the-counter pain remedies

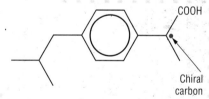

Figure 5 Structure of ibuprofen, used in many medicines to relieve pain

Questions

1 What do you understand by the term *chiral centre*?

2 Optical isomers are examples of stereoisomers. What do you understand by the term *stereoisomer*?

3 Discuss the possible disadvantages of producing a drug that contains more than one stereoisomer.

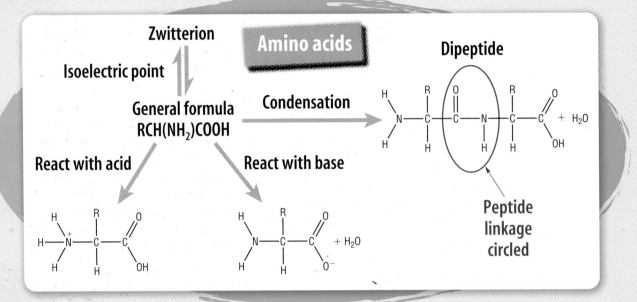

Amino acids

Zwitterion

Isoelectric point

General formula
$RCH(NH_2)COOH$

Condensation

React with acid

React with base

Dipeptide

Peptide linkage circled

+ H_2O

+ H_2O

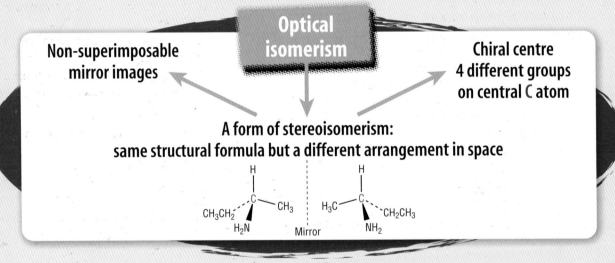

Optical isomerism

Non-superimposable mirror images

Chiral centre
4 different groups on central C atom

A form of stereoisomerism:
same structural formula but a different arrangement in space

Mirror

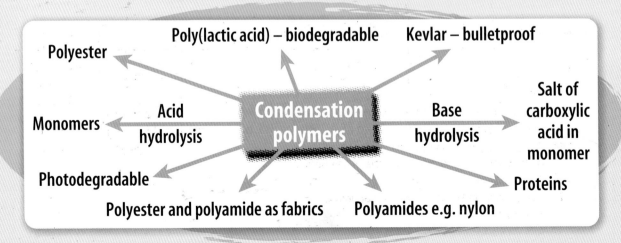

Condensation polymers

Poly(lactic acid) – biodegradable

Kevlar – bulletproof

Polyester

Monomers

Acid hydrolysis

Base hydrolysis

Salt of carboxylic acid in monomer

Photodegradable

Proteins

Polyester and polyamide as fabrics

Polyamides e.g. nylon

Practice questions

1 The structures of two amino acids are shown below:

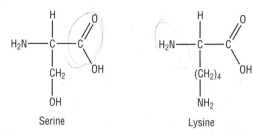

Serine   Lysine

(a) Which functional groups are involved in the formation of the peptide bond?

(b) Draw a diagram to show how these groups are linked in a protein chain.

(c) Amino acids form zwitterions.

   (i) What do you understand by the term *zwitterion*?

   (ii) What name is given to the pH at which an amino acid exists as a zwitterion?

   (iii) Draw the zwitterion form of serine.

(d) Draw the structure of lysine at pH 13.

2 Glutamic acid is an amino acid, $H_2NCH(CH_2CH_2COOH)COOH$.

(a) Glutamic acid forms two stereoisomers. Describe the structural feature of glutamic acid that enables the formation of these two isomers. Illustrate your answer with suitable diagrams of these isomers.

(b) At pH 3, glutamic acid exists in its zwitterion form as shown in the diagram to the right.

Show the structure of glutamic acid at:

   (i) pH 1;

   (ii) pH 13.

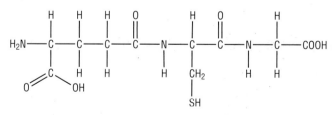

(c) Glutathione is a naturally occurring molecule made from glutamic acid and two other amino acids. The structure of glutathione is shown below.

   (i) Show the structures of the other two amino acids used to make glutathione.

   (ii) State the type of reaction that takes place when glutathione is made from its three constituent amino acids.

   (iii) How can glutathione be hydrolysed in the laboratory?

3 (a) What is meant by the term *chiral*?

(b) Draw the full displayed formula for each of the following molecules:

   (i) pentan-1-ol;   (ii) pent-2-ene;

   (iii) 2-butylamine;   (iv) 1-chloroethanol.

(c) Indicate with a * the chiral carbon atoms in any of the above molecules.

4 The polymer Terylene is made from the monomers benzene-1,4-dioic acid and ethane-1,2-diol.

(a) What type of polymerisation is involved in the making of Terylene?

(b) What functional group is present in the repeat unit of Terylene?

(c) (i) Draw each of the monomers used to make Terylene.

   (ii) Show one repeat unit of Terylene.

5 Nylon-6,6 is made from polymerisation of 1,6-diaminohexane and hexane-1,6-dioic acid.

(a) Draw a section of nylon-6,6 showing the repeat unit clearly.

(b) State one way in which the structure of this synthetic polymer is:

   (i) similar to the structure of a protein;

   (ii) different from the structure of a protein.

6 Draw and name the monomers used to make the following three polymers.

(a)

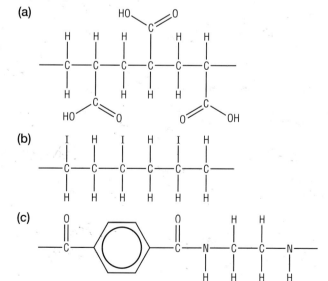

(b)

(c)

1   Compound A is currently being tested as a possible anti-allergy drug.

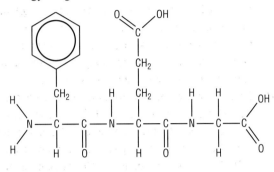

Compound **A**

(a) Compound **A** can be hydrolysed to form three organic products.
   (i)   Name a suitable reagent and conditions for the hydrolysis of compound **A**.   [2]
   (ii) The three organic products all belong to the same class of compound. State the general name for this class of organic compound.   [1]
   (iii) Draw the structure of one of the organic products from the hydrolysis of **A** using the reagent you have given in (a)(i) above.   [2]
   (iv) Explain what is meant by the term *hydrolysis*. Use this reaction to illustrate your answer.   [2]
(b) Compound **A** can exist as a number of stereoisomers, but only one of them is pharmacologically active as the anti-allergy drug.
   (i)   Explain what causes stereoisomerism in compounds such as **A**.   [3]
   (ii) Explain why there are four different stereoisomers of compound **A**.   [2]
   (iii) Suggest how a drug company could synthesise compound **A** so that the drug contains only the pharmacologically active stereoisomer.   [1]
   (iv) Sometimes it is difficult to manufacture a drug containing only the one pharmacologically active stereoisomer. Describe two possible disadvantages of producing a drug containing a mixture of several stereoisomers.   [2]

[Total: 15]
(OCR 2814 Jun05)

2   Leucine (2-amino-4-methylpentanoic acid) is a naturally occurring α-amino acid that is often used in protein supplements. Leucine has a structural formula of $(CH_3)_2CHCH_2CH(NH_2)COOH$.
(a) (i)   State the general formula of an α-amino acid.   [1]
   (ii) Draw a displayed formula of leucine.   [1]

(b) Leucine can exist as a *zwitterion*.
   (i)   State what is meant by the term *zwitterion*.   [1]
   (ii) Explain with the aid of a diagram how the zwitterion is formed from the functional groups in leucine.   [2]
(c) Leucine can be obtained from a source of protein such as meat.
   (i)   State suitable reagents and conditions to break down a protein into amino acids.   [2]
   (ii) State the type of reaction occurring.   [1]
(d) Leucine can also be synthesised in the laboratory from simpler compounds.
   (i)   One reaction in this synthesis converts 2-chloro-4-methylpentanoic acid, $(CH_3)_2CHCH_2CHClCOOH$, into leucine. State the reagents and conditions needed for this reaction.   [1]
   (ii) Explain how a purified sample of leucine synthesised in the laboratory would differ from a sample of leucine purified from meat.   [3]

[Total: 12]
(OCR 2814 Jun06)

3   Polymers can be made either from a single monomer or from more than one monomer. Two polymers, **L** and **M**, are shown below.

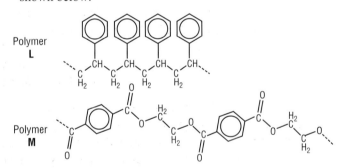

(a) Deduce the structures of the monomers from which **L** and **M** could be obtained.   [3]
(b) Polymer **N** can be made from the monomer **P** only, shown below. Suggest a structure for polymer **N**, showing three repeat units.

Monomer
P   HO—⬡—COOH
   [2]

(c) Polymers **M** and **N** are made by the same type of polymerisation. Name this type of polymerisation and describe its characteristic features.   [2]
(d) State a major use for polymers such as **M**.   [1]

[Total: 8]
(OCR 2814 Jan03)

4 (a) A section of polymer has the structure shown below.

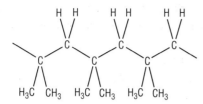

(i) Circle a repeat unit of this polymer on the diagram above. [1]

(ii) Deduce the empirical formula of this polymer. [1]

(iii) Draw a structure for a monomer from which this polymer could be made. Your structure should show any multiple bonds. [1]

(b) Proteins are natural polymers made from α-amino acids, such as glycine, H_2NCH_2COOH.

(i) Name the functional group made during amino acid polymerisation and draw its displayed formula. [2]

(ii) Name this type of polymerisation reaction. [1]

(iii) Draw a displayed and a skeletal formula for the dipeptide **H**. $C_4H_8N_2O_3$, made from glycine, H_2NCH_2COOH. [2]

(iv) A student made 1.10 g of dipeptide **H** starting from 1.40 g of glycine. Calculate the percentage yield obtained. Give your answer to 3 significant figures. [4]

(v) When glycine is treated with hydrochloric acid a compound **J**, $C_2H_6ClNO_2$, is formed. Draw a structure for compound **J**. [2]

[Total: 14]
(OCR 2814 Jun02)

5 Clear plastic drink bottles are usually made from either HDPE (high-density polythene) or from PET (polyethylene terephthalate). HDPE is an **addition polymer** and is used for softer bottles, whereas PET is a **condensation polymer** used for bottles needing a greater rigidity. PET is also known as *Terylene* when it is used to make fibres. The monomers used to make these two polymers are shown below.

Ethene

Ethane-1,2-diol

Benzene-1,4-dicarboxylic acid

(a) In this question, one mark is available for the quality of written communication.

Using HDPE and PET as your examples, explain the difference between *addition* and *condensation* polymerisation. Include equations to show the polymerisation reaction forming each polymer and clearly show the repeat unit in HDPE and PET. [8]

(b) Bottles made from PET are often used to contain carbonated drink. However, PET allows the gas to escape slowly. A layer of another polymer can be included to decrease the gas leakage. One such polymer is PVOH (polyvinyl alcohol). This is manufactured in two steps as shown below.

(i) Draw the structure of the monomer used in **step 1**. [1]

(ii) **Step 2** involves hydrolysis of the ester in the side chain. State the reagents and conditions required for **step 2**. [3]

(iii) Another compound is formed in **step 2**. Identify this compound. [1]

(c) Research is being carried out to find alternatives to PET that do not allow gases to escape. The formula of the repeat unit of one possible condensation polymer is shown. Deduce the structures of the **two** monomers from which this polymer could be made.

[2]

[Total: 15]
(OCR 2814 Jun04)

6 The 'Nylon Rope Trick' is a well-known laboratory demonstration for the formation of a condensation polymer from its monomers. Two solutions, one containing each monomer, are placed in a beaker as shown. In this reaction, the more reactive hexanedioyl dichloride is used instead of hexanedicarboxylic acid. The nylon forms where the two immiscible liquids join, and can be pulled out from between the layers in a continuous strand.

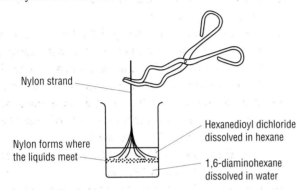

Nylon strand

Nylon forms where the liquids meet

Hexanedioyl dichloride dissolved in hexane

1,6-diaminohexane dissolved in water

(a) Complete the equation below to show the formation of the nylon polymer from its monomers. Show a repeat unit for the polymer and the other product formed.

$$\underset{\substack{Cl}}{\overset{O}{\parallel}}{C} - (CH_2)_4 - \underset{\substack{Cl}}{\overset{O}{\parallel}}{C} \quad + \quad \text{1,6-diaminohexane}$$

Hexanedioyl dichloride

↓

Repeat unit of polymer   +   Other product   [2]

(b) Nylon is sometimes used for electrical insulation. However, if there is a risk of high temperatures then a polymer such as Nomex®, with a higher melting point, is used. The repeat unit of Nomex® is shown below.

$$\left[\underset{}{\overset{O}{\parallel}}{C} \quad \underset{}{\overset{O}{\parallel}}{C} - \underset{H}{N} \quad \underset{H}{N} \right]_n$$

(i) Draw the structures of two monomers that could be used to form Nomex®.   [2]

(ii) Suggest a reason why the melting point of Nomex® is higher than that of nylon.   [1]

[Total: 5]
(OCR 2814 Jun07)

7 From the information given, draw the structural formula for each organic compound.

(a) This compound is made by reaction of benzene with concentrated nitric acid in the presence of concentrated sulfuric acid.   [1]

(b) These two compounds react together in the presence of concentrated sulfuric acid to make methyl ethanoate, CH_3COOCH_3.   [2]

(c) These two different compounds can be made by reaction of $C_6H_5CH(NH_2)COOH$ with $CH_3CH(NH_2)COOH$.   [2]

[Total: 5]
(OCR 2814 Jun03)

8 A commercial synthesis of the ester **G** is shown below.

$$C_6H_5CH_2Cl \xrightarrow[\text{Stage I}]{} C_6H_5CH_2OH \xrightarrow[\substack{\text{conc. } H_2SO_4}]{CH_3COOH,} CH_3COOCH_2C_6H_5$$

$$\quad \mathbf{E} \qquad\qquad \mathbf{F} \qquad\qquad\qquad \mathbf{G}$$

(a) Stage I:
  (i) Suggest a suitable reagent.   [1]
  (ii) State the type of reaction occurring.   [2]
  (iii) Write the equation for this reaction.   [1]

(b) Stage II:
  (i) Draw the displayed formula for the ester **G**.   [1]
  (ii) Write the equation.   [1]
  (iii) Suggest a general use for esters such as **G**.   [1]
  (iv) **G** can also be made directly from **E** by the reaction with $CH_3COO^-Na^+$. Suggest a possible mechanism for this reaction.   [3]

[Total: 10]
(OCR 2814 Jun02)

9 Compound **A** has the structure shown.

$$HO-CH_2-CH_2-O-\underset{}{\overset{O}{\parallel}}{C}- \bigcirc -\underset{}{\overset{O}{\parallel}}{C}-OH$$

(a) Deduce the molecular formula, $C_XH_YO_Z$, of compound **A**.   [1]

(b) Compound **A** can undergo **(i)** oxidation, **(ii)** hydrolysis and **(iii)** neutralisation. Complete the schemes below by stating the reagent(s) and conditions, if any. Draw the structure for each organic product.   [9]
  (i) Oxidation of compound **A**.
  (ii) Hydrolysis of compound **A**.
  (iii) Neutralisation of compound **A**.

[Total: 10]
(4821 Jun01)

10 A diagram of a section of nylon-6,6 is shown below.

$$-CO(CH_2)_4CONH(CH_2)_6NHCO(CH_2)_4CONH(CH_2)_6NH-$$

  (a) Identify the monomer(s) from which nylon-6,6 is obtained. [2]

  (b) State and explain the type of polymerisation reaction which gives nylon-6,6. [2]

  (c) Proteins and polypeptides are polymers which have been described as being similar to nylon-6,6. Suggest with the aid of diagrams and equations
- one structural similarity
- one chemical similarity
- one important difference

    (*In this question, 1 mark is available for the quality of written communication.*) [7]

                                    [Total: 11]

                              (OCR 2814 Jan02)

11 There are two major types of polymerisation: addition polymerisation and condensation polymerisation.

  (a) **(i)** Propene undergoes addition polymerisation. Give a balanced equation for this polymerisation, using structural formulae. [2]

      (ii) Explain the differences between addition polymerisation and condensation polymerisation. [2]

  (b) Polymer **G** is also formed by addition polymerisation. A section of polymer **G** is shown below.

    Deduce the structure of a monomer from which **G** could be made. [1]

  (c) The monomer shown below can form a condensation polymer, **H**.

    (i) Suggest a structure for the polymer, showing two repeat units. [2]

    (ii) Concentrated aqueous NaOH solution can be transported in containers made of poly(propene) but not in containers made of polymer **H**. Suggest reasons for this difference. [3]

                                 [Total: 10]

                          (OCR 2814 Jun03)

12 Mordant orange 1 is an azo dye with the structure shown below.

  (a) Draw a circle around the azo group that identifies this molecule as an azo dye. [1]

  (b) Describe how mordant orange 1 can be made starting with 4-nitrophenylamine and other suitable reagents. You should include full structures for any intermediate formed and state essential reagents and conditions. [6]

  (c) Mordant orange 1 is yellow in acidic conditions, but changes to a red colour in the presence of strong bases, such as aqueous sodium hydroxide.
Draw the structure of Mordant Orange 1 after reaction with excess sodium hydroxide. [2]

  (d) The colour of the dye can be changed by reducing the nitro group.

    (i) State a suitable reducing mixture to reduce the nitro group. [1]

    (ii) Write an equation for the reduction of mordant orange using [H] to represent the reducing agent. [2]

                                 [Total: 12]

                          (OCR 2814 Jan08)

Module 3
Analysis

Introduction

As scientific knowledge has developed, traditional methods of analysis, such as titrations and test tube reactions, have been supplemented by powerful instrumental techniques such as gas chromatography, infrared spectroscopy, mass spectrometry and NMR spectroscopy.

The use of wet chemical tests and spectral analysis enables the analytical chemist to analyse milk for traces of pesticides, to determine the percentage of impurities in steel and to check the quality of food products. Modern analytical techniques are at the forefront of law enforcement, being used by Customs to detect forgeries in imported goods and by Trading Standards to ensure that products on sale contain no dangerous or banned chemicals.

In this module, you will examine the theory behind some of the more important analytical techniques and gain the skills required to interpret analytical data and spectra in order to identify organic molecules.

Test yourself

1 State two applications of mass spectrometry.
2 What information is given by the molecular ion peak?
3 What is meant by *fragmentation*?
4 What is chromatography used for?

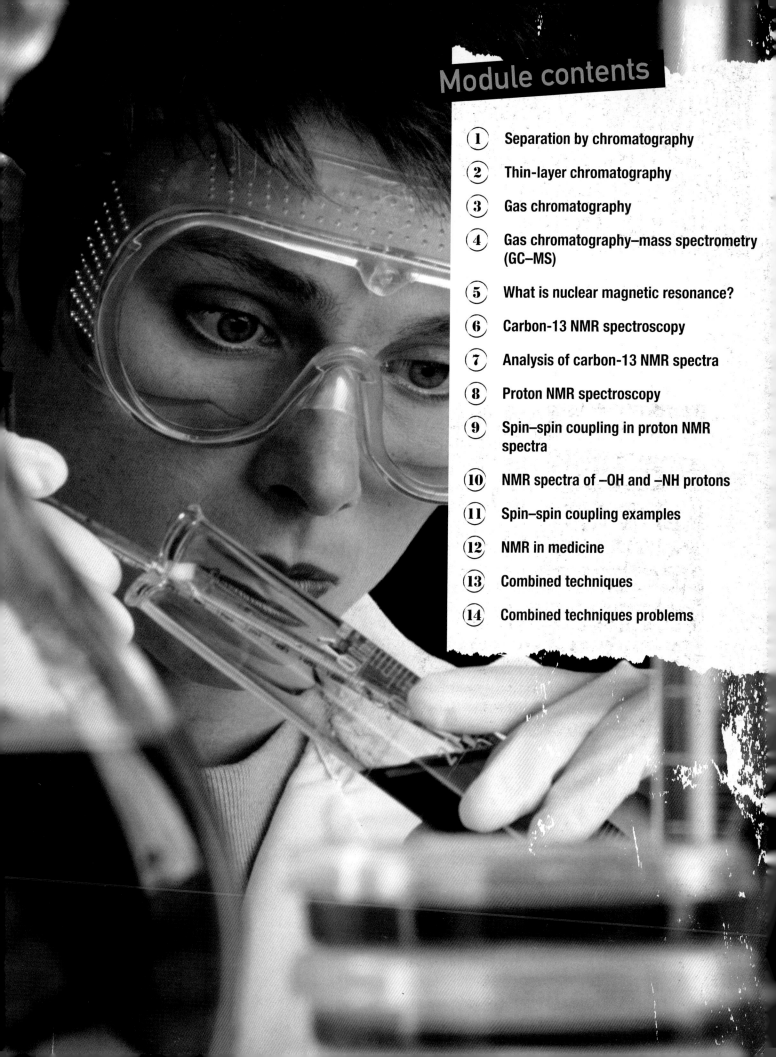

Module contents

By the end of this spread, you should be able to ...

* Describe chromatography as an analytical technique that separates components in a mixture.
* State that the mobile phase may be a liquid or a gas.
* State that the stationary phase may be a solid, or either a liquid or solid on a solid support.
* State that a solid stationary phase separates by adsorption.
* State that a liquid stationary phase separates by relative solubility.

What is chromatography?

Chromatography is an analytical technique used to separate the components in a mixture. In the early 1900s, the Russian botanist Mikhail Semyonovich Tsvet invented chromatography as a technique to separate pigments in the leaves of plants. He invented the word 'chromatography' to describe his new analytical technique. The name comes from the Greek 'chroma' meaning *colour* and 'graphein' meaning *to write*. Figure 1 shows the result of separation of plant pigments by chromatography.

Figure 2 The technique of column chromatography: the column is filled with a porous gel. Two bands of colour can be seen in the column; these are different components of a mixture that is separated by the different rates at which its components travel down the column. The technique can be used to purify and isolate a compound

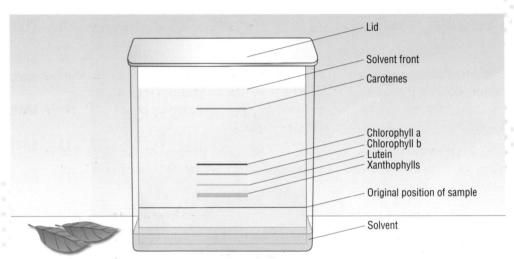

Figure 1 Separation of plant pigments by thin-layer chromatography

The chromatography plate is carefully placed in the chromatography tank with the sample line toward the bottom. The plate is in contact with the solvent, but it is important that the solvent level is below where the sample has been placed on the plate. The tank is covered and the plate is left until the solvent front has reached near to the top of the plate. The plate is then removed and the position of the solvent front is immediately marked. The solvent is then allowed to evaporate and the plate is analysed.

Chromatography is a very versatile method of analysis, but its power lies with its ability to separate out components with similar physical properties in very complex mixtures.

Chromatography is used in the analysis of drugs, plastics, flavourings, foods, pesticides, forensic evidence, fuels, air samples, water samples, etc. Only small sample sizes are required and, after separation, the pure components can be analysed precisely using other analytical methods.

How does chromatography work?

Phases

In A2 chemistry you will study two different types of chromatography:

- thin-layer chromatography, TLC
- gas chromatography, GC.

During chromatography, a **mobile phase** sweeps a mixture over a **stationary phase**:

- the stationary phase is fixed in place
- the mobile phase moves in a definite direction.

Chromatography works on the basis that different components have different affinities for a stationary phase and for a mobile phase.

- The stationary phase interacts with the components in the mixture, slowing them down. The greater the interaction, the more the components are slowed down.
- This allows different components to flow over the stationary phase at different speeds, separating the components.

In thin-layer chromatography:

- the stationary phase is a solid, and the mobile phase is a liquid.

In gas chromatography:

- the stationary phase is liquid or solid on a solid support, and the mobile phase is a gas.

Separation

Efficient separation depends on two main factors. A solid stationary phase separates by **adsorption**. As the mobile phase passes over the stationary phase, component molecules are able to bind to the solid making up the stationary phase. This happens at the *surface* of the stationary phase – see Figure 3.

A liquid stationary phase separates by relative solubility. Some components dissolve more readily in the liquid and these are slowed down more than components with lower solubility – see Figure 3.

> ### Key definitions
>
> A **phase** is a physically distinctive form of a substance, such as the solid, liquid, and gaseous states of ordinary matter.
>
> The **mobile phase** is the phase that moves in chromatography.
>
> The **stationary phase** is the phase that does not move in chromatography.

> ### Key definition
>
> **Adsorption** is the process by which a solid holds molecules of a gas or liquid or solute as a thin film on the surface of a solid or, more rarely, a liquid.

> ### Examiner tip
>
> Be careful not to confuse adsorption with absorption.
>
> *Adsorption* means interacting at the surface of a material. *Absorption* means soaking into a material.

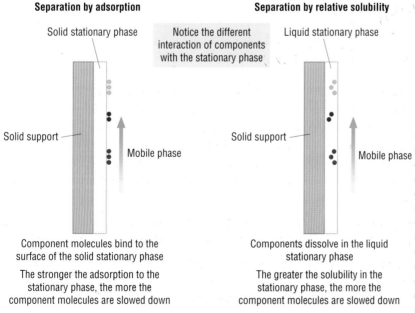

Separation by adsorption

Solid stationary phase

Notice the different interaction of components with the stationary phase

Solid support

Mobile phase

Component molecules bind to the surface of the solid stationary phase

The stronger the adsorption to the stationary phase, the more the component molecules are slowed down

Separation by relative solubility

Liquid stationary phase

Solid support

Mobile phase

Components dissolve in the liquid stationary phase

The greater the solubility in the stationary phase, the more the component molecules are slowed down

Figure 3 Separation by adsorption and by relative solubility

Questions

1 In chromatography, what is meant by the terms *mobile phase* and *stationary phase*?
2 How is separation achieved with
   (a) a solid stationary phase;   **(b)** a liquid stationary phase?

By the end of this spread, you should be able to ...

✱ State that the mobile phase in TLC is a liquid and that the stationary phase is a solid on a solid support.

✱ State that the solid stationary phase in TLC separates by adsorption.

✱ Explain the term R_f value, and interpret chromatograms in terms of R_f values.

Thin-layer chromatography

TLC is a simple and quick method used to monitor the extent of a chemical reaction or to check the purity of compounds. It is particularly useful in organic chemistry.

* The stationary phase is a thin layer of an *adsorbent* such as silica gel (SiO_2) or alumina (Al_2O_3) coated on a flat, inert support, usually a sheet of glass or plastic – this is called the *TLC plate*.

* The mobile phase is a liquid solvent, which moves vertically up the TLC plate.

Producing the chromatogram

* A small sample of the mixture is dissolved.

* A small spot of the sample solution is placed on the TLC plate a short distance from one end of the plate. It is useful to first draw a pencil line as a guide for the sample.

* The sample spot is allowed to dry and the TLC plate is then placed in a jar containing a shallow layer of solvent. The solvent must be below the sample spot or the solvent will just wash the sample off the plate. The jar is then sealed to saturate the space in the jar with solvent vapour. This slows down the evaporation of solvent from the TLC plate and prevents solvent loss.

* As the solvent rises, it meets the sample, and the components in the mixture are swept upwards with the solvent. Separation is achieved by *adsorption* of the components in the mixture by binding to the surface of the solid stationary phase (see spread 1.3.1). Some components bind to the adsorbent strongly, others more weakly. The result is that different components in the mixture are separated as they travel different distances up the TLC plate.

* To achieve maximum separation of the components, the solvent is left to rise until it nearly reaches the top of the TLC plate. The plate is then taken out of the solvent. The position of the solvent front is marked with a pencil line and the solvent is allowed to evaporate. The result of the chromatography is called the **chromatogram**.

* Each separated component appears as a spot on the TLC plate. If the components are coloured, then they can be seen easily, but if the components are colourless you may need to use a locating agent to 'develop' the chromatogram. Alternatively, UV radiation is often used to show up each component by fluorescence.

Figure 1 shows how components are separated during the course of chromatography.

Key definition

A **chromatogram** is a visible record showing the result of separation of the components of a mixture by chromatography.

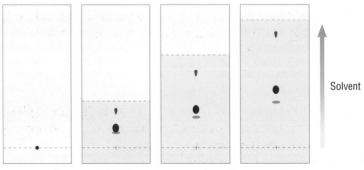

Figure 1 Running a TLC chromatogram – how the components are separated during the course of the chromatography

R_f values

TLC chromatograms can be interpreted in terms of **R_f** values.

An R_f value shows how far a component has moved compared with the solvent front:

$$R_f = \frac{\text{distance moved by component}}{\text{distance moved by solvent front}}$$

Figure 2 shows a chromatogram in which the solvent front has travelled 4.85 cm.

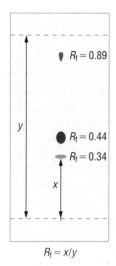

$R_f = 0.89$

y

$R_f = 0.44$

$R_f = 0.34$

x

$R_f = x/y$

Figure 2 Measuring R_f values – for the green spot, x = 1.65 cm and y = 4.85 cm. ∴ the R_f value = 1.65/4.85 = 0.34

- The green spot has travelled 1.65 cm from the base line, so its R_f value = $\frac{1.65}{4.85}$ = 0.34

- The pink spot has travelled 2.15 cm from the base line, so its R_f value = $\frac{2.15}{4.85}$ = 0.44

- The blue spot has travelled 4.30 cm from the base line, so its R_f value = $\frac{4.30}{4.85}$ = 0.89

If this chromatogram were to be run again using the same conditions and the same solvent, then the R_f values for the three components would be the same.

R_f values of components can also be compared with those of pure compounds to aid in the identification of an unknown substance. A chromatogram can be run with samples of known substances alongside the mixture. It is then easy to identify a component by comparing the distances travelled by the component and the known substance.

Limitations

Analysis by thin-layer chromatography has limitations.
- Similar compounds often have similar R_f values.
- Unknown compounds have no reference R_f for comparison.
- It may be difficult to find a solvent that separates all the components in a mixture. If the components are very soluble in the solvent, they will just be washed up the TLC plate with the solvent front. If the components have little solubility, they will hardly move. Trial and error may be required before a suitable solvent, or mixture of solvents, is discovered.

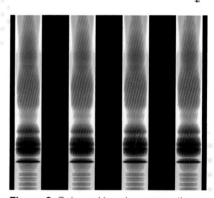

Figure 3 Coloured bands representing the separation of different chemicals by chromatography. The process uses a solvent to move the sample across a substrate such as paper. The different physical properties of different chemicals will cause them to move at different rates and thus separate; when they are coloured, as they are here, the separation can be clearly observed. There are many variants of the technique, which is used in chemical and biochemical analyses ranging from isolating a new chemical to DNA sequencing to amino acid identification in protein analyses

Questions

1 In TLC, what is used as the stationary phase and what as the mobile phase?
2 On a TLC chromatogram, the solvent front has moved 8.6 cm. Component **A** has moved 3.3 cm. Component **B** has moved 0.62 cm.
What are the R_f values of each component?

③ Gas chromatography, GC

By the end of this spread, you should be able to . . .

* **Explain the term** *retention time*.
* **Interpret gas chromatograms in terms of retention times and the approximate proportions of the components of a mixture.**
* **Explain that analysis by gas chromatography has limitations.**

Gas chromatography

Gas chromatography (GC) is a technique used to separate volatile components in a mixture. GC is particularly useful for many organic compounds that have a low boiling point and evaporate easily.

Gas chromatography takes place in an instrument called a gas chromatograph (Figures 1 and 2).

* The stationary phase is a thin layer of a liquid or solid, coated on the inside of some capillary tubing, which acts as an inert solid support (see Figure 1). The tubing may be up to 30 m long and is wound into a coil so that it fits inside a thermostatically controlled oven. The tubing is called the chromatography *column*. The liquid for the stationary phase is often a long-chain alkane with a high boiling point. Suitable solid stationary phases include silicone polymers. For separation of different types of compound, columns containing different liquids and solids are used.
* The mobile phase is a carrier gas which moves through the column. An inert or unreactive gas such as helium or nitrogen is usually used.

Figure 1 Stationary phase in a gas chromatography column

Solid support
Stationary phase
(Not to scale)

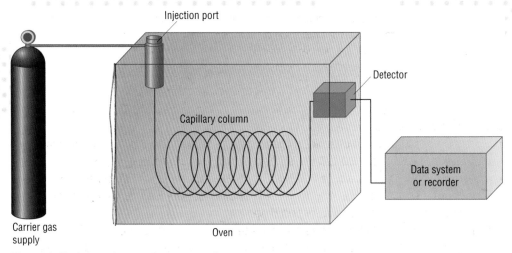

Figure 2 Equipment for gas chromatography

Injection port
Detector
Capillary column
Data system or recorder
Carrier gas supply
Oven

Producing the chromatogram

* The mixture is injected into the gas chromatograph, where it is vaporised.
* The mobile carrier gas flushes the mixture though the column.
* As the mixture moves through the column, components slow down as they interact with the stationary phase lining the column. If the lining is a liquid, components may dissolve; if the lining is a solid, components may become adsorbed to the surface. The carrier gas flushes the components further along the tube. The greater the solubility or adsorption, the more the components are slowed down. Different components are slowed down by different amounts, so they separate. Separation can be improved by using different oven temperatures and different flow rates for the carrier gas.
* Each component leaves the column at a different time and is detected as it leaves the column. A computer processes the results to display a gas chromatogram.

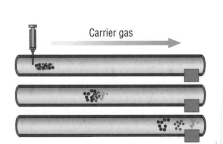

Carrier gas

Figure 3 Gas chromatograph separating three components

Retention time

Retention time is the time for a component to pass from the column inlet to the detector.

- Different compounds have different retention times. A retention time can be compared with those of known compounds to help in the identification of an unknown substance.
- The area under each peak is proportional to the amount of a compound in the sample.

Figure 4 shows a gas chromatogram of alcohols and carbonyl compounds in the blood. In this chromatogram, the retention time of the main component, ethanol, is 1.1 min. The approximate peak areas can be estimated by inspection.

Modern chromatographs automatically record peak areas accurately.

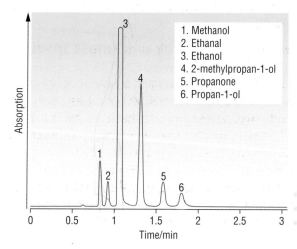

1. Methanol
2. Ethanal
3. Ethanol
4. 2-methylpropan-1-ol
5. Propanone
6. Propan-1-ol

Figure 4 Gas chromatogram of blood alcohols and related compounds showing retention times. You can estimate the relative concentrations of each compound by comparing the peak areas

Limitations of gas chromatography

GC analysis has several limitations.
- Potentially, thousands of chemicals may have the same retention time, peak shape and detector response. So GC does not positively identify most components.
- Not all substances in a sample will necessarily be separated and detected. A small amount of a substance can 'hide' beneath another substance that has a higher concentration and the same retention time.
- Unknown compounds have no reference retention times for comparison. To make results meaningful, analysts need to know which components are to be expected.

To improve reliability, results must be checked against a GC analysis of a reference sample containing only the suspected substance. GC results are often used in the law courts, and any doubts about the scientific method will cast doubt on the evidence and may make a conviction more difficult to secure. Partly for these reasons, gas chromatography is often combined with mass spectrometry – this technique is discussed in detail in spread 1.3.4.

Questions

1 In gas chromatography, what is used as the stationary phase and what as the mobile phase?
2 **(a)** In the GC chromatogram in Figure 4, what are the retention times of each separated component?
  (b) Suggest a likely use for the chromatogram in Figure 4.

By the end of this spread, you should be able to ...

* Explain that mass spectrometry can be combined with gas chromatography in GC–MS to provide a far more powerful analytical tool than gas chromatography alone.
* Explain that the mass spectra generated can be analysed or compared with spectral databases for positive identification of a component.
* State the use of GC–MS in analysis in forensics, environmental analysis, airport security and space probes.

Combining gas chromatography with mass spectrometry

Gas chromatography and mass spectrometry have different strengths.
* GC can separate components in a mixture, but cannot identify them conclusively.
* MS can provide detailed structural information on most compounds for exact identification, but it cannot separate them.

By combining the two techniques we obtain a far more powerful analytical tool than from each separate technique. The coupled technique is called gas chromatography–mass spectrometry or GC–MS.
* The components in a mixture are first separated using GC. Retention time may provide a preliminary identification.
* The separated components are directed into a mass spectrometer, where they are detected.
* The generated mass spectrum can be analysed or compared with a spectral database by computer.
* A mass spectrum is unique and definitive for a single compound, so MS allows positive identification of each component in the original mixture.

Uses for mass spectrometry

Applications of GC–MS include forensic and drug analysis, environmental analysis, airport security and space probes.

GC–MS in forensics

GC–MS can analyse minute particles found at the scene of a crime. The mass spectrometer positively identifies the presence of particular substances in a forensic sample – this often provides critical evidence during a trial in the law courts.

Figure 1 shows the stages in the GC–MS process. Figure 2 shows a gas chromatograph (left) connected to a mass spectrometer (right) in a forensic laboratory. This equipment is sensitive enough to detect minute quantities of illegal drugs in the hair of a suspect weeks after any drugs have been taken. A sample injection robot transfers samples from

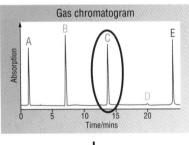

Sample: mixture of volatile liquids

Gas chromatograph

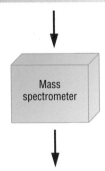

Gas chromatogram

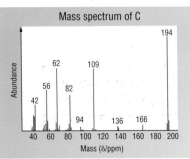

Mass spectrometer

Mass spectrum of C

Figure 1 Stages in GC–MS

Figure 2 Modern GC–MS for forensic drug detection

small vials to the gas chromatograph, where the samples are evaporated and separated. The mass spectrometer identifies even minute traces of chemicals as they emerge from the chromatography tube.

GC–MS in environmental analysis

GC–MS is used for monitoring and analysing organic pollutants in the environment, including the quality of waste water and drinking water, and detecting the presence of pesticides in food. The rise of GC–MS comes at the expense of GC because many labs are replacing GCs with GC–MS systems as a single technique.

GC–MS in airport security

Since the attacks on the World Trade Center in New York in 2001, improved security systems have been put in place at airports to detect explosives in luggage and on human beings. Many security systems are based on GC–MS.

GC–MS in space probes

GC–MS has even found its way into outer space. In the 1970s, two GC–MSs were sent to Mars to collect and analyse material. Another has visited Venus and has analysed the planet's atmosphere. In 2005, the Huygens probe landed a GC–MS on Titan, Saturn's largest moon. This probe found evidence for liquid methane on the moon's surface.

Figure 3 GC–MS goes to Saturn's largest moon, Titan

Problems with drug testing

Nandrolone is a member of a class of drug called anabolic steroids – muscle-building chemicals which occur naturally in the human body in tiny quantities. Use of these drugs is prohibited in sports, including athletics.

In the body, nandrolone is converted into 19-norandrosterone. It is this compound that is detected by the drug test. 19-norandrosterone is excreted from the body in urine, making it easy to obtain samples. The International Olympic Committee has set a limit of two nanograms (0.000 000 002 g) per cm^3 in urine samples. If this concentration is exceeded then the drug test is considered to be 'positive'. GC–MS is the critical analytical technique used to detect these drugs.

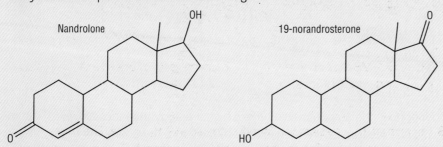

Figure 4 Nandrolone and 19-norandrosterone – note that the names both end in 'one': in each of these structures, you should be able to find the ketone functional group that gives the '-one' ending to the names

In recent years, a number of big-name athletes have been banned from competition after being found to have nandrolone in their system. Nandrolone has been implicated in hundreds of recent doping cases involving athletes in just about every sport – especially athletics, cycling, football and tennis.

Some scientists believe that the number of positive tests is surprisingly high. They suspect that nandrolone may be an ingredient in improperly labelled nutritional supplements popular with high-performance athletes.

A likely theory has recently emerged from an investigation being carried out at Aberdeen University. The findings suggest that a combination of dietary supplements and exercise can result in increased concentrations of nandrolone. Scientists are uncertain why this might happen and more work is required before the 'nandrolone mystery' can be solved.

Questions

1 How does GC–MS enable individual components in a mixture to be identified?
2 State four uses of GC–MS.
3 Carry out some research to find out what object in space should be visited by a GC–MS in 2014.

By the end of this spread, you should be able to . . .

* State that NMR spectroscopy involves interaction of materials with low-energy radio-frequency radiation.
* Describe the use of tetramethylsilane, TMS, as the standard for chemical shift measurement.
* State the need for deuterated solvents, e.g. $CDCl_3$, when running an NMR spectrum.

Nuclear magnetic resonance

Nuclear magnetic resonance, NMR, has been developed as an analytical technique over the last 70 years. NMR is a truly fantastic technique for examining molecular structure in detail. The isotopes most commonly studied using NMR spectroscopy are ^1H, ^{13}C, ^{19}F and ^{31}P. In organic chemistry, proton (^1H) NMR and carbon-13 (^{13}C) NMR spectroscopy are the most useful general-purpose techniques, but the basic principles are the same for NMR spectroscopy of other isotopes.

The requirements for NMR spectroscopy are:
* a strong magnetic field applied using an electromagnet
* low-energy radio-frequency radiation.

Nuclear spin

Chemists use a model of the atom in which the nucleus is made up of protons and neutrons (each called a 'nucleon').
* Protons and neutrons have a property called *spin*, which can be in one of two opposite directions – rather like electrons can have opposite spins.
* In the nucleus, opposite spins pair up. However, the nuclei of some isotopes have an uneven number of nucleons. These isotopes have an unpaired nucleon, which produces a small residual nuclear spin. This generates a magnetic field.

Nuclei with a magnetic spin can be thought of as tiny magnets. In a strong magnetic field, these nuclei can line up in one of two directions:
* with the field
* opposed to the field.

Nuclei that oppose the magnetic field have a higher energy than those aligned with the magnetic field. The stronger the magnetic field, the larger the energy gap, ΔE, between these energies – as shown in Figure 1.

Resonance

A nucleus in its low-energy spin state can be promoted to its upper-energy spin state by providing energy that is exactly equal to the energy gap, ΔE – the process is called 'excitation'.
* In an NMR spectrometer, this energy is supplied by low-energy radio-frequency radiation. ΔE is proportional to frequency so the larger the energy gap, the higher the frequency required.
* The excited nucleus will later drop back to its low-energy ground state, emitting exactly the same amount of energy in the form of radiation – this process is called 'relaxation'.

The cycle of excitation and relaxation of the nucleus is called *resonance*. Resonance continues so long as the frequency of the radiation supplied has an energy that exactly matches the energy gap between the spin states. The name *nuclear magnetic resonance* describes this process aptly.

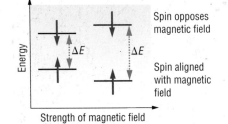

Figure 1 Different nuclear spins in an external magnetic field. The stronger the magnetic field, the larger the energy gap, ΔE, between the nuclear spins states. The green dotted lines show the energy gap between spin states at different magnetic field strengths

Key definition

Chemical shift, δ, is a scale that compares the frequency of an NMR absorption with the frequency of the reference peak of TMS at $\delta = 0$ ppm.

Nuclear shielding and chemical shift

Nuclear shielding

The magnetic field felt by a nucleus depends on two factors:

- the applied strong magnetic field
- a weaker magnetic field generated from electrons surrounding the nucleus and in nearby atoms.

An atom's electrons shield the nucleus from some of the applied magnetic field – this is called *nuclear shielding*. The extent of nuclear shielding depends on the electron density in nearby atoms or groups. This alters the environment of a nucleus, changing the energy gap. The result is that atoms in different environments have different resonance frequencies – see Figure 2.

In an NMR spectrum any differences in environment will show up as different signals and these can be identified. The interpretation of NMR spectra is discussed in detail in later spreads.

Chemical shift

Chemical shift, δ, is the place in an NMR spectrum at which a nucleus absorbs energy. NMR spectra are calibrated using the δ (delta) scale, measured in units of parts per million, or ppm. The δ scale is based on the frequencies used by the NMR spectrometer.

Chemical shift is measured relative to a reference signal from a standard compound called tetramethylsilane (TMS), $(CH_3)_4Si$.

Each molecule of TMS has 12 equivalent protons and these give rise to a single, sharp NMR signal which can be identified easily. The chemical shift of TMS is defined as $\delta = 0$ ppm.

When NMR spectra are recorded, a small amount of TMS is usually added so that the spectrometer can be calibrated against the TMS reference peak. TMS is chemically unreactive and volatile so it can easily be removed from a sample after running an NMR spectrum.

Figure 3 Tetramethylsilane (TMS), $(CH_3)_4Si$

Solvents for NMR spectroscopy

NMR analysis of a sample is usually carried out in solution. A solvent is required, but organic solvents contain carbon and hydrogen and these atoms themselves produce a signal.

The use of deuterated solvents solves this problem. Deuterium, D, is an isotope of hydrogen with one proton and one neutron in the nucleus. Deuterium has an even number of nucleons and produces no signal in an NMR spectrum. Solvents such as $CDCl_3$ are commonly used when running both proton and carbon-13 NMR spectra. For carbon-13 spectra, the carbon peak from $CDCl_3$ is usually removed from the spectrum.

After running the NMR spectrum, the sample can be recovered by evaporating off the $CDCl_3$ solvent.

A solution of the sample is prepared in a deuterated solvent with a small amount of TMS added. The turntable above the magnet (Figure 5) inserts samples into the spectrometer, allowing spectra to be run automatically.

^1H nucleus

Deuterium nucleus, ^2H

Figure 4. Isotopes of hydrogen, ^1H and ^2H; ^2H is commonly called 'deuterium'

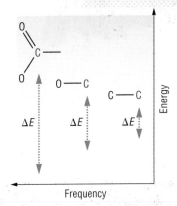

Figure 2 Nuclear shielding – electrons modify the magnetic field experienced by the nucleus. Here, the bonded oxygen attracts electrons away from the carbon atom increasing the energy gap and resonance frequency. In NMR spectra, increasing frequency is shown to the left on a scale called chemical shift

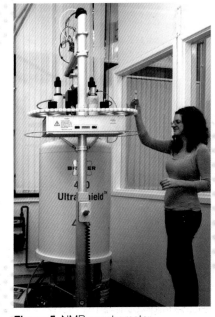

Figure 5 NMR spectrometer. A superconducting magnet is cooled to 4 K by liquid helium. The large cylinder is a giant thermos flask used to keep the helium below 4 K

Questions

1 What compound is used as the standard reference for measuring NMR chemical shifts?

2 Give an example of a deuterated solvent and explain why deuterated solvents are used in NMR spectroscopy.

By the end of this spread, you should be able to ...

＊ Analyse carbon-13 NMR spectra to make predictions about the different types of carbon atoms present.

＊ Predict the chemical shifts of carbons in a given molecule.

Carbon-13 NMR spectroscopy

Carbon has a relative atomic mass of 12.011. Almost all carbon atoms are carbon-12 with just a small proportion (1.1%) being carbon-13. Carbon-13 has an odd number of nucleons (13), resulting in a residual magnetic spin. This property is exploited in carbon-13 NMR spectroscopy and allows the identification of carbon atoms in an organic molecule. This provides an important tool in the determination of structure in organic chemistry.

Typical carbon-13 chemical shifts

In carbon-13 NMR, chemical shifts indicate the chemical environments of the carbon atoms present. The presence of an electronegative atom or group causes a significant chemical shift, δ.

Carbon-13 is sensitive to nuclear shielding and gives a large range of chemical shift values, typically 0–230 ppm. With such a large range, unless carbon atoms are present in an identical environment each carbon atom is likely to show as a separate signal.

Figure 1 shows typical chemical shift values for carbon-13 in different types of environment. However, these ranges are approximate and may not cover all carbon atoms of a particular type. The actual values can vary slightly depending on the solvent, concentration and substituents.

> **Examiner tip**
>
> A 'different type of environment' means that an atom is bonded to different atoms or groups of atoms.
>
> Equivalent carbons can be described in an identical way.

> **Examiner tip**
>
> You don't need to learn these chemical shifts – the data is provided in exams on a *Data Sheet*.

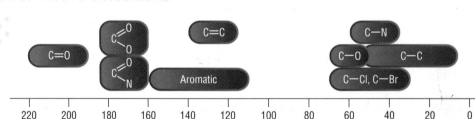

Figure 1 Carbon-13 chemical shifts

Interpreting carbon-13 NMR spectra

From a carbon-13 NMR spectrum, the most important information is:
• the *number* of different carbon environments – from the number of peaks
• the *types* of carbon environment – from the chemical shifts.

Unfortunately, the *size* of the peak tells us nothing about the number of carbon atoms responsible. This is in contrast to proton NMR spectroscopy, discussed in later spreads.

Propan-1-ol and propan-2-ol are structural isomers of C_3H_8O. Their carbon-13 NMR spectra are shown in Figure 2.

How many carbon environments?

The number of peaks gives the number of different carbon environments.
• The NMR spectrum of propan-1-ol has 3 peaks: 3 **C** atoms in 3 environments
• The NMR spectrum of propan-2-ol has 2 peaks: 3 **C** atoms in 2 environments

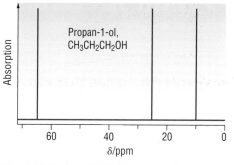
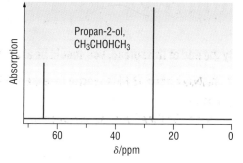

Figure 2 Carbon-13 NMR spectra of propan-1-ol and propan-2-ol

What type of carbon environments?

We find the environment of each carbon atom from chemical shift values (see Figure 1).

The NMR spectrum of propan-1-ol, $CH_3CH_2CH_2OH$ has 3 peaks so there are 3 carbon environments:

- 1 peak at δ = 64 ppm for **C**–O
- 1 peak at δ = 15 ppm for **C**–C.
- 1 peak at δ = 27 ppm for **C**–C

The NMR spectrum of propan-2-ol, $CH_3CHOHCH_3$, has 2 peaks so there are 2 carbon environments:

- 1 peak at δ = 64 ppm for **C**–O
- 1 peak at δ = 28 ppm for **C**–C.

In propan-2-ol, the two CH_3 carbon atoms are in the same environment. Each CH_3 group has its carbon atom bonded to $-CHOHCH_3$. The two CH_3 carbon atoms are equivalent – they are the 'same type of carbon'.

The NMR spectra have been repeated in Figure 3 below, showing the carbon atoms responsible for each peak.

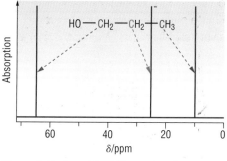
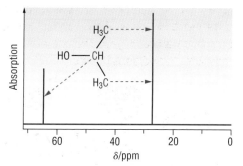

Figure 3 Carbon-13 NMR spectra of propan-1-ol and propan-2-ol showing peak assignments

Questions

Use the ^{13}C chemical shifts in Figure 1 to help you answer these questions.

1 For each compound below, predict the number of peaks and the chemical shift of each peak:
   (a) CH_3CH_2OH; **(b)** $CH_3CH_2NH_2$; **(c)** $CH_3COCH_2CH_3$.
2 For each ^{13}C spectrum below, identify the carbon atom(s) responsible for each labelled peak. Note that the δ scale is different in (a) and (b).

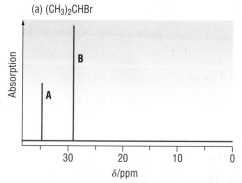
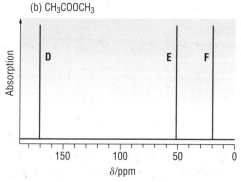

Figure 4 Carbon-13 NMR spectra of (a) $CH_3CHBrCH_3$ and (b) CH_3COOCH_3

By the end of this spread, you should be able to . . .

* Analyse carbon-13 NMR spectra to make predictions about possible structures for an unknown compound.

Making predictions

The strength of carbon-13 NMR spectroscopy lies in the analysis of unknown compounds and the confirmation of a suspected organic structure. This spread uses worked examples to look closely at these aspects of ^{13}C NMR spectroscopy. You will need to refer back to the chemical shifts given in the previous spread.

Worked example 1

A carbonyl compound has the molecular formula C_3H_6O. Using the carbon-13 NMR spectrum of the compound below, identify the compound.

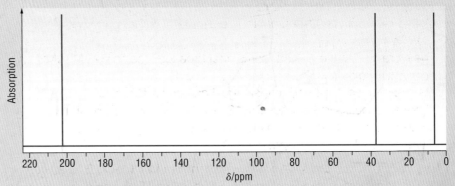

Figure 1 Carbon-13 NMR spectra of isomer of C_3H_6O

There are two possible isomers of C_3H_6O that are carbonyl compounds: propanal, CH_3CH_2CHO, and propanone, CH_3COCH_3.

The carbon environments in CH_3CH_2CHO and CH_3COCH_3 are shown in Figure 2.

Figure 2 Structural isomers of the carbonyl isomers of C_3H_6O showing different carbon environments

- Propanal has three carbon environments.
- Propanone has two carbon environments.

The carbon-13 NMR spectrum above must have been produced by propanal because there are 3 peaks for 3 different carbon environments:
- 1 peak (C^3) at $\delta = 205$ ppm for $C=O$
- 1 peak (C^2) at $\delta = 37$ ppm for $C–C$
- 1 peak (C^1) at $\delta = 6$ ppm for $C–C$.

Worked example 2

An aromatic compound has the molecular formula C_8H_{10}. Use the carbon-13 NMR spectrum of the compound in Figure 3 to identify the compound.

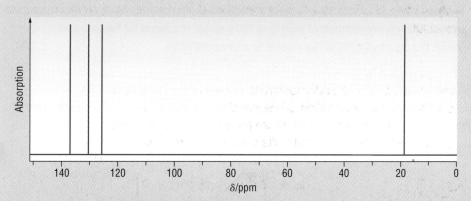

Figure 3 Carbon-13 NMR spectra of an aromatic compound with molecular formula C_8H_{10}

- The carbon-13 NMR spectrum in Figure 3 shows 4 carbon environments with 3 aromatic carbon environments.
- Only isomer A has 4 carbon environments with 3 aromatic environments. (See Figure 4.)
- Therefore compound A is the isomer that produced the carbon-13 NMR spectrum above.
- There are 4 peaks in total, so there are 4 carbon environments.
- There are 3 peaks in the range $\delta = 125–137$ ppm. These show aromatic carbon atoms in three different environments.
- There is 1 peak at $\delta = 18$ ppm for C–C.

There are four possible aromatic structural isomers of C_8H_{10} – A, B, C and D. These have different numbers of carbon environments. See Figure 4.

Isomer	Carbon environments	Aromatic environments
A 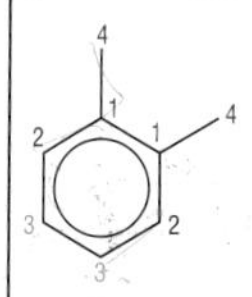	4	3
B 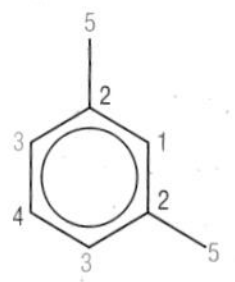	5	4
C 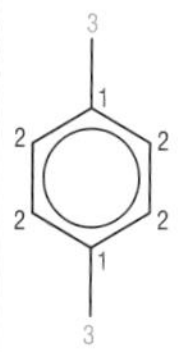	3	2
D 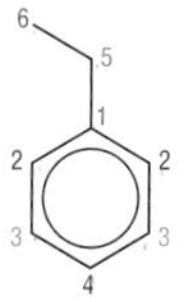	6	4

Figure 4 Structural isomers of the aromatic isomers of C_8H_{10}

Question

1 A carbonyl compound has the molecular formula C_4H_8O. Its carbon-13 NMR spectrum is shown in Figure 5.

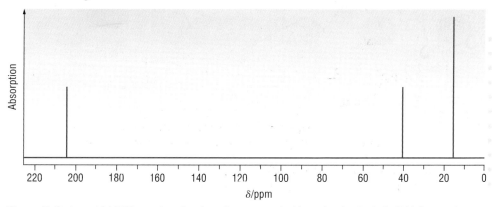

Figure 5 Carbon-13 NMR spectra of carbonyl compound with molecular formula C_4H_8O

(a) Show the possible structural isomers of the carbonyl compound C_4H_8O.
(b) How many carbon environments does each isomer have?
(c) Identify the isomer of C_4H_8O that produced the carbon-13 spectrum in Figure 5.
(d) Explain whether or not the other isomers of C_4H_8O that are carbonyl compounds could be identified from their carbon-13 NMR spectra.

⑧ Proton NMR spectroscopy

By the end of this spread, you should be able to …

* Analyse a proton NMR spectrum to make predictions about the different types of proton present, the relative numbers of each type and possible structures for the molecule.
* Predict the chemical shifts of the protons in a given molecule.

Proton NMR

Proton nuclear magnetic resonance spectra are run routinely and quickly in the analysis of organic compounds.

A ^1H nucleus has just a single proton in its nucleus and has a residual magnetic spin. ^1H is the commonest isotope of hydrogen, with a relative abundance of 99.9%. This is in contrast to ^{13}C, which makes up just 1.1% of carbon atoms. Because of the greater abundance of ^1H nuclei, proton NMR requires far less material than ^{13}C NMR.

A proton NMR spectrum is interpreted in a similar way to a carbon-13 NMR spectrum:
- the number of peaks gives the number of proton environments
- the chemical shifts give the type of proton environment of each peak.

A proton NMR spectrum also gives additional information:
- the relative peak areas give the proportions of protons in each environment
- spin–spin coupling gives information about adjacent protons.

These additional features are introduced over the next two spreads.

Typical chemical shifts

Examiner tip

You don't need to learn these chemical shifts – they are provided on exam papers.

Chemical shifts indicate the type of proton environments and the functional groups present. Figure 1 shows typical chemical shift values for protons in different chemical environments.

A proton NMR spectrum has a much narrower range of chemical shifts than ^{13}C and this can lead to overlapping of signals.

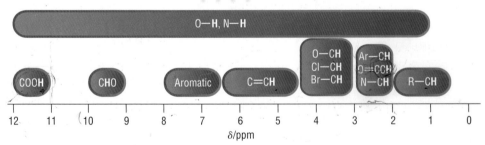

Figure 1 Proton chemical shifts

The actual chemical shifts may be slightly different depending on the environment of the protons.
- The chemical shift for O–**H** and N–**H** can vary considerably, depending on concentration, solvent and other factors.

Integration traces

In a proton NMR spectrum, the area under each peak is in direct proportion to the number of protons responsible for the absorption.

- The spectrometer measures this area. This information can be displayed as an extra line on the NMR spectrum called an **integration trace**.
- The integration trace shows a series of steps, increasing in height as it passes each peak. The ratio of the step heights is equal to the ratio of each type of proton. The values for the integration steps can also be displayed on the spectrum, labelled 'Integration', as shown in the worked example below.

Worked example

The proton NMR spectrum of a compound with the molecular formula $C_3H_6O_2$ is shown in Figure 2. The integration data are displayed at the right-hand corner of the spectrum.

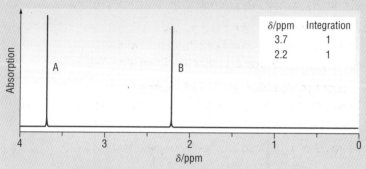

δ/ppm	Integration
3.7	1
2.2	1

Figure 2 Proton NMR spectrum of $C_3H_6O_2$ with integration data

The molecular formula is $C_3H_6O_2$:
- there are 6 protons in total.

There are 2 peaks:
- there are 2 proton environments
- with 6 protons in total, some protons must be equivalent.

Integration data:
- the ratio of protons in each environment is 1 : 1
- with 6 protons in total, each peak must represent 3 protons.

Chemical shifts:
- peak A at δ = 3.7 ppm matches O–C**H** on Figure 1 – this must be O–C**H**$_3$
- peak A at δ = 2.2 ppm matches OC–C**H** on Figure 1 – this must be OC–C**H**$_3$.

Putting all this information together, the compound is CH_3COOCH_3.

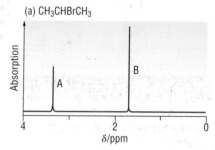

(a) $CH_3CHBrCH_3$

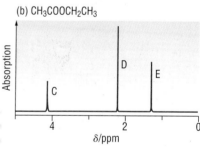

(b) $CH_3COOCH_2CH_3$

Figure 3 Proton NMR spectra of
(a) $CH_3CHBrCH_3$ and
(b) $CH_3COOCH_2CH_3$

Questions

Use the proton chemical shifts in Figure 1 to help you answer these questions.

1. For each compound below predict the number of peaks, the relative peak areas and the chemical shifts in a proton NMR spectrum:
   - **(a)** CH_3CH_2CHO;
   - **(b)** CH_3COCH_3;
   - **(c)** $HCOOCH_2CH_3$;
   - **(d)** $(CH_3)_2CHCHO$;
   - **(e)** $C_6H_5CH_2CH_3$.

2. For each proton NMR spectrum in Figure 3, identify the proton(s) responsible for each labelled peak.

Note

In an actual NMR spectrum, the peaks in these questions may be split by spin–spin coupling (discussed in spread 1.3.9). Don't worry about this at the moment.

After studying the next spread, you should be able to come back to these questions and predict any splitting patterns.

Spin–spin coupling in proton NMR spectra

By the end of this spread, you should be able to ...

✳ **Analyse a proton NMR spectrum to make predictions about the number of non-equivalent protons and possible structures for the molecule.**

✳ **Predict the splitting patterns of the protons in a given molecule.**

Spin–spin coupling

Some signals in an NMR spectrum may be split into distinctive patterns.

- This splitting is called **spin–spin coupling** and arises from interactions with the spin states of protons on adjacent carbon atoms.
- The splitting pattern can be used to find the number of protons on an adjacent carbon atom.

In an external magnetic field, each proton on an adjacent carbon has a small magnetic field of its own, which can align with or against the external field.

Adjacent protons can align their spin states in different ways. Different numbers of adjacent protons may generate different magnetic fields. The possible combinations are shown in Figure 1.

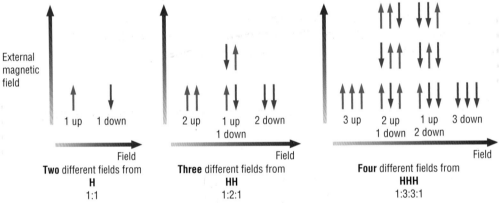

Figure 1 Different combinations of spin states for protons in adjacent H atoms

In Figure 1:

- there is always one more field than the number of adjacent protons
- the strength of a field depends on the number of ways that the spins can be aligned.

The different magnetic fields split the NMR signal of a proton into a series of sub-peaks. The number of sub-peaks matches the number of magnetic fields.

In NMR spectroscopy, we predict the splitting pattern using the **$n + 1$ rule**:

- for n protons on an adjacent carbon atom, the number of peaks in a splitting pattern equals $n + 1$.

The different splitting patterns are summarised in Table 1.

n	$n + 1$	Multiplet	Ratio of peak areas within multiplet
0	$0 + 1 = 1$	Singlet	1
1	$1 + 1 = 2$	Doublet	$1 : 1$
2	$2 + 1 = 3$	Triplet	$1 : 2 : 1$
3	$3 + 1 = 4$	Quartet	$1 : 3 : 3 : 1$

Table 1 Spin–spin coupling patterns and the $n + 1$ rule

The NMR spectrum of methyl propanoate

The proton NMR spectrum of methyl propanoate, $CH_3CH_2COOCH_3$, is shown in Figure 2.

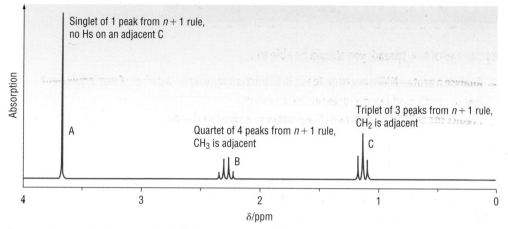

Figure 2 Proton NMR spectrum of methyl propanoate showing splitting patterns

Spin–spin coupling shows interactions between the protons on adjacent carbon atoms. You will usually see a pair of signals with one being split by the other. This gives part of the structural formula.

- The three CH_3 protons cause splitting of the adjacent CH_2 protons into a quartet.
- The two CH_2 protons cause splitting of the adjacent CH_3 protons into a triplet.
- CH_3 is adjacent to CH_2 and the bonding sequence is CH_3–CH_2–.

Spin–spin coupling happens only between non-equivalent protons.
- The three protons in CH_3 have the same chemical shift and are equivalent.
- The three protons in CH_3 do not couple with one another.

Examiner tip

Spin–spin coupling usually gives pairs of signals. The pattern gives part of the structural formula in the molecule.

The 'triplet–quartet' pairing is very common and is a giveaway for the sequence CH_3CH_2– in a molecule.

Questions

1 In the spectra below, identify the sequence in the molecule that produces the splitting patterns.

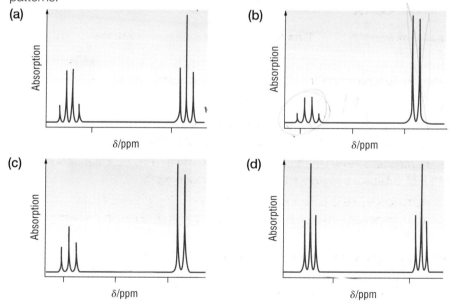

2 Predict the splitting patterns in the following molecules:
   (a) CH_3CH_2Cl;   (b) CH_3CHO;   (c) $CH_3COOCH_2CH_2Cl$

By the end of this spread, you should be able to . . .

✱ Describe the identification of O–H and N–H protons by proton exchange using D₂O.

NMR spectra of compounds with –OH and –NH protons

We have not looked at proton NMR spectra of compounds containing protons in hydroxyl groups, –OH, or amino groups, –NH.

It can be difficult to identify –OH and –NH protons:
- peaks can appear over a wide range of different chemical shift values, depending on the solvent used and the concentration of the sample
- the signals are often broad
- there is usually no splitting pattern.

These factors not only make it difficult to assign an absorption to an –OH or –NH proton, but also may lead to confusion with other peaks. Consequently an –OH or –NH absorption may mistakenly be assigned to another type of proton on the basis of an unpredictable chemical shift value.

What is needed is a way of filtering out absorptions due to –OH and –NH protons – and this is where deuterium oxide, D_2O, comes in.

Use of D₂O

In spread 1.3.5, you met the use of $CDCl_3$ as a solvent in NMR spectroscopy. Deuterium oxide, D_2O, is the same chemical compound as water, H_2O, but with the isotope deuterium, ^2H or D, in place of ^1H, the common isotope of hydrogen. This is why the common name for deuterium oxide is 'heavy water'. Unlike ^1H, deuterium does not produce an NMR signal (remember that an odd number of nucleons is needed for NMR).

D_2O is used by following these stages:
- first a proton NMR spectrum is run
- a small amount of D_2O is added to the sample solution and the mixture is shaken
- then a second proton NMR spectrum is run, and any peak due to –OH or –NH protons disappears.

This works because the deuterium in D_2O exchanges with H present in –OH and –NH.

For example, when D_2O is added to ethanol, CH_3CH_2OH, deuterium exchanges with the –OH proton:

$$CH_3CH_2OH + D_2O \rightleftharpoons CH_3CH_2OD + HOD$$

When the second NMR spectrum is run, the material being analysed is now CH_3CH_2OD rather than CH_3CH_2OH. In the absence of the –OH proton there will obviously be no –OH signal. So absorptions due to –OH and –NH protons can be easily identified by simply comparing the two spectra.

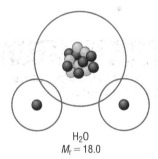

H_2O
$M_r = 18.0$

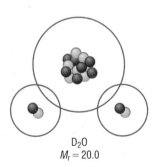

D_2O
$M_r = 20.0$

Figure 1 Water, H_2O, and heavy water, D_2O. Water and heavy water react identically in chemical reactions, but they have different physical properties. D_2O has an M_r value of 20.0 and a density of 1.11 g cm⁻³, so it is truly heavier than normal water. D_2O melts at 3.8 °C and boils at 101.4 °C. Most importantly for NMR, the deuterium nucleus does not produce a signal in a proton NMR spectrum

Figure 2 NMR sample tubes

Worked example

Compare the spectra of ethanol, C_2H_5OH, in Figure 3(a) with no D_2O and (b) with D_2O added.

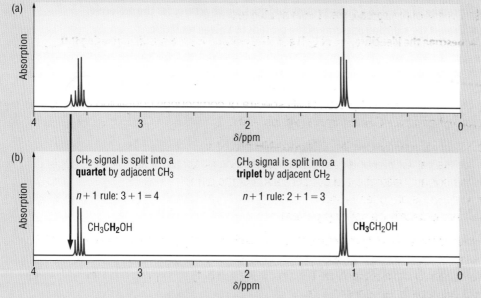

(b) CH_2 signal is split into a **quartet** by adjacent CH_3

$n + 1$ rule: $3 + 1 = 4$

CH_3CH_2OH

CH_3 signal is split into a **triplet** by adjacent CH_2

$n + 1$ rule: $2 + 1 = 3$

CH_3CH_2OH

Figure 3 Proton NMR spectrum of ethanol, C_2H_5OH: **(a)** without D_2O; **(b)** with D_2O

After D_2O is added, the –OH peak disappears.

Splitting from –OH and –NH protons

As a general rule, NMR peaks for –OH or –NH protons are not split. An –OH or –NH peak usually shows as a singlet, which may be broad.

- It is difficult to get solvents that are absolutely dry – traces of water in the solvent form hydrogen bonds with –OH and –NH protons in the compound being analysed. This results in the broadening of –OH or –NH signals.
- Protons on adjacent carbon atoms are not split by the –OH or –NH; neither is the –OH or –NH proton split itself.

In the NMR spectrum of ethanol in Figure 3, the protons in adjacent CH_3 and CH_2 groups are split:

- the CH_3 signal at $\delta = 1.1$ ppm is split into a triplet by the adjacent CH_2 ($n + 1$ rule: $2 + 1 = $ **3**).
- the CH_2 signal at $\delta = 3.6$ ppm is split into a quartet by the adjacent CH_3 ($n + 1$ rule: $3 + 1 = $ **4**).
  The adjacent OH group does not split this signal.

Questions

1 What peaks would you expect in the proton NMR spectra of the following compounds run without D_2O and with D_2O?
   (a) glycolic acid, $HOCH_2COOH$;
   (b) glycine, H_2NCH_2COOH;
   (c) lactic acid, $CH_3CHOHCOOH$.
2 The proton NMR spectrum in Figure 4 is for a compound with molecular formula $C_3H_7NO_2$. Analyse the spectrum to suggest a structure for this compound.

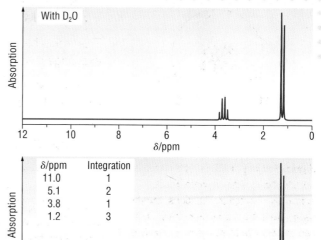

δ/ppm	Integration
11.0	1
5.1	2
3.8	1
1.2	3

Figure 4 Compound with molecular formula $C_3H_7NO_2$. The upper spectrum has been run after addition of D_2O

By the end of this spread, you should be able to ...

* Analyse a proton NMR spectrum to make predictions about the number of non-equivalent protons adjacent to a given proton, and possible structures for the molecule.

* Predict the splitting patterns of the protons in a given molecule.

In this spread, we will see how spin–spin coupling can distinguish between molecules that are structurally very similar. We will also see that chemical shifts might also be needed to aid identification.

Examiner tip

Spin–spin coupling usually comes in pairs, with one type of proton splitting the other type of proton.

The pairings are for protons on adjacent carbon atoms.

Worked example 1: using splitting

Two structural isomers of $C_3H_5ClO_2$ are $CH_3CHClCOOH$ and $ClCH_2CH_2COOH$.

The proton NMR spectra for the two isomers, run in D_2O, are shown in Figure 1.

Because D_2O has been used, we don't need to worry about the carboxyl proton, COOH.

* The upper spectrum has a quartet and doublet for CH_3CH.
* The lower spectrum has two triplets for CH_2CH_2.

By just using the splitting patterns, we can distinguish between the spectra:

* the upper spectrum is for $CH_3CHClCOOH$
* the lower spectrum is for $ClCH_2CH_2COOH$.

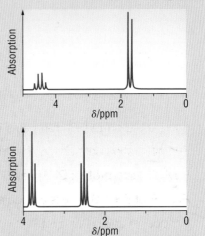

Figure 1 Proton NMR spectra of $CH_3CHClCOOH$ and $ClCH_2CH_2COOH$, both in D_2O

Worked example 2: using splitting and chemical shifts

There are four esters that have the molecular formula $C_4H_8O_2$. Their structures are:

A: $CH_3CH_2COOCH_3$; B: $CH_3COOCH_2CH_3$;
C: $HCOOCH_2CH_2CH_3$; D: $HCOOCH(CH_3)_2$

The proton NMR spectra for two of these esters are shown on the right. Match the spectra with structures.

Both spectra have the same splitting pattern:

* a quartet and triplet for CH_3CH_2
* a singlet for protons on a carbon with no adjacent protons.

Structure A and structure B are the only esters with the CH_3CH_2 sequence. This means that the singlet in each spectrum must be a CH_3 ... but which spectrum is which?

* To finish the identification, we need to consider chemical shifts.
* If you compare the spectra with the chemical shifts, you can see that the positions of the quartet and singlet are swapped over. Table 1 summarises the proton environments in A and B.
* Using the chemical shift data from Table 1, you can see that the upper spectrum is for ester A $CH_3CH_2COOCH_3$ and the lower spectrum is for ester B $CH_3COOCH_2CH_3$.

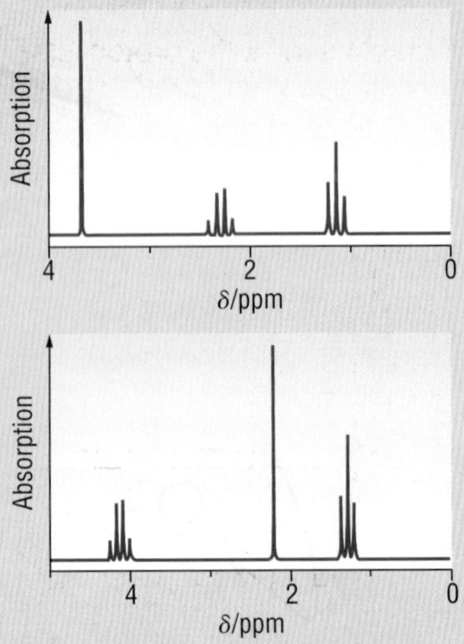

Figure 2 Proton NMR spectra of two esters of $C_4H_8O_2$

A	B	δ/ppm
CH_3CH_2	CH_3CH_2	0.7–1.6
CH_2CO	CH_3CO	2.0–2.9
OCH_3	OCH_2	3.3–4.9

Table 1

Worked example 3: what about protons on both sides?

If a CH or CH_2 group has non-equivalent protons on either side, you need to be careful. The number of adjacent carbon atoms has to be counted in both directions!

Figure 3 shows the proton NMR spectrum of 2-chloropropane, $CH_3CHClCH_3$.

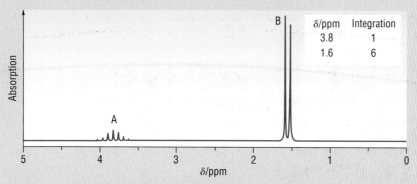

δ/ppm	Integration
3.8	1
1.6	6

Figure 3 Proton NMR spectrum of 2-chloropropane, $CH_3CHClCH_3$

There are 2 proton environments for 7 protons.

The integration data shows that the ratio of protons is A : B = 1 : 6

Peak A has 1 proton: CH; peak B has 6 protons: $2 \times CH_3$

Peak B at δ = 1.6 ppm is a doublet – there must be one adjacent proton as CH.

For peak A at δ = 3.8 ppm there is a CH_3 group on either side of the CHCl.
- There are 6 adjacent protons in total, which gives a heptet (pattern of 7 sub-peaks). With so much splitting, it can be very difficult to see the outer peaks in the split. Look carefully on the spectrum above and you can see all seven peaks in the splitting pattern at δ = 3.8 ppm.

Worked example 4: equivalent protons are not split

Figure 4 shows the proton NMR spectrum of 1,2-dichloroethane, $ClCH_2CH_2Cl$.

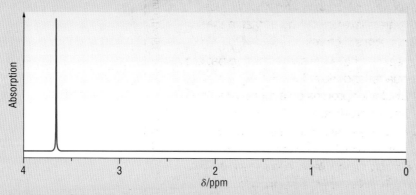

Figure 4 Proton NMR spectrum of 1,2-dichloroethane, $ClCH_2CH_2Cl$

Although there is the sequence CH_2CH_2, the NMR spectrum shows a singlet only.
- Both CH_2 groups are in the same environment and all protons are equivalent.
- Equivalent protons do not split one another.

STRETCH and CHALLENGE

The relative intensities of the lines in NMR splitting follow a natural pattern called Pascal's triangle. To work out the intensity, just add the two adjacent numbers in the row above:

```
            1
          1   1
        1   2   1
      1   3   3   1
    1   4   6   4   1
```

- a doublet has two lines with relative intensities 1:1
- a triplet has three lines with relative intensities 1:2:1
- a quartet has four lines with relative intensities 1:3:3:1.

Have a look at the splitting patterns in the NMR spectra shown on these spreads and you will see this relationship.

If you look closer, you will see that the gaps between the lines of two pairs of split patterns are the same – almost a signal to us that the peaks are interacting with one another.

The pattern in Pascal's triangle crops up in many natural series. You can find out more about this fascinating mathematical pattern on the internet. See if you can discover the number patterns that are hidden within the triangle.

Question

1 Look back at the other two esters in worked example 2. Sketch the likely NMR spectrum of each ester, showing the splitting patterns clearly.

By the end of this spread, you should be able to . . .

* Explain that NMR spectroscopy is the same technology as that used in magnetic resonance imaging (MRI).

NMR in analysis

Drugs and medicines are used to relieve pain and to combat disease. Pharmaceutical companies have developed many modern drugs and their benefits have been felt by people all over the world.

In 2006, pharmaceutical companies in the UK spent £3.95 billion on pharmaceutical research and development. This makes up about 65% of the total UK capital spent on medical research. The development and testing of a new medicine takes an average of 12 years and costs over £500 million.

Pharmaceutical firms use NMR as an important analytical technique in checking the composition of organic compounds during the synthesis of drugs.

NMR in body scanning

Nuclear magnetic resonance has been developed into the medical scanning technique known as magnetic resonance imaging, MRI. 'Nuclear magnetic resonance' was thought to be the wrong name for medical use because patients might associate 'nuclear' with the use of radioactive elements. So 'nuclear' was removed to give 'MRI'. The first MRI body scanners were available in the early 1980s. Since then, MRI has developed rapidly and is now used routinely for the diagnosis of medical conditions. Worldwide, more than 60 million MRI investigations are carried out each year.

MRI was made possible by much scientific work, particularly the efforts of Paul C Lauterbur from Illinois University in the USA, and Peter Mansfield from Nottingham University. In 2003, these two scientists received recognition for their discoveries when they were jointly awarded the Nobel Prize in Physiology or Medicine.

The MRI scan

An MRI scan is a similar technique to that used in the NMR analysis of organic compounds. However, the patient is now 'the sample' being analysed. The patient is placed inside a large, cylindrical electromagnet and radio waves are then sent through the body. Protons align with or against the strong magnetic field and they resonate in response to pulses of radio-frequency radiation. Despite this resonance, patients receiving an MRI scan report that they do not feel anything, even though protons in their bodies are resonating in response to the magnetic field and radio waves.

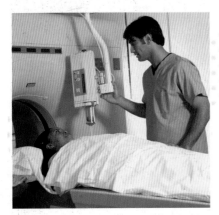

Figure 1 MRI scanning – a patient is prepared for a scan by a nurse in a hospital

Hazards

A great advantage of MRI is that it is harmless and non-invasive, ruling out the need for exploratory operations. In contrast to high-energy X-rays, MRI uses harmless low-energy radio-frequency radiation. Despite these advantages, there are some patients who should not be subjected to an MRI scan. These include patients with ferromagnetic metal (Fe, Co, Ni) implants that would be attracted towards the magnet within the scanner. Heart pacemakers are affected by the strong magnetic field and people with these implants are not even allowed to enter an MRI area. The MRI electromagnetic magnet is so strong that anything ferromagnetic – pens, knives, glasses and furniture – may fly across the room! This is potentially very hazardous and these items are not allowed in an MRI scanning area. It has even been claimed that the magnetic field is strong enough to pick up a car or to wipe a credit card!

Diagnosis of medical problems

In MRI, the image is essentially a three-dimensional NMR spectrum of protons in water and other hydrogen-containing molecules. Many diseases change the water content within tissues and organs, and the scanner detects these differences. A computer processes the information from the scan by taking slices over a period of time. This makes it possible to build up a three-dimensional image that reflects the chemical structure of the tissues.

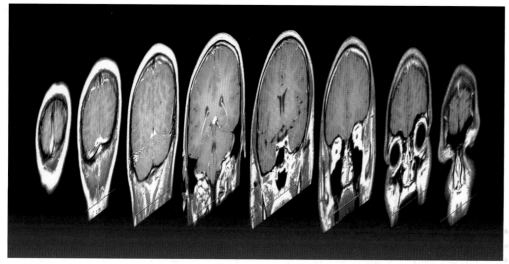

Figure 2 Eight vertical sections of an MRI head scan of a healthy 16-year-old male. The images are seen in perspective to show the slicing nature of MRI imaging

Magnetic resonance images can be obtained from almost all the parts of the body. Bones contain the fewest hydrogen atoms and produce a dark image. Tissues with many hydrogen atoms, such as fatty tissue, produce a much brighter image. An MRI scan can produce clear pictures of parts of the body that are surrounded by bone tissue, making the technique particularly useful when examining the brain and spinal cord. Most brain disorders result in changes to water content and this is reflected in the MR image.

MRI examinations are also very important in the diagnosis, treatment and follow-up of cancer. Images can reveal the exact limits of a tumour, providing important information to the surgeon for precise surgery and radiotherapy. MRI can also help to show whether a tumour has infiltrated surrounding tissue and to inform decisions on the best method of treatment.

MRI in sport

Seemingly every week, there are reports of injured sportsmen and sportswomen 'having a scan'. An MRI scan is commonly used to identify the extent of injuries such as tears in muscles, tendons and ligaments. In Figure 3 an MRI scan reveals a ruptured Achilles tendon. The Achilles tendon is the largest and strongest tendon in the human body, connecting the heel bone to the calf muscles. It can be injured during exercise, or by a trip or a fall, resulting in pain and swelling.

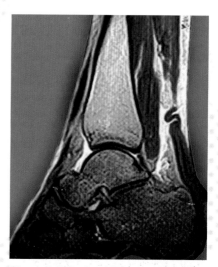

Figure 3 Side view MRI scan of a ruptured Achilles tendon. The tendon is the dark blue line to the right of the ankle – the rupture is midway along its length. Treatment is rest and physiotherapy, although in extreme cases surgery may be needed

Questions

1 Why was the name 'nuclear magnetic resonance' not used in medicine?
2 Which patients should not have an MRI scan?
3 How is a three-dimensional MR image built up?
4 Why is an MRI scan particularly good for diagnosing spinal and brain disorders?

By the end of this spread, you should be able to ...

* Analyse absorptions in an infrared spectrum to identify the presence of functional groups in an organic compound.
* Analyse molecular ion peaks and fragmentation peaks in a mass spectrum to identify parts of an organic structures.
* Combine evidence from NMR, IR and mass spectra to deduce organic structures.

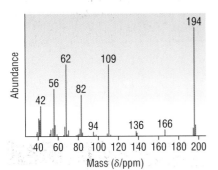

Figure 1 A mass spectrum

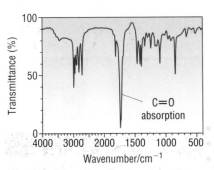

Figure 2 An IR spectrum

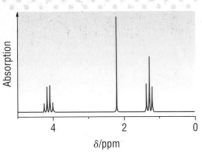

Figure 3 An NMR spectrum

Mass spectrometry

Chemical analysis of an unknown compound can provide the percentage by mass of each element. The empirical formula can be determined from the results. A mass spectrum gives the relative molecular mass of a compound. Using the empirical formula, the molecular formula can then be found.

Unfortunately, the same M_r value could apply to more than one compound with the same molecular formula – for example, CH_3COOH and $HCOOCH_3$ both have an M_r value of 60.0.

Fragmentation patterns also give clues about the carbon skeleton in the molecules of a compound.

Infrared spectroscopy

Infrared spectroscopy gives information about the bonds present in a molecule and the likely functional groups present.

However, different members of a homologous series have the same functional groups – for example, different alcohols have C–O and O–H absorptions in their IR spectra (see *AS Chemistry* spread 2.2.11).

NMR spectroscopy

A carbon-13 NMR spectrum gives information on the number and types of carbon environments in a molecule.

Proton NMR gives additional information about the number of each type of proton environment and the number of protons on adjacent carbon atoms.

Table 1 compares the information that can be obtained from carbon-13 and proton NMR spectra.

Feature	Carbon-13 NMR	Proton NMR
number of peaks	number of types of C	number of types of H
chemical shift	type of C δ range 0–220 ppm	type of H δ range 0–12 ppm
peak area	–	relative number of protons of each type
spin–spin coupling	–	number of protons on adjacent carbon

Table 1 Comparison of information from ^{13}C and ^{1}H NMR spectra

Often, a single technique is inconclusive and a combination of techniques is used in practice. This spread looks at how combining several different techniques can lead to the identification of an organic compound. Spread 1.3.14 offers practice for working on these combined techniques.

Overall, NMR gives more information than MS or IR.

Worked example

The empirical formula of an unknown compound is C_2H_4O. Use the three spectra in Figure 4 to suggest a possible structure for the compound.

To solve the problem, evidence has been annotated on each spectrum.

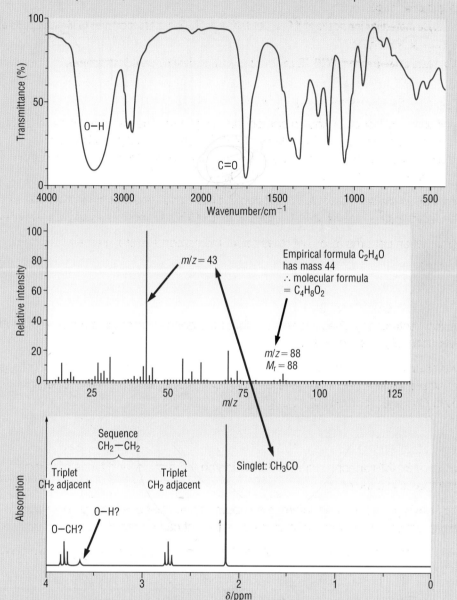

Figure 4 IR spectrum, mass spectrum and proton NMR spectrum for an unknown compound

Tying together the information is like doing a jigsaw puzzle.

- Look at the different pieces of information and get a structure that fits.
- The proton NMR gives the most information, but D_2O would have helped to confirm the O–H.
- CH_3CO from the NMR is backed up by the fragment ion in the mass spectrum at m/z = 43, suggesting CH_3CO^+.

Linking everything suggests the structure in Figure 5.

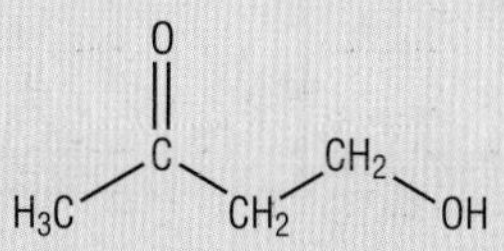

Figure 5 Suggested structure for the unknown compound

Examiner tip

In an IR spectrum, look for
C=O around 1700 cm^{-1}
O–H around 3200–3500 cm^{-1}
O–H (carboxylic acid) around 2500–3300 cm^{-1}.

Examiner tip

In a mass spectrum, first find the peak with the highest m/z value to get the mass.

Then look for fragment ions to identify parts of the skeleton of the molecule.

Examiner tip

In an NMR spectrum, look for δ values for types of proton or carbon; look for splitting for protons on adjacent carbon atoms; use the integration trace for number of each type of proton.

Question

1 Suggest values for the relative peak areas for the protons in the NMR spectrum in Figure 4.

Questions

1 The empirical formula of an unknown compound is $C_5H_{10}O_2$.

Use the four spectra in Figure 1 to suggest a possible structure for the compound.

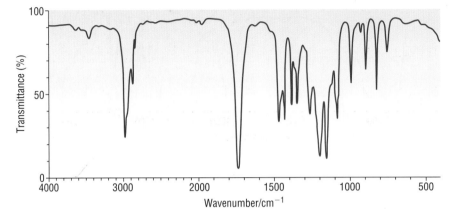

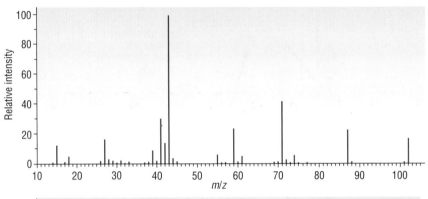

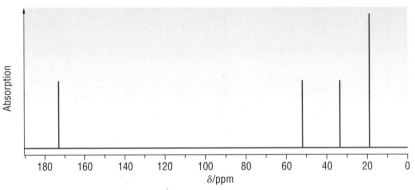

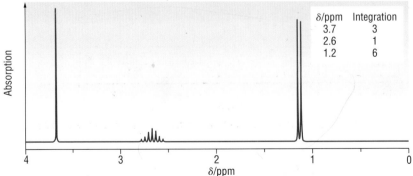

δ/ppm	Integration
3.7	3
2.6	1
1.2	6

Figure 1 IR spectrum, mass spectrum, ^{13}C and ^{1}H NMR spectra of an unknown compound

2 The empirical formula of an unknown compound is $C_9H_{10}O$.
Use the four spectra in Figure 2 to suggest a possible structure for the compound.

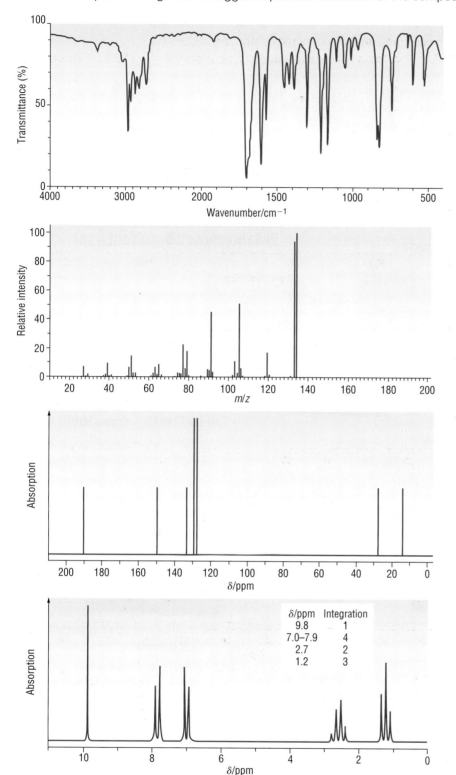

Examiner tip

Any absorption at 3200–3550 cm^{-1}?

Any absorption at 2500–3300 cm^{-1}?

Any absorption at ~1700 cm^{-1}?

Examiner tip

What is the m/z value of the molecular ion peak?

What is the molecular formula?

Check the fragment ions once you have more information.

Examiner tip

How many types of C?

How many different types of aromatic C?

What type of C absorbs at $\delta = 190$ ppm?

Examiner tip

How many types of H? (Hint – the peaks from $\delta = 7.5$–7.9 ppm are all aromatic protons.)

What are the types of H?

What does the splitting tell you?

Does the splitting of the aromatic protons give you any clues?

δ/ppm	Integration
9.8	1
7.0–7.9	4
2.7	2
1.2	3

Figure 2 IR spectrum, mass spectrum, carbon-13 and proton NMR spectra of an unknown compound

Chromatography

The mobile phase moves
- in TLC, mobile phase is a solvent
- in GC, mobile phase is carrier gas

Stationary phase does not move
- solid stationary phase separates by adsorption
- a liquid stationary phase separates by relative solubility

Retention time = time for component to pass through column

$$R_f = \frac{\text{Distance moved by component}}{\text{Distance moved by solvent front}}$$

GC–MS **GC separation** → **MS analysis**

Carbon-13 and proton NMR

		Carbon-13 NMR	Proton NMR
The number of peaks:	number of environments	✓	✓
Chemical shifts:	type of environment	✓	✓
Relative peak areas:	proportions of protons in each environment		✓
Spin–spin coupling:	protons on adjacent carbon		✓
Use of D_2O	OH and NH detection		✓

Carbon-13 NMR

Isomers of C_3H_6O

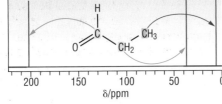

CH_3CH_2CHO – **three** carbon environments

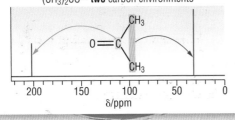

$(CH_3)_2CO$ – **two** carbon environments

Proton NMR

Splitting: number of H atoms (n) on adjacent C: **$n + 1$ rule**

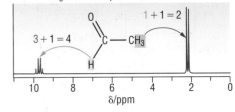

CH_3CHO – **two** proton environments

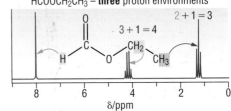

$HCOOCH_2CH_3$ – **three** proton environments

Practice questions

1 (a) In chromatography, what is meant by the terms *mobile phase* and *stationary phase*?

(b) What is used for the stationary and mobile phase in:
(i) TLC; **(ii)** GC?

(c) How is separation achieved with:
(i) a solid stationary phase;
(ii) a liquid stationary phase?

2 (a) On a TLC chromatogram, the solvent front has moved 7.5 cm. Component **A** has moved 2.1 cm, component **B** has moved 4.6 cm and component **C** has moved 5.7 cm.
What are the R_f values of each component?

(b) How does GC–MS enable components in a mixture to be identified?

3 (a) Predict the number of peaks in a carbon-13 NMR spectrum of the following carbonyl isomers of C_5H_8O:
(i) $CH_3CH_2CH_2CH_2CHO$;
(ii) $(CH_3)_3CCHO$;
(iii) $CH_3COCH(CH_3)_2$

(b) The three structural isomers, **A**, **B** and **C**, below can be distinguished using their carbon-13 NMR spectra.

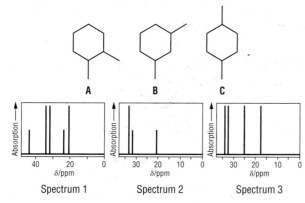

A   B   C

Spectrum 1   Spectrum 2   Spectrum 3

Match each spectrum to the compound that produces it.

4 Use the proton chemical shifts in Figure 1 in spread 1.3.8 to help you to answer these questions.

(a) For each compound below, predict the number of peaks, the relative peak areas and the chemical shifts in a proton NMR spectrum:
(i) $CH_3CH_2CH_2OH$; **(ii)** $CH_3CHOHCH_3$;
(iii) $CH_3CH_2OCH_3$

(b) The four spectra below show a splitting pattern present in the proton NMR spectrum of each of four compounds, **A–D**.
Compound **A**: $ClCH_2CH_2OH$
Compound **B**: $HCOOCH_2CH_3$
Compound **C**: CH_3CHNH_2COOH
Compound **D**: Cl_2CHCH_2OH

Identify the compound that has produced each splitting pattern.

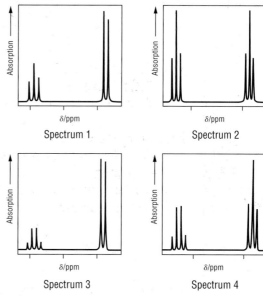

Spectrum 1   Spectrum 2

Spectrum 3   Spectrum 4

5 Outline the use of $CDCl_3$, TMS and D_2O in proton NMR spectroscopy.

6 Compound **E** is a carboxylic acid with the molecular formula $C_6H_{12}O_2$. The proton NMR spectrum of compound **E** has three singlet peaks. Identify compound **E**.

7 Compounds **F** and **G** are both structural isomers of $C_4H_8Br_2$. The peaks in their proton NMR spectra are summarised below – the relative peak areas are shown in brackets after each peak description.

Compound	Molecular formula	NMR peaks
F	$C_4H_8Br_2$	singlet (3), triplet (3), quartet (2)
G	$C_4H_8Br_2$	singlet (3), singlet (1)

Analyse this information to identify **F** and **G**.

8 Compound **H** has the following composition by mass: C, 66.67%; H, 11.11%; O, 22.22%. The mass spectrum of compound **H** has a molecular ion peak at $m/z = 72$ and a large fragment peak at $m/z = 43$. The proton NMR spectrum of compound **H** is shown below.

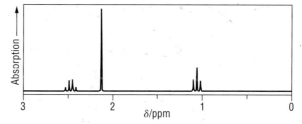

Identify compound **H**.
Explain your reasoning.

1 Chromatography is a versatile technique that may be used to separate and identify compounds.
  (a) (i) Name a type of chromatography that is used to separate and identify dissolved substances. [1]
      (ii) State what quantitative value may be determined from the chromatogram to identify the substances present in the solution. [1]
      (iii) Sketch a chromatogram to show how the value in (ii) is determined. [1]
  (b) Gas chromatography is used to separate and identify gases and liquids.
      (i) State what quantitative value is normally used to identify the components in this type of chromatography. [1]
      (ii) Sketch a chromatogram to show how the value in (i) is determined. [1]
  (c) (i) State the physical process on which the separation used in gas chromatography depends. [1]
      (ii) Describe briefly how the separation works. [4]
      [Total: 10]
      (OCR 2815/4 Jan05)

2 The spectra (shown below) and the data (shown at the top of the next column) were all produced from compound **X**, which is an ester. In this question, you will need to use this information to identify **X**.

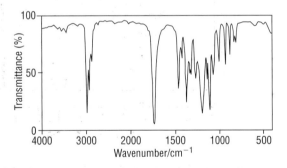

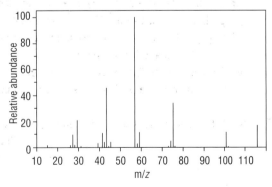

The NMR shows four peaks as detailed below.

δ/ppm	Relative peak area	Splitting pattern
0.9	3	Triplet
1.2	6	Doublet
2.3	2	Quartet
4.1	1	Multiplet (septet)

  (a) Use the infrared spectrum to confirm that compound **X** could be an ester. [2]
  (b) Identify the molecular formula and the structural formula of compound **X**.
      Clearly explain how you make your deductions from the spectra and the data. [8]
      [Total: 10]
      (OCR 2815/4 Jun07)

3 An unknown organic compound **Y** was analysed as follows.
  (a) Elemental analysis showed that compound **Y** had an empirical formula of $C_3H_6O_2$. The mass spectrum of **Y** is shown below.

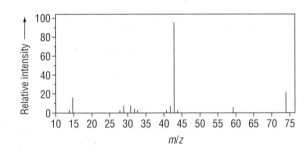

Show how the molecular formula of compound **Y** can be deduced from the information above. Explain your reasoning. Circle the peak you have used on the spectrum. [3]
  (b) (i) Compound **Y** did not give a precipitate with 2,4-dinitrophenylhydrazine. State what this tells you about compound **Y**. [1]
      (ii) Compound **Y** did not decolourise bromine water. State what this tells you about compound **Y**. [1]
  (c) Three structures for compound **Y** were suggested.

Structure 1   Structure 2   Structure 3

State the names of structures 1 and 3. [2]

(d) The infrared spectrum of compound **Y** is shown below.

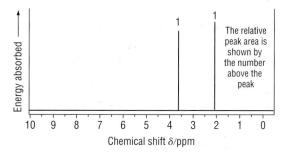

For each of the structures 1 to 3, state whether this spectrum supports its identity as compound **Y** or not. Explain your reasoning by commenting on the presence or absence of the relevant peaks. [4]

(e) In this question, one mark is available for the quality of use and organisation of scientific terms.

The NMR spectrum of compound **Y** is shown below.

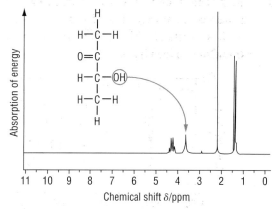

Use this spectrum to conclusively identify the structure of compound **Y**. Explain your reasoning clearly by showing why the structure you have suggested would give rise to this spectrum. Refer to the δ value of each peak, the relative peak areas and the lack of peak splitting. [6]

[Total: 17]

(OCR 2814 Jun05)

4 *Thua nao* is a traditional sauce made in northern Thailand by fermenting cooked soybeans. Its unique flavour is due to a range of volatile compounds formed during the fermentation.

One of these compounds is 3-hydroxybutanone.

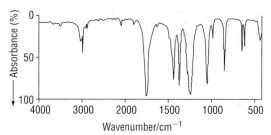

3-hydroxybutanone

(a) State the meaning of the term *volatile*. [1]

(b) Several hydroxyketones with similar boiling points can be separated from the fermentation mixture. Describe a method, which does not involve spectroscopy, that could be used to distinguish 3-hydroxybutanone from the other hydroxyketones. [4]

(c) 3-hydroxybutanone can also be identified using the NMR spectrum, shown below.

(i) Use the structure shown on the spectrum to label the parts of the molecule that are responsible for each of the peaks. One has been done for you.

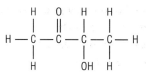

[2]

(ii) Explain how you could confirm that the labelled peak on the spectrum is the hydroxyl peak. [2]

(iii) Explain the splitting patterns shown by the peaks at δ = 1.4 ppm and δ = 4.3 ppm. [2]

(iv) For each of the four peaks on the spectrum above, write the relative peak area expected for that peak. [1]

(v) What determines the relative area of a peak? [1]

[Total: 13]

(OCR 2814 Jan06)

5 One of the final stages in winemaking involves the fermentation of malic acid to lactic acid. An equation for the reaction is shown below.

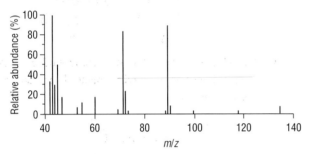

Malic acid   Lactic acid

(a) Both acids contain a chiral centre.
- **(i)** Identify the chiral centre on the structure of malic acid above using an asterisk (*). [1]
- **(ii)** Draw a diagram to show the 3D arrangement of groups around the chiral centre in malic acid. [1]

(b) Wine can be analysed for the presence of these acids by mass spectrometry.

The mass spectrum of malic acid is shown below.

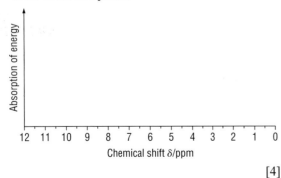

- **(i)** Draw a circle around the molecular ion peak on this spectrum of malic acid. [1]
- **(ii)** Deduce the m/z value of the molecular ion peak of lactic acid. [1]

(c) Lactic acid produces an NMR spectrum in D_2O with peaks at chemical shift values of $\delta = 1.4$ ppm and 4.3 ppm.
- **(i)** On the axes below, sketch the NMR spectrum of lactic acid in D_2O.

Show any splitting pattern and state the relative areas of the two peaks.

(blank axes: Absorption of energy vs Chemical shift δ/ppm, 12 to 0)

[4]

- **(ii)** How many peaks would you expect if the NMR spectrum of lactic acid was run in an inert solvent rather than in D_2O? Explain your answer. [2]

(d) As a wine ages, some of the acids slowly react with ethanol in the wine to produce esters.
- **(i)** Draw a displayed formula to show the structure of the ester formed when lactic acid reacts with ethanol. [1]
- **(ii)** Suggest what effect this process might have on the flavour of the wine. Explain your reasoning. [1]

[Total: 12]
(OCR 2814 Jun06)

6 An ester **D** with the formula $CH_3CH_2COOCH_2CH(CH_3)_2$ is used in rum flavouring.

(a) Draw a displayed formula of ester **D**. [2]

(b) Outline how you could obtain a sample of ester **D**, starting with a named carboxylic acid and a named alcohol. Include any essential reaction conditions and write an equation for the reaction. You do not need to include any details of the separation or purification of the ester. [6]

(c) State the spectroscopic method that could be used to confirm that a sample of ester **D** has a molecular mass of 130.

Explain how you would obtain the molecular mass of **D** from the spectrum. [2]

(d) (i) In this question, one mark is available for the quality of the use and organisation of scientific terms.

Describe and explain the different ways that a NMR spectrum can give information about a molecule. [8]

- **(ii)** The NMR spectrum of ester **D**, $CH_3CH_2COOCH_2CH(CH_3)_2$, is shown below.

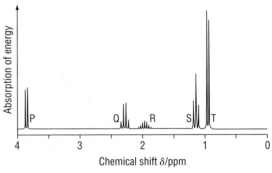

Fill in the boxes below to identify which protons in ester **D** are responsible for peaks labelled P to T on the spectrum. Peak R has been identified for you.

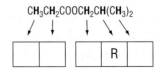

[3]

[Total: 21]
(OCR 2814 Jun07)

7   Forest fires release a large number of organic compounds into the atmosphere. These include alcohols and carboxylic acids. An environmental chemist is trying to identify one of these compounds in a sample of air.
The unknown compound **E** is thought to be a carboxylic acid with empirical formula $C_2H_3O_2$.

(a) Mass spectrometry is used to help deduce the molecular formula of compound **E**.

   (i)   Describe how the mass spectrum of compound **E** is used to determine its relative molecular mass.   [2]

   (ii) The relative molecular mass of compound **E** is shown to be 118.
   Explain how this relative molecular mass and the empirical formula are used to deduce that the molecular formula of compound **E** is $C_4H_6O_4$. Show any working.   [2]

(b) The two dicarboxylic acids with molecular formula $C_4H_6O_4$ are shown below.

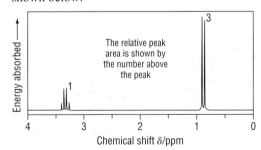

NMR spectroscopy is used to deduce which of these is the unknown compound.
The environmental chemist obtains an NMR spectrum of compound **E** and then adds some D_2O and obtains a second NMR spectrum.

   (i)   What difference would you expect between these two NMR spectra?   [1]

   (ii) The NMR spectrum of compound **E** in D_2O is shown below.

The relative peak area is shown by the number above the peak

Energy absorbed →

Chemical shift δ/ppm

Use this spectrum to identify which of the two carboxylic acids is compound **E**.
Draw the correct structure.
Use labels to show which protons on the structure are responsible for each peak on the spectrum.
Explain your reasoning by referring to the relative peak areas and the spin–spin splitting patterns of the peaks.   [6]

(iii) Predict the number of peaks and any spin–spin splitting expected on the NMR spectrum of a solution in D_2O of the other acid with formula $C_4H_6O_4$.
Explain your reasoning.   [3]

[Total: 14]
(OCR 2814 Jan05)

8   Spectroscopic techniques can be used to distinguish between compounds **X** and **Y**, shown below.

(a) In this question, one mark is available for the quality of use and organisation of scientific terms.
Describe the similarities and differences in the infrared and mass spectra of compounds **X** and **Y**, shown below.   [6]

X

Y

(b) The NMR spectrum of another compound, **Z**, shows three different proton environments. The structure of **Z** is shown below.

$H_3C - CH_2 - C$

$O - CH_3$

Z

Complete the table below for the protons in compound **Z**.

Proton environment	Chemical shift, δ/ppm	Splitting pattern	Relative peak area
$-O-CH_3$			
$-CH_2-CO$			
CH_3-C-			

[6]
[Total: 12]
(OCR 2815/4 Jan08)

Module 1
Rates, equilibrium and pH

Introduction

How do we predict whether a reaction will take place or not and, once started, how do we know how far and how fast the reaction will go? Does a chemical reaction take place in one step or does it involve many steps? What does an equilibrium constant tell us about the relative proportions of reactants and products in an industrial process?

By studying chemical kinetics and equilibrium, we can answer many of these questions and also understand the quantitative effect of temperature, pressure and concentration on chemical processes.

Acid and base reactions in solution are some of the most important processes in both chemical and biological systems. Weak acids exist in oranges and lemons, vinegar and many insect bites. Strong acids such as sulfuric and hydrochloric acid are used extensively as catalysts and in industrial processes.

In this module, you will study the complex relationships between kinetics and equilibrium. You will discover what the pH scale really tells us about acids and bases and how acidity is linked to chemical equilibrium.

Test yourself

1 What is meant by an *acid*?
2 State le Chatelier's principle.
3 Define the term *activation energy*.
4 Give three characteristics of an equilibrium system.
5 Name two industrial processes which use a catalyst.
6 Sketch a Boltzmann distribution for a gas at constant temperature.

Module contents

By the end of this spread, you should be able to ...

* Explain and use the term rate of reaction.
* Deduce the rate of a reaction from a concentration–time graph.

Introduction

In AS chemistry, you learned that the rate of a reaction depends on collisions between particles (see *AS Chemistry*, spread 2.3.10). Collisions must have more energy than the activation energy for a reaction to take place. The proportion of molecules that exceed E_a can be changed altering the temperature or in the presence of a catalyst.

Rate of reaction

The rate of a chemical reaction measures:

* how fast reactants are being used up
* how fast products are being made.

Rate of reaction is usually measured as the change in concentration of a reactant or product with time:

$$rate = \frac{\text{change in concentration of reactant or product}}{\text{time for the change to take place}}$$

$$\text{Units of rate} = \frac{\text{mol dm}^{-3}}{\text{s}} = \underbrace{\text{mol dm}^{-3}\text{ s}^{-1}}_{\text{concentration/time}}$$

Sometimes it is more convenient to use a different physical quantity that is proportional to concentration, or a different time measurement. For example, you might measure the volume of carbon dioxide gas produced during a reaction in units of cm^3 min^{-1}.

Representing concentrations

Chemists use square brackets [. . . .] to represent the concentration, in mol dm^{-3}, of a reactant or product. For example, [CO_2] means the 'concentration of CO_2 in mol dm^{-3}'.

Measuring rates

As a reaction proceeds, the concentrations of the reactants decrease:

* fewer collisions take place per second between reactant particles
* the rate slows down.

The rate of a reaction can be determined by measuring the concentration of a reactant, or product, at time intervals during the course of a reaction. The experimental results are processed and a graph is plotted of *concentration* against *time*.

* At any instant of time, the rate is equal to the slope of the curve.
* The slope is measured by drawing a tangent to the curve at this time.
* The gradient, or slope, of the tangent is then calculated.

Measuring the decrease in the concentration of a reactant

The concentration of a reactant falls during the course of a reaction – see Figure 1.

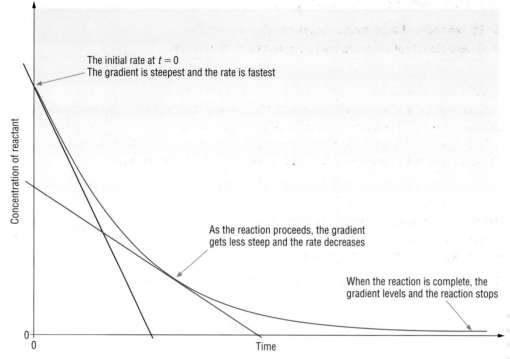

Figure 1 Concentration change of a reactant during a reaction

The following labels appear on Figure 1:
- The initial rate at $t = 0$. The gradient is steepest and the rate is fastest
- As the reaction proceeds, the gradient gets less steep and the rate decreases
- When the reaction is complete, the gradient levels and the reaction stops

The axes are labelled *Concentration of reactant* (vertical) and *Time* (horizontal).

- The reactants are being used up, so we are measuring the rate of *decrease* in concentration.
- The value of the gradient at the start of the reaction ($t = 0$) gives the **initial rate**.

Measuring the increase in the concentration of a product

The rate could also be followed by measuring the rate of *increase* in concentration of one of the products of this reaction.

Figure 2 shows how you can measure the initial rate from a concentration–time graph of a product against time.

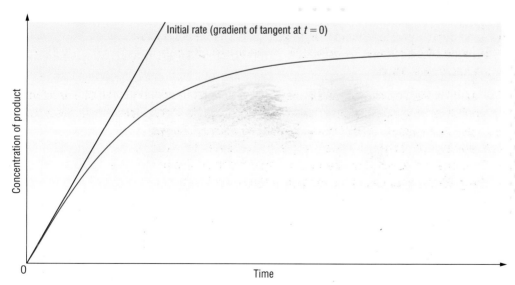

Figure 2 Measuring an initial rate

The label on Figure 2 reads: Initial rate (gradient of tangent at $t = 0$). The axes are labelled *Concentration of product* (vertical) and *Time* (horizontal).

Key definition

Initial rate of reaction is the change in concentration of a reactant, or product, per unit time at the start of the reaction when $t = 0$.

Questions

1 What are the units of rate of reaction?
2 How is a rate of reaction measured from a concentration–time graph?
3 What is meant by the *initial rate* of reaction?

By the end of this spread, you should be able to ...

❋ Plot a concentration–time graph from experimental results.
❋ Deduce the rate of a reaction from a concentration–time graph.

Obtaining data for a concentration–time graph

We can monitor reaction rates using many experimental techniques – some are outlined below.

For reactions involving acids or bases, we can measure:
• pH changes by carrying out titrations
• pH changes by using a pH meter.

For reactions that produce gases, we can measure:
• the change in volume or pressure
• the loss in mass of reactants.

For reactions that produce visual changes, we can observe:
• the formation of a precipitate
• a colour change.

Using a colorimeter to follow colour changes

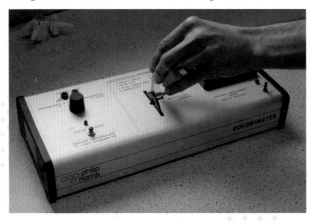

Figure 1 A colorimeter measures the intensity of different colours in a liquid – the colorimeter passes filtered light through the sample and the intensity of the colour is recorded

Bromine reacts with methanoic acid as shown in Figure 2.

$$Br_2(aq) + HCOOH(aq) \longrightarrow 2Br^-(aq) + 2H^+(aq) + CO_2(g)$$

During the course of the reaction, the orange colour of the bromine becomes less intense as its concentration falls. The intensity of the colour depends on the concentration of bromine, so the reaction rate can be followed by colorimetry.

Figure 2 Change in colour of bromine during reaction with methanoic acid

Measuring rate of reaction from a concentration–time graph

Sulfur dichloride dioxide, SO_2Cl_2, decomposes according to the equation:

$$SO_2Cl_2(g) \longrightarrow SO_2(g) + Cl_2(g)$$

In an experiment, the concentration of the reactant SO_2Cl_2 was measured over a period of time.

The results are shown below.

Time, t/s	0	500	1000	2000	3000	4000
[$SO_2Cl_2(g)$]/mol dm^{-3}	0.50	0.43	0.37	0.27	0.20	0.15

A concentration–time graph of [$SO_2Cl_2(g)$] against time, t, is shown in Figure 3.

Tangents have been drawn after 0 s (the initial rate, in red) and 3000 s (in green).

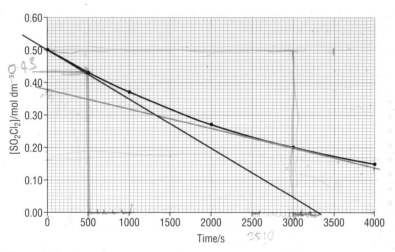

Figure 3 Drawing tangents to a concentration–time graph

The value of the gradient of each tangent is measured to give the rate at each time, t.

After $t = 0$ s, initial rate $= \dfrac{\text{change in concentration of } SO_2Cl_2}{\text{time for the change to take place}} = \dfrac{0.50 - 0.00}{3300 - 0}$ ← red line

$= 1.5 \times 10^{-4}$ mol dm^{-3} s^{-1}

After $t = 3000$ s, rate $= \dfrac{\text{change in concentration of } SO_2Cl_2}{\text{time for the change to take place}} = \dfrac{0.38 - 0.14}{4000 - 0}$ ← green line

$= 6.0 \times 10^{-5}$ mol dm^{-3} s^{-1}

Questions

1 Hydrogen peroxide, H_2O_2, decomposes spontaneously:

$$2H_2O_2(aq) \longrightarrow 2H_2O(l) + O_2(g)$$

In an experiment, the concentration of the H_2O_2 was measured over a period of time. The results are shown in the table.

Time/s	0	15	30	60	100	180
[$H_2O_2(aq)$]/mol dm^{-3}	0.40	0.28	0.19	0.07	0.03	0.01

(a) Plot a graph of [$H_2O_2(aq)$] on the y-axis against time on the x-axis.
(b) Draw tangents at the start of the reaction ($t = 0$), after 20 s and after 90 s.
(c) Measure the gradient of each tangent to find the rate of the reaction at these three times.

By the end of this spread, you should be able to . . .

* ✳ Explain and use the terms order and rate constant.
* ✳ Deduce a rate equation from orders.

Orders

In a chemical reaction, some particles have an energy greater than the activation energy. It is only these particles that have sufficient energy to react.

The greater the concentration:
* the larger the number of collisions per second
* the faster the reaction.

By carrying out experiments, we can find out how the concentration of each reactant affects the reaction rate. Mathematically this effect is called the **order** with respect to a reactant.

For a reactant A:
* $rate \propto [A]^x$
* where x = order with respect to A.

Order is always defined in terms of reactant concentrations. You will come across three common orders.

Zero order

If the order is 0 with respect to a reactant A, then $rate \propto [A]^0$.
* The rate is unaffected by changing the concentration of A.
  (Note that any number to the power 0 is equal to 1.)

First order

If the order is 1 with respect to a reactant B, then $rate \propto [B]^1$.
* If [B] increases by 2 times, the rate also increases by 2 times.
* If [B] increases by 3 times, the rate also increases by 3 times.

Second order

If the order is 2 with respect to a reactant C, then $rate \propto [C]^2$.
* If [C] increases by 2 times, the rate increases by $2^2 = 4$ times.
* If [C] increases by 3 times, the rate increases by $3^2 = 9$ times.

The rate equation and the rate constant, k

The rate equation shows the relationship between the rate of a reaction and the concentrations of the reactants raised to the powers of their orders.

Take the reaction A + B + C $\longrightarrow$ products.
Let the orders with respect to each reactant be A = 0, B = 1 and C = 2:

  $rate \propto [A]^0$, $rate \propto [B]^1$ and $rate \propto [C]^2$

Combining these expressions gives $rate \propto [A]^0[B]^1[C]^2$

The **rate constant, k**, is the constant that links the rate of reaction with the concentrations of the reactants.

The **rate equation** is $rate = k[A]^0[B]^1[C]^2$.
But remember that any number to the power zero = 1, so:

  $rate = k[B]^1[C]^2$

Key definitions

The **rate equation** for a reaction A + B → C is given by $rate = k[A]^m[B]^n$.

m is the **order of reaction** with respect to A.

n is the **order of reaction** with respect to B.

The **order** with respect to a reactant is the power to which the concentration of the reactant is raised in the rate equation.

The **overall order** of reaction is the sum of the individual orders, $m + n$.

The **rate constant, k**, is the constant that links the rate of reaction with the concentrations of the reactants raised to the powers of their orders in the rate equation.

Module 1
Rates, equilibrium and pH
Orders and the rate equation

Overall order

The **overall order** of a reaction is the sum of the individual orders.

In the example above, $rate = k[B]^1[C]^2$ and the overall order is $1 + 2$, which is 3.

The rate equation can be determined *only* from experimental results. The orders are *not* the same as the balancing numbers in the overall equation. The reasons for this are discussed in spread 2.1.7.

Units of rate constants

The units of k depend on the overall order of the rate reaction. The units of k are determined by substituting units for rate and concentration into the rate equation.

Zero order, $rate = k[A]^0 = k$ 　　Units of $k = mol\ dm^{-3}\ s^{-1}$ 　　$= mol\ dm^{-3}\ s^{-1}$

First order, $rate = k[A]$ 　$k = \dfrac{rate}{[A]}$ 　Units of $k = \dfrac{(mol\ dm^{-3}\ s^{-1})}{(mol\ dm^{-3})}$ 　$= s^{-1}$

Second order, $rate = k[A]^2$ 　$k = \dfrac{rate}{[A]^2}$ 　Units of $k = \dfrac{(mol\ dm^{-3}\ s^{-1})}{(mol\ dm^{-3})^2}$ 　$= dm^3\ mol^{-1}\ s^{-1}$

Third order, $rate = k[A]^2[B]$ 　$k = \dfrac{rate}{[A]^2[B]}$ 　Units of $k = \dfrac{(mol\ dm^{-3}\ s^{-1})}{(mol\ dm^{-3})^2(mol\ dm^{-3})}$ 　$= dm^6\ mol^{-2}\ s^{-1}$

Worked example

The reaction between hydrogen, H_2, and nitrogen monoxide, NO has the following rate equation.

$rate = k[H_2(g)][NO(g)]^2$

$6.0 \times 10^{-3}\ mol\ dm^{-3}\ H_2(g)$ was reacted with $3.0 \times 10^{-3}\ mol\ dm^{-3}\ NO(g)$. The initial rate of this reaction was $4.5 \times 10^{-3}\ mol\ dm^{-3}\ s^{-1}$.

Calculate the rate constant, k, for this reaction and state its units.

Rearrange the equation: 　　$k = \dfrac{rate}{[H_2(g)]\,[NO(g)]^2}$

Substitute values and calculate k: 　$k = \dfrac{4.5 \times 10^{-3}}{6.0 \times 10^{-3} \times (3.0 \times 10^{-3})^2} = 8.3 \times 10^4$

Work out the units. 　Units of $k = \dfrac{(mol\ dm^{-3}\ s^{-1})}{(mol\ dm^{-3})\,(mol\ dm^{-3})^2} = dm^6\ mol^{-2}\ s^{-1}$

Answer: $k = 8.3 \times 10^4\ dm^6\ mol^{-2}\ s^{-1}$

Questions

1 The rate equation for a reaction between P and Q is $rate = k[P]^2[Q]$.
  How will the rate change if:
  (a) the concentration of P is doubled;
  (b) the concentration of Q is tripled;
  (c) the concentration of P and Q are both tripled?

2 In a reaction between R, S and T:
  • when the concentration of R is doubled, the rate is unchanged
  • when the concentration of S is quadrupled, the rate increases by 16 times
  • when the concentration of T is halved, the rate is halved.
  (a) Deduce the orders with respect to R, S and T.
  (b) Hence write down the rate equation for the reaction.

3 The reaction between ozone, $O_3(g)$, and ethene, $C_2H_4(g)$ has the following rate equation.

  $rate = k[O_3(g)][C_2H_4(g)]$

  $5.0 \times 10^{-8}\ mol\ dm^{-3}\ O_3(g)$ was reacted with $1.0 \times 10^{-8}\ mol\ dm^{-3}\ C_2H_4(g)$. The initial rate of this reaction was $1.0 \times 10^{-12}\ mol\ dm^{-3}\ s^{-1}$.
  Calculate the rate constant, k, for this reaction and state its units.

By the end of this spread, you should be able to ...

* ✳ Explain and use the terms order and half-life.
* ✳ Deduce the half-life of a first-order reaction from a concentration–time graph.
* ✳ State that the half-life of a first-order reaction is independent of the concentration.

Concentration–time graphs

The concentration of a reactant is measured at time intervals during the course of a reaction.

* From the results, a concentration–time graph is plotted.
* You can determine the order with respect to the reactant from the shape of the concentration–time graph.

Half-life

The **half-life** of a reactant is an important feature of a concentration–time graph. Half-life is the time for the concentration of a reactant to reduce by half.

Concentration–time graphs for a first-order reaction

A first-order reaction can easily be identified from a concentration–time graph.

* A first-order reaction has a constant half-life.
* The half-life is the same no matter what the concentration is.

When heated strongly, nitrous oxide decomposes into its elements:

$$2N_2O(g) \longrightarrow 2N_2(g) + O_2(g)$$

This reaction is first order with respect to N_2O.

The graph in Figure 1 shows how nitrous oxide decomposes with time at constant temperature.

> **Key definition**
>
> The **half-life** of a reactant is the time taken for the concentration of the reactant to reduce by half.

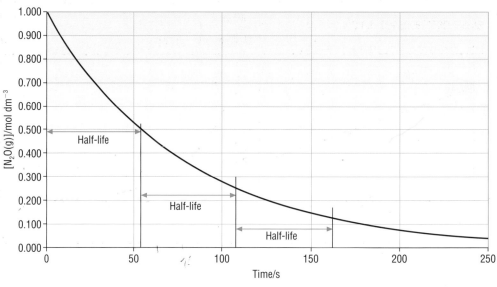

Figure 1 Measuring half-lives

The half-life is 54 s.

The initial concentration of N_2O is 1.000 mol dm⁻³.

* After 54 s (one half-life), $[N_2O]$ has fallen to 0.500 mol dm⁻³, ½ the original value.
* After 108 s (two half-lives), $[N_2O]$ has fallen to 0.250 mol dm⁻³, ¼ the original value.
* After 162 s (three half-lives), $[N_2O]$ has fallen to 0.125 mol dm⁻³, ⅛ the original value.

STRETCH and CHALLENGE

A half-life is a logarithmic relationship in terms of the exponential constant *e*. If you study maths, find out how you could use concentration and time to plot a straight line instead of a curve for a first-order reaction.

Exponential decay

Many natural processes have a constant half-life, with a value halving every half-life. This is called **exponential decay**.

Radioactive decay

This is the most well-known example of exponential decay. The constant half-life of a radioactive element is the time taken for half the element to decay. Radioactive half-lives are used to measure the stability of a radioactive element over time and are used in radiocarbon dating to find the age of historical objects.

Removal of a drug from the body

The biological half-life of a substance is the time required for half of that substance to be removed from an organism by either physical or chemical processes. Many drugs behave in this way in the body.

- Salbutamol, a drug used for treatment of asthma, has a half-life of 7 hours.
- The anti-cancer drug *cis*-platin has a half-life of 20 to 30 minutes.

Removal of charge from a capacitor

Capacitors store electric charge. When being discharged, the charge decays exponentially with a constant half-life.

Figure 2 Image-processed photo of part of the Turin Shroud, which has been kept at Turin for centuries and has been claimed to be the death-shroud of Jesus by the Christian church. Radiocarbon dating has suggested that the shroud dates from about 500 years ago rather than the time of Jesus

Other concentration–time graphs

In addition to the constant half-life for first-order reactions, concentration–time graphs for zero-order and second-order reactions also have special shapes – these are summarised in the table.

Zero order	First order	Second order
Concentration decreases at a constant rate Half-life decreases with time	The concentration halves in equal time intervals Half-life is constant	Concentration decreases rapidly, but the rate of decrease then slows down Half-life increases with time
[graph: Concentration vs Time, straight declining line] **Figure 3**	*[graph: Concentration vs Time, exponential decay curve]* **Figure 4**	*[graph: Concentration vs Time, steep then slow decay curve]* **Figure 5**

Questions

1 Hydrogen peroxide, H_2O_2, reacts in a first-order reaction with a half-life of 27 seconds. If the initial concentration of H_2O_2 is 1.60 mol dm^{-3}, what is the concentration after 81 seconds?

2 Paracetomol has a biological half life of 380 s. How long will it take for the level of paracetamol in the body to fall to one-sixteenth of its original value?

By the end of this spread, you should be able to ...

* ✳ Deduce the order (0, 1 or 2) with respect to a reactant from a rate–concentration graph.
* ✳ Determine, using the initial rates method, the order (0, 1 or 2) with respect to a reactant.

Orders and rate–concentration graphs

You can determine the order with respect to a reactant by plotting a rate–concentration graph.

Different orders give graphs with different shapes.

Zero order	First order	Second order
If the order is 0 with respect to a reactant A, then *rate* $\propto [A]^0$. • The rate is unaffected by changes in concentration.	If the order is 1 with respect to a reactant B, then *rate* $\propto [B]^1$. • If [B] increases by 2 times, the rate also increases by 2 times. • If [B] increases by 3 times, the rate also increases by 3 times.	If the order is 2 with respect to a reactant C, then *rate* $\propto [C]^2$. • If [C] increases by 2 times, the rate increases by $2^2 = 4$ times. • If [C] increases by 3 times, the rate increases by $3^2 = 9$ times.
Figure 1	Figure 2	Figure 3

Initial rates

Initial rate can be determined from a concentration–time graph. A tangent is drawn at $t = 0$ and the rate is the gradient of this tangent (see spreads 2.1.1 and 2.1.2).

Another method of obtaining a value for the initial rate is to measure the time for a certain amount of product to be formed. *Clock reactions* are ideal for this purpose – a clock reaction measures the time from the start of the reaction until there is a visual change (preferably sudden) such as:

* appearance of a precipitate
* disappearance of a solid
* a change in colour.

During a clock reaction, you are measuring the time, t, for the initial part of the reaction to take place. To a good approximation, the initial rate is proportional to $1/t$.

You carry out a series of clock reactions, varying each reactant in turn. You then plot a graph of initial rate against initial concentration for each reactant.

STRETCH and CHALLENGE

A second-order relationship can be confirmed by plotting a graph of rate against $[X]^2$, which gives a straight line. If you study maths, you can work out why.

Worked example

Sodium thiosulfate, $Na_2S_2O_3$, reacts with hydrochloric acid, HCl:

$$Na_2S_2O_3(aq) + 2HCl(aq) \longrightarrow 2NaCl(s) + S(s) + SO_2(aq) + H_2O(l)$$

Sulfur forms as a cloudy precipitate. The reaction is timed until a certain amount of precipitate has formed.

Two series of experiments are carried out using different initial concentrations of the reactants.
- *Series 1* – the concentration of $Na_2S_2O_3$ is changed while the concentration of HCl is kept constant.
- *Series 2* – the concentration of HCl is changed while the concentration of $Na_2S_2O_3$ is kept constant.

If t is the time from the start of the reaction until a certain amount of the sulfur precipitate is formed,
- the *initial rate* of the reaction is proportional to $1/t$.

The results of this investigation are shown in Figure 5 and Figure 6 as two graphs. The orders can be determined by matching the shapes to those in spread 2.1.4 (Figures 3–5).

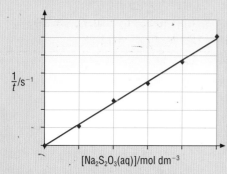

Figure 5

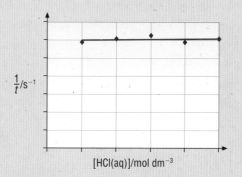

Figure 6

From the shapes of the graphs:
- the reaction is first order with respect to $Na_2S_2O_3$
- rate $\propto [Na_2S_2O_3]^1$

- the reaction is zero order with respect to HCl
- rate $\propto [HCl]^0$

From these results, the rate equation for the reaction can be written as:

$$rate = k[Na_2S_2O_3]^1[HCl]^0$$

or more simply as:

$$rate = k[Na_2S_2O_3].$$

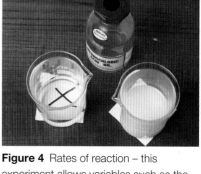

Figure 4 Rates of reaction – this experiment allows variables such as the concentration of a reactant to be investigated.

Aqueous sodium thiosulfate (left) is a colourless solution, which reacts with hydrochloric acid (upper centre) to form a cloudy solution (right).

A cross drawn on paper is placed under the glassware and the experiment is timed from when the reactants are mixed until when the cross disappears from sight

Examiner tip

Remember, you do not need to include a reactant that is zero order in a rate equation. Any value to the power '0' is equal to '1'. Here $[HCl]^0 = 1$.

Questions

1 A reaction between two reactants, P and Q, gave the rate–concentration graphs shown in Figure 7.
   (a) Determine the order with respect to P and Q.
   (b) Hence write the rate equation for the reaction.
2 The rate equation for a reaction between R and S is $rate = k[R]^2[S]$.
   (a) Determine the order with respect to R and S.
   (b) Sketch the rate–concentration graphs for reactants R and S.

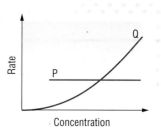

Figure 7

By the end of this spread, you should be able to ...

✳ Deduce, from orders, a rate equation of the form $rate = k[A]^m[B]^n$.

✳ Calculate the rate constant, k, from a rate equation.

✳ Explain qualitatively the effect of temperature change on a rate constant and the rate of a reaction.

Determination of orders by inspection

You may be provided with data showing the initial rates of reaction with different initial concentrations of reactants. You can find the order with respect to each reactant by comparing how the rate changes with changes in concentration. The worked example below shows how to solve this type of problem.

Worked example

Nitrogen dioxide is one of the major pollutants in air, formed by reaction of nitrogen monoxide with oxygen:

$$2NO(g) + O_2(g) \longrightarrow 2NO_2(g)$$

Three experiments are carried out to investigate the rate of this reaction. After the first experiment, the concentration of each reactant is varied. The results are shown in the table.

Experiment	Initial [NO]/mol dm⁻³	Initial [O₂]/mol dm⁻³	Initial rate/mol dm⁻³ s⁻¹
1	0.001 00	0.001 00	1.82×10^{-6}
2	0.001 00	0.003 00	5.46×10^{-6}
3	0.002 00	0.001 00	7.28×10^{-6}

Step 1: Determine the orders
Comparing experiments 1 and 2:
• [NO] has been kept constant
• [O₂] has increased by 3 times; the rate has also increased by 3 times.
• so the order with respect to $O_2 = 1$.

Comparing experiments 1 and 3:
• [O₂] has been kept constant
• [NO] has increased by 2 times; the rate has increased by $4 = 2^2$ times.
• so the order with respect to NO = 2.

Step 2: Use the orders to write the rate equation
• $rate = k[NO]^2[O_2]$
• the overall order of this reaction is $(2 + 1)$ = third order.

Step 3: Calculate the rate constant for the reaction
The rate equation is rearranged:

• The rate constant, $k = \dfrac{rate}{[NO]^2[O_2]}$

• The rate constant, k, is calculated using values from one of the experiments. Using the results from experiment 2:

$$k = \frac{(5.46 \times 10^{-6})}{(0.001\,00)^2(0.003\,00)} = 1\,820 \text{ dm}^6 \text{ mol}^{-2} \text{ s}^{-1}$$

Examiner tip

Refer back to spread 2.1.3 for details of how to work out the units of a rate constant, k.

The rate constant, *k*

Reaction rate depends on both the rate constant and the concentrations of the reactants present in the rate equation:

$$rate = k\,[NO]^2[O_2]$$

rate constant   concentrations

The larger the value of *k*, the faster the reaction:
- a fast reaction has a large value of *k*
- a slow reaction has a small value of *k*.

The effect of temperature on the rate constant, *k*

In AS chemistry, you learned that an increase in temperature gives more energy to the molecules. There are then more frequent collisions, and more of the collisions exceed the activation energy of the reaction. This was explained in terms of the Boltzmann distribution (see *AS Chemistry* spread 2.3.13).

The key factor affecting the reaction rate is the number of collisions that exceed the activation energy. This means that rate increases with temperature by much more than can be explained solely from any increased frequency of collisions.

Look at the rate equation above – if the rate increases with increasing temperature when the concentrations are the same, then the rate constant must increase with temperature.
- Raising the temperature speeds up the rate of most reactions by increasing the rate constant, *k*.
- For many reactions, the rate doubles for each 10 °C increase in temperature. This reflects the greater number of reacting particles that exceed the activation energy.

In spread 2.1.5, we looked at a reaction of hydrochloric acid and sodium thiosulfate:

$$Na_2S_2O_3(aq) + 2HCl(aq) \longrightarrow 2NaCl(s) + S(s) + SO_2(aq) + H_2O(l)$$

The rate of this reaction can be studied at different temperatures, and its rate increases rapidly with increasing temperature. This reflects the increase in the rate constant, *k*, of the reaction. Figure 1 shows how *k* increases with temperature for this reaction – this relationship is typical of many reactions.

Questions

1 Look back at the table of results in the worked example in this spread.
Work out the value of *k* for the other two experiments. You should get the same numerical value for each calculation.
2 A, B and C react together. Four experiments are carried out to investigate the kinetics of this reaction. The results are shown in the table.

Experiment	Initial [A]/ mol dm^{-3}	Initial [B]/ mol dm^{-3}	Initial [C]/ mol dm^{-3}	Initial rate/ mol dm^{-3} s^{-1}
1	0.001 00	0.003 00	0.006 00	2.16×10^{-7}
2	0.001 00	0.006 00	0.006 00	8.64×10^{-7}
3	0.000 50	0.006 00	0.006 00	4.32×10^{-7}
4	0.003 00	0.003 00	0.003 00	6.48×10^{-7}

(a) Deduce the order with respect to A, B and C. Show your reasoning.
(b) Write the rate equation.
(c) Calculate the rate constant, *k*, for the reaction.

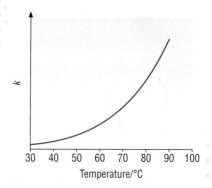

Figure 1 The rate constant, *k*, increases with increasing temperature

By the end of this spread, you should be able to ...

* Propose a rate equation that is consistent with the rate-determining step.
* Propose the steps in a reaction mechanism from the rate equation and the balanced equation for the overall reaction.

The rate-determining step

Chemical reactions often take place in a series of steps – this is called the **reaction mechanism** of the reaction.

Multi-step reactions often have one step that is much slower than the others.
* A slow step becomes an obstacle to the whole process – reactants can become products only as fast as they can get through this slow step.
* The overall reaction can be no faster than the slowest step.

The rate of the reaction is dominated by this slow step, called the **rate-determining step.**

Predicting reaction mechanisms from rate equations

Nitrogen dioxide, NO_2, reacts with carbon monoxide, CO, to form nitrogen monoxide, NO, and carbon dioxide, CO_2:

$$NO_2(g) + CO(g) \longrightarrow NO(g) + CO_2(g)$$

The overall balanced equation tells you that:
* 1 mol $NO_2(g)$ reacts with 1 mol CO(g) to produce 1 mol NO(g) and 1 mol $CO_2(g)$.

The overall equation does *not* tell you anything about the reaction mechanism. For this we have to carry out some rate experiments.

The results of rate experiments carried out on this reaction show that the reaction is:
* second order with respect to NO_2
* zero order with respect to CO.

This gives the rate equation, $rate = k[NO_2]^2$.

Worked example

If a reactant appears in the rate equation, that reactant is involved in the rate-determining step.

The order with respect to the reactant tells you how many particles of the reactant are involved in the rate-determining step.

$$rate = k\ [NO_2]^2$$

Rate-determining step involves **two molecules of NO_2**

Figure 1

The two molecules of NO_2 go on the left-hand side of the equation for the rate-determining step.

$$NO_2 + NO_2 \longrightarrow \quad \text{slow, rate-determining step}$$

Two molecules in the rate-determining step

- The rate-determining step must be followed by *further fast* steps.
- Together, the sum of all the steps must add up to give the overall equation.

We will propose a two-step mechanism for this reaction. We first summarise what we know so far.

1st step $\quad\quad$ $NO_2 + NO_2 \longrightarrow$ $\quad\quad$ slow, rate-determining step
2nd step $\quad\quad\quad\quad\quad\longrightarrow$ $\quad\quad\quad\quad$ fast
Overall equation $\quad NO_2 + CO \longrightarrow NO + CO_2$

Let's see what we must do:
- CO and CO_2 must be involved in the second step because they are in the *overall equation*.
- NO_2 must be a product of the second step because one molecule of NO_2 is in the *overall equation*.

1st step $\quad\quad$ $NO_2 + NO_2 \longrightarrow$ $\quad\quad$ slow, rate-determining step
2nd step $\quad\quad\quad\quad + CO \longrightarrow NO_2 + CO_2$ $\quad$ fast
Overall equation $\quad NO_2 + CO \longrightarrow NO + CO_2$

We are nearly there.
- The other reactant in the second step must be NO_3.
- This now gives us the products of the first step, NO and NO_3.

The completed mechanism is:
1st step $\quad\quad$ $NO_2 + NO_2 \longrightarrow NO + NO_3$ $\quad$ slow, rate-determining step
2nd step $\quad\quad$ $NO_3 + CO \longrightarrow NO_2 + CO_2$ $\quad$ fast
Overall equation $\quad NO_2 + CO \longrightarrow NO + CO_2$

NO_3 is called an **intermediate**:
- NO_3 is generated in the first step and is consumed in the second step
- NO_3 does not appear in the overall equation.

Intermediates are typically short-lived.

The tentative nature of science

A reaction mechanism is a detailed description of the way that a reaction occurs. It is a sequence of steps which lead from reactants to products.

Chemists propose mechanisms using rate equations derived from experimental data and balanced equations for the overall reaction.

Mechanisms can be proven wrong, though! They are at best 'educated guesses' and tentative explanations of chemistry that fit the experimental evidence.

Questions

1 The rate-determining step of a reaction between X and Y is:

$2X \longrightarrow Z$

Predict the rate equation for this reaction.

2 Nitrogen monoxide and oxygen react together as in the overall equation below:

$2NO(g) + O_2(g) \longrightarrow 2NO_2(g)$

The rate equation for this reaction is $rate = k[NO]^2$.
(a) Explain what is meant by *rate-determining step*.
(b) What does the rate equation tells you about the rate-determining step?
(c) Suggest a possible two-step mechanism for this reaction.

Examiner tip

It is very unlikely that more than two molecules will be involved in each step. Remember that a collision may lead to a reaction – the chances of more than two molecules colliding simultaneously are very slim!

So try to restrict each step to either one molecule (which may decompose) or two molecules (which may react on collision).

Examiner tip

The overall equation is the sum of the equations from each step in the reaction mechanism.

Key definition

An **intermediate** is a species formed in one step of a multi-step reaction that is used up in a subsequent step, and is not seen as either a reactant or a product of the overall equation.

By the end of this spread, you should be able to . . .

✳ Deduce expressions for the equilibrium constant, K_c, for homogeneous reactions.

✳ Determine the units for K_c.

Introduction

In AS chemistry, you learned that dynamic equilibrium is established in a closed system when:

- the rate of the forward reaction is the same as the rate of the reverse reaction
- the concentrations of the reactants and the products remain the same.

You used le Chatelier's principle to predict how changes in conditions may alter the equilibrium position. (See *AS Chemistry* spreads 2.3.14–2.3.15)

The equilibrium law

The equilibrium law tells us the relative proportions of reactants and products present at equilibrium.

The overall equation for a general reversible equation is shown below.

$$aA + bB \rightleftharpoons cC + dD$$

- [A], [B], [C] and [D] are the *equilibrium* concentrations of the reactants and products.
- *a*, *b*, *c* and *d* are the balancing numbers in the overall equation.

For this equilibrium, the **equilibrium law** states:

$$K_c = \frac{[C]^c[D]^d}{[A]^a[B]^b}$$

- K_c is the equilibrium constant in terms of concentrations.
- The equilibrium concentrations of the *products* are multiplied together on the *top* of the fraction.
- The equilibrium concentrations of the *reactants* are multiplied together on the *bottom* of the fraction.
- The equilibrium concentration of each product and reactant is raised to the *power* of its *balancing number* in the overall equation.

Approaching equilibrium

Dinitrogen tetroxide, N_2O_4, is a colourless gas. Nitrogen dioxide, NO_2, is a dark brown gas.

N_2O_4 and NO_2 are linked by a reversible reaction.

- A molecule of N_2O_4 can decompose into two NO_2 molecules:

$$N_2O_4(g) \longrightarrow 2NO_2(g)$$

- Two molecules of NO_2 can combine to form a molecule of N_2O_4 molecule, the reverse reaction:

$$2NO_2(g) \longrightarrow N_2O_4(g)$$

If $N_2O_4(g)$ is placed in a sealed container:

- N_2O_4 molecules start to decompose into $NO_2(g)$ molecules – because there will be very few NO_2 molecules present, the reverse reaction can happen only very slowly.

As the reaction proceeds:

- there are fewer N_2O_4 molecules available to decompose and the N_2O_4 concentration decreases – the rate of the forward reaction decreases
- there are more NO_2 molecules to recombine and the NO_2 concentration increases – the rate of the reverse reaction increases.

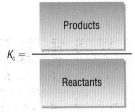

$$K_c = \frac{\text{Products}}{\text{Reactants}}$$

Figure 1

Key definition

The **equilibrium law** states that for the equilibrium

$$aA + bB \rightleftharpoons cC + dD$$

$$K_c = \frac{[C]^c[D]^d}{[A]^a[B]^b}$$

Eventually, a **dynamic equilibrium** is established:
- the forward reaction takes place at exactly the same rate as the reverse reaction
- the concentrations of the reactants and products are constant.

The dynamic equilibrium is indicated by using the '$\rightleftharpoons$' sign in the equilibrium equation:

$$N_2O_4(g) \rightleftharpoons 2NO_2(g)$$

The graph in Figure 2 shows how the concentrations of N_2O_4 and NO_2 would change until equilibrium is achieved at constant temperature.

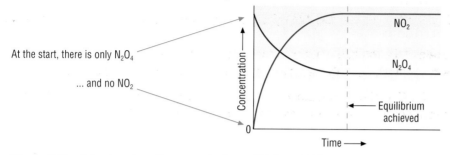

Figure 2 Graph to show how the concentrations of N_2O_4 and NO_2 change until equilibrium is achieved

Writing expressions for K_c

At equilibrium, the concentrations of $NO_2(g)$ and $N_2O_4(g)$ are constant. Their equilibrium concentrations are linked by the equilibrium constant, K_c, of the equilibrium reaction.

Applying the equilibrium law,

$$N_2O_4(g) \rightleftharpoons 2NO_2(g) \qquad K_c = \frac{[NO_2(g)]^2}{[N_2O_4(g)]}$$

Two further examples of dynamic equilibria are shown below, together with their K_c expressions.

$$2SO_2(g) + O_2(g) \rightleftharpoons 2SO_3(g) \qquad K_c = \frac{[SO_3(g)]^2}{[SO_2(g)]^2[O_2(g)]}$$

$$H_2(g) + I_2(g) \rightleftharpoons 2HI(g) \qquad K_c = \frac{[HI(g)]^2}{[H_2(g)][I_2(g)]}$$

These two examples are both **homogeneous equilibria** – all species have the same physical state, in these examples the gaseous state. A **heterogeneous equilibrium** would have a mixture of physical states.

Units of K_c

The units for K_c expressions have to be worked out afresh for each K_c expression. Each concentration term in the K_c expression is replaced by its units.

The units for each of the three equilibria above are shown below.

$$K_c = \frac{[NO_2(g)]^2}{[N_2O_4(g)]} \qquad \text{Units: } \frac{(\text{mol dm}^{-3})^2}{(\text{mol dm}^{-3})} = \text{mol dm}^{-3}$$

One of the concentration terms, mol dm^{-3}, cancels.

$$K_c = \frac{[SO_3(g)]^2}{[SO_2(g)]^2[O_2(g)]} \qquad \text{Units: } \frac{(\text{mol dm}^{-3})^2}{(\text{mol dm}^{-3})^2(\text{mol dm}^{-3})} = \text{dm}^3 \text{ mol}^{-1}$$

Note that positive indices are placed before negative indices – so the units are dm^3 mol^{-1} rather than mol^{-1} dm^3.

$$K_c = \frac{[HI(g)]^2}{[H_2(g)][I_2(g)]} \qquad \text{Units: } \frac{(\text{mol dm}^{-3})^2}{(\text{mol dm}^{-3})(\text{mol dm}^{-3})} = \text{no units}$$

All the units cancel.

Module 1
Rates, equilibrium and pH
The equilibrium constant, K_c

Key definition

A **dynamic equilibrium** exists in a closed system when the rate of the forward reaction is equal to the rate of the reverse reaction.

Examiner tip

By convention, 'reactants' are shown on the left-hand side of the equilibrium expression. 'Products' are on the right-hand side of the equilibrium expression.

A K_c expression always relates to a particular equilibrium equation.

Examiner tip

For A level, you will be asked only about homogeneous equilibria. These will be either gaseous, in solution or a mixture of liquids.

Key definitions

A **homogeneous equilibrium** is an equilibrium in which all the species making up the reactants and products are in the same physical state.

A **heterogeneous equilibrium** is an equilibrium in which species making up the reactants and products are in different physical states.

Question

1 For each of the following equilibria, write down the expression for K_c and state the units:
   (a) $2NO(g) + O_2(g) \rightleftharpoons 2NO_2(g)$
   (b) $H_2(g) + Br_2(g) \rightleftharpoons 2HBr(g)$
   (c) $N_2(g) + 3H_2(g) \rightleftharpoons 2NH_3(g)$

By the end of this spread, you should be able to ...

* Calculate the value of an equilibrium constant, K_c, including its units.
* Calculate the concentration or quantities of substances present at equilibrium.

Worked examples to determine the equilibrium constant, K_c

The worked examples in this spread show you how to work out the value of K_c from equilibrium concentrations. They also show you how to work out the amounts and concentrations present at equilibrium from the results of experiments.

Determination of K_c from equilibrium concentrations

Notes

This is the simplest form of calculation. You are given the equilibrium concentrations of all the components in the equilibrium mixture. You need to write the expression for K_c and then substitute the equilibrium concentrations – don't forget the units.

Examiner tip

Notice that all the values have been given to three significant figures, so express your answer to three significant figures.

Worked example 1

$H_2(g)$, $I_2(g)$ and $HI(g)$ exist in equilibrium in a closed system:

$$H_2(g) + I_2(g) \rightleftharpoons 2HI(g)$$

The equilibrium concentrations are:

$H_2(g)$ 0.140 mol dm^{-3}; $I_2(g)$ 0.0400 mol dm^{-3}; $HI(g)$ 0.320 mol dm^{-3}

Calculate K_c under these conditions.

Write the expression for K_c and work out the units:

$$K_c = \frac{[HI(g)]^2}{[H_2(g)][I_2(g)]} \quad \text{Units:} \quad \frac{(\text{mol dm}^{-3})^2}{(\text{mol dm}^{-3})(\text{mol dm}^{-3})} = \text{no units (all cancel)}$$

Use the equilibrium concentrations to calculate a value for K_c:

$$K_c = \frac{0.320^2}{0.140 \times 0.0400} = 18.3$$

Notes

This example has a twist. You are given *amounts*, in mol, and not concentrations. You are also given the total volume of the equilibrium mixture. So you first need to work out the equilibrium concentrations in mol dm^{-3} using:

$$\text{concentration} = \frac{\text{amount in mol}}{\text{volume in dm}^3}$$

Worked example 2

$N_2O_4(g)$ and $NO_2(g)$ exist in equilibrium in a closed container with a volume of 2.00 dm^3.

$$N_2O_4(g) \rightleftharpoons 2NO_2(g)$$

The equilibrium amounts present are 0.400 mol $N_2O_4(g)$ and 3.20 mol $NO_2(g)$. Calculate K_c under these conditions.

Take care. In this example, you are given *amounts*, in mol, and not concentrations.

First you need to work out the equilibrium concentrations.

$$[N_2O_4(g)] = \frac{\text{amount of } N_2O_4 \text{ in mol}}{\text{volume in dm}^3} = \frac{0.400}{2.00} = 0.200 \text{ mol dm}^{-3};$$

$$[NO_2(g)] = \frac{3.20}{2.00} = 1.60 \text{ mol dm}^{-3}$$

Write the expression for K_c and work out the units:

$$K_c = \frac{[NO_2(g)]^2}{[N_2O_4(g)]} \quad \text{Units:} \quad \frac{(\text{mol dm}^{-3})^2}{(\text{mol dm}^{-3})} = \text{mol dm}^{-3}$$

Use the equilibrium concentrations to calculate a value for K_c:

$$K_c = \frac{1.60^2}{0.200} = 12.8 \text{ mol dm}^{-3}$$

Examiner tip

If you have a question in which K_c has no units (with the same total moles on both sides of the equilibrium), then the volume cancels. If you have a problem like this, then use V to represent the volume. So don't panic if you haven't been given the volume in such questions.

Calculating the quantities and concentrations present at equilibrium

In AS chemistry, you carried out calculations involving stoichiometry and reacting quantities.

From the balanced equation you can find:
- the reacting quantities needed to prepare a required quantity of a product
- the quantities of products formed by reacting together known quantities of reactants.

But reactions that go to completion, you can assume that all the reactants are converted into products. This is not the case in equilibrium reactions. We still use the balanced equation, but we need to do a little more work to find the amounts of reactants and products present at equilibrium.

The equilibrium constant, K_c, can be determined from experimental results. The examples below show how to determine the equilibrium concentrations of the components in an equilibrium mixture.

Worked example

$H_2(g)$, $I_2(g)$ and $HI(g)$ exist in equilibrium.

$$H_2(g) + I_2(g) \rightleftharpoons 2HI(g)$$

0.60 mol of $H_2(g)$ was mixed with 0.40 mol of $I_2(g)$ in a sealed container with a volume of 1 dm^3. The mixture was allowed to reach equilibrium and it was found that 0.28 mol of $H_2(g)$ remained.

Find the concentrations of all three components in the equilibrium mixture.

It is useful to summarise the information in a table. At the start, only $H_2(g)$ and $I_2(g)$ are present and these values have been added to the table below in blue. You are also told how much $H_2(g)$ is present at equilibrium and this value has been added to the second row of the table in green.

Component	$H_2(g)$	$I_2(g)$	$HI(g)$
Initial amount/mol	0.60	0.40	0
Equilibrium amount/mol	0.28		

We now need to find the equilibrium amounts of I_2 and HI.

You can use the information in the table to find out how much $H_2(g)$ has reacted.

Amount of H_2 reacted = initial amount – equilibrium amount = 0.60 – 0.28 = 0.32 mol

You now use the balanced equation to find out how much I_2 has reacted and how much HI has formed.

From the equation, 1 mol H_2 + 1 mol I_2 $\longrightarrow$ 2 mol HI

$\therefore$ 0.32 mol H_2 + 0.32 mol I_2 $\longrightarrow$ 0.64 mol HI

You can now work out the equilibrium amounts of I_2 and HI using the same method as for H_2 above.

Amount of I_2 reacted = initial amount – equilibrium amount = 0.40 – 0.32 = 0.08 mol

Amount of HI formed = initial amount + equilibrium amount = 0 + 0.64 = 0.64 mol

Finally, convert the amounts to concentrations. Here the volume of the container is 1.00 dm^3, so we already have the amount in 1 dm^3. The concentration numbers are the same as the equilibrium amounts.

$\therefore$ $[H_2(g)]$ = 0.28 mol dm^{-3}; $[I_2(g)]$ = 0.08 mol dm^{-3}; $[HI(g)]$ = 0.64 mol dm^{-3}

Questions

1 In a closed system, $N_2(g)$, $H_2(g)$ and $NH_3(g)$ exist in equilibrium:

$$N_2(g) + 3H_2(g) \rightleftharpoons 2NH_3(g)$$

The equilibrium concentrations are $N_2(g)$ 1.20 mol dm^{-3}, $H_2(g)$ 2.00 mol dm^{-3} and $NH_3(g)$ 0.876 mol dm^{-3} Calculate the value for K_c under these conditions.

2 CH_3COOH, C_2H_5OH, $CH_3COOC_2H_5$ and H_2O exist in equilibrium:

$$CH_3COOH + C_2H_5OH \rightleftharpoons CH_3COOC_2H_5 + H_2O$$

2.00 moles of CH_3COOH were mixed with 3.00 moles of C_2H_5OH and the mixture allowed to reach equilibrium when 0.43 moles of CH_3COOH remained.

(a) How many moles of CH_3COOH have reacted?

(b) Determine the equilibrium concentrations of CH_3COOH, C_2H_5OH, $CH_3COOC_2H_5$ and H_2O. (You will need to use V to represent the volume, but this will cancel out in your calculation.)

(c) Hence calculate K_c.

By the end of this spread, you should be able to . . .

* Understand the significance of K_c values.
* Explain the effect of changing temperature on the value of K_c.

What is the significance of a K_c value?

The magnitude of K_c indicates the extent of a chemical reaction.

A K_c value of 1 would indicate that the position of equilibrium is halfway between reactants and products.

When $K_c \gg 1$:
* the reaction is *product-favoured*
* the products on the right-hand side predominate at equilibrium.

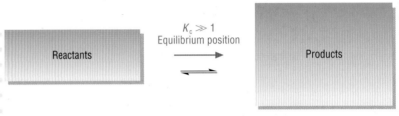

Figure 1 When $K_c \gg 1$ equilibrium lies to the right

When $K_c \ll 1$:
* the reaction is *reactant-favoured*
* the reactants on the left-hand side predominate at equilibrium.

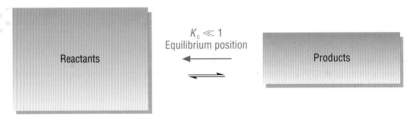

Figure 2 When $K_c \ll 1$ equilibrium lies to the left

How do changes in temperature affect K_c?

In AS chemistry, you used le Chatelier's principle (*AS Chemistry* spread 2.3.14) to predict how changes in temperature affect the position of equilibrium. You used the sign of ΔH to make predictions.
* An *increase* in temperature shifts the position of equilibrium in the *endothermic* direction (ΔH +ve).
* A *decrease* in temperature shifts the position of equilibrium in the *exothermic* direction (ΔH −ve).
* The ΔH values for the forward and reverse reactions in an equilibrium have the same magnitude, but they have opposite signs.

K_c rules

Shifts in the position of equilibrium are actually controlled by K_c, which changes its value *only* with changes in temperature. But the way that K_c changes is linked to ΔH. These changes are shown on the next page.

Module 1
Rates, equilibrium and pH
The equilibrium position and K_c

Forward reaction endothermic: ΔH +ve

When the forward reaction is endothermic as temperature increases, K_c *increases*:
- the equilibrium yield of the products on the right-hand side *increases*
- the equilibrium yield of the reactants on the left-hand side *decreases*.

Increase temperature –
K_c increases,

EXOTHERMIC
$\Delta H = -58\,\text{kJ}\,\text{mol}^{-1}$

equilibrium shifts
$N_2O_4(g) \rightleftharpoons 2NO_2(g)$
Colourless   Dark brown

ENDOTHERMIC
$\Delta H = +58\,\text{kJ}\,\text{mol}^{-1}$

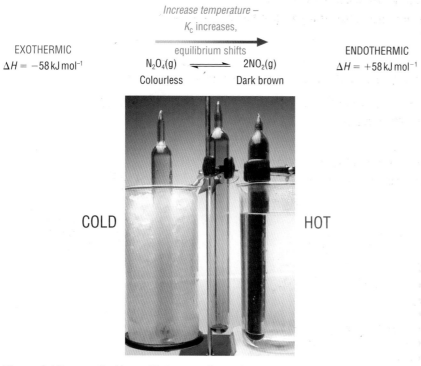

COLD   HOT

Figure 3 Nitrogen dioxide equilibrium – a change in temperature in tubes of N_2O_4 and NO_2 gases shifts the equilibrium between the two species; when more NO_2 is produced, the gas mixture inside the tube becomes darker

Forward reaction exothermic: ΔH –ve

When the forward reaction is exothermic as temperature increases, K_c *decreases*:
- equilibrium yield of the products on the right-hand side *decreases*
- the equilibrium yield of the reactants on the left-hand side *increases*.

Increase temperature –
K_c decreases,

ENDOTHERMIC
$\Delta H = +92\,\text{kJ}\,\text{mol}^{-1}$

equilibrium shifts
$N_2(g) + 3H_2(g) \rightleftharpoons 2NH_3(g)$

EXOTHERMIC
$\Delta H = -92\,\text{kJ}\,\text{mol}^{-1}$

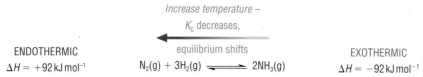

Figure 4 Ammonia equilibrium – when the forward reaction is exothermic, an increase in temperature decreases the value of K_c

Question

1 Predict whether the two reactions, A and B, are exothermic or endothermic. Explain your answer.

Temperature/K	Numerical value of K_c	
	Reaction A	Reaction B
500	5000	4.5×10^{-2}
800	0.3	2.8×10^3

The equilibrium constant, K_c, and the rate constant, k

By the end of this spread, you should be able to ...

* State that the value of K_c is unaffected by changes in concentration, pressure or the presence of a catalyst.
* Make predictions for shifts in equilibrium position from concentration and pressure changes.
* Understand that compromise conditions rely on a balance between K_c and k.

How do changes in concentration and pressure affect K_c?

In AS chemistry, you used le Chatelier's principle (*AS Chemistry* spread 2.3.14) to predict how changes in concentration and pressure affect the position of equilibrium. The reasons for these changes lie with K_c.

Although K_c is altered by changes in temperature, the value of K_c is unaffected by changes in concentration and pressure.

Changes in concentration

For the equilibrium $N_2O_4(g) \rightleftharpoons 2NO_2(g)$, the equilibrium concentrations are:

* $[NO_2(g)] = 1.60$ mol dm^{-3}
* $[N_2O_4(g)] = 0.200$ mol dm^{-3}.

$$\therefore K_c = \frac{[NO_2(g)]^2}{[N_2O_4(g)]} = \frac{1.60^2}{0.200} = 12.8 \text{ mol dm}^{-3}$$

If $[N_2O_4(g)]$ is doubled from 0.200 mol dm^{-3} to 0.400 mol dm^{-3}:

* the concentrations in the K_c expression now give the ratio $\dfrac{1.60^2}{0.400} = 6.4$ mol dm^{-3}

The system is now no longer in equilibrium:

* the equilibrium position must shift to restore this ratio back to the K_c value of 12.8 mol dm^{-3}.

The system must:

* *increase* $[NO_2(g)]$ on the top
* *decrease* $[N_2O_4(g)]$ on the bottom.

This results in a shift in the equilibrium position from *left* to *right*.

Changes in pressure

If the pressure is doubled, the concentrations of *both* $[NO_2(g)]$ and $[N_2O_4(g)]$ will also double.

* The concentrations in the K_c expression now give the ratio

$$\frac{(2 \times 1.60)^2}{2 \times 0.200} = \frac{3.20^2}{0.400} = 25.6 \text{ mol dm}^{-3}$$

The system is now no longer in equilibrium:

* the equilibrium position must shift to restore this ratio back to the K_c value of 12.8 mol dm^{-3}.

The system must:

* *decrease* $[NO_2(g)]$ on the top
* *increase* $[N_2O_4(g)]$ on the bottom.

This results in a shift in the equilibrium position from *right* to *left*.

Notes

By le Chatelier's principle, we would predict a shift in the equilibrium position from left to right to decrease $[N_2O_4(g)]$.

le Chatelier's principle works because K_c controls the relative concentrations of reactants and products.

Notes

By le Chatelier's principle, we would predict a shift in the equilibrium position from right to left towards the side with fewer gaseous moles.

This shift has been directed by the value of K_c being restored.

How does the presence of a catalyst affect K_c?

The short answer to this is 'It doesn't'.

Catalysts affect the *rate* of a chemical reaction, but not the *position* of equilibrium. Catalysts speed up *both* the forward and reverse reactions in the equilibrium by the same factor. We certainly reach equilibrium more quickly, but the equilibrium position is unchanged by the action of a catalyst.

The equilibrium constant, K_c, and the rate constant, k

In this module, you have met two of the big ideas in chemistry – equilibrium ('how far') and rates ('how fast') – and the constants that go with those ideas.

The equilibrium constant, K_c

The equilibrium constant in terms of concentrations, K_c, compares the concentrations of reactants and products present at equilibrium – this is the 'how far?'

- A large value for K_c means that the position of equilibrium lies well to the right-hand side, in favour of the products.
- If the forward reaction is endothermic, K_c increases with an increase in temperature.
- If the forward reaction is exothermic, K_c decreases with an increase in temperature.
- You can write a K_c equilibrium expression solely from a balanced equation for the equilibrium.

The rate constant, k

A rate constant, k, measures the rate of a reaction – this is the 'how fast?'

- A large value for k means a fast rate of reaction.
- k increases with an increase in temperature, so rate increases with increasing temperature.
- You can only measure k experimentally. Reactions may take several steps to reach the required products and the rate equation, from which k is determined, is a snapshot of the slowest or rate-determining step (see spread 2.1.7).

The importance of compromise

During AS chemistry, you looked at the importance of compromise in industrial processes. We discussed equilibrium in terms of le Chatelier's principle (see *AS Chemistry* spreads 2.3.14 and 2.3.15).

The equilibrium below shows the formation of ammonia from nitrogen and hydrogen:

$$N_2(g) + 3H_2(g) \rightleftharpoons 2NH_3(g) \qquad \Delta H = -92 \text{ kJ mol}^{-1}$$

Using le Chatelier's principle, we can say that increasing the temperature moves the equilibrium position to the left-hand side (the endothermic direction) so that the system compensates for the increase in energy.

In terms of K_c and k:

- at low temperatures, K_c is large (reverse reaction is endothermic) but k is small (low temperature)
- at high temperatures, K_c is small (forward reaction is exothermic) but k is large (high temperature).

The necessary compromise is achieved by getting the right balance between k and K_c:

- at low temperatures we obtain a high equilibrium yield of ammonia, but slowly
- at high temperatures we reach equilibrium quickly, but obtain little ammonia because the position of equilibrium moves far to the left-hand side.

Using compromise conditions increases the value of k enough to get a reasonable yield soon, without K_c decreasing the equilibrium yield of ammonia by too much.

Examiner tip

Rate constants are calculated from experimental data only.

Equilibrium constants are calculated from the overall balanced equation for a reaction.

Read these two lines repeatedly until ingrained in your head!

Questions

1 In a closed system, $NO(g)$, $O_2(g)$ and $NO_2(g)$ exist in equilibrium:

$$2NO(g) + O_2(g) \rightleftharpoons 2NO_2(g)$$

(a) Write the expression for K_c.

(b) The concentration of $NO(g)$ is increased. Explain in terms of K_c what happens to the equilibrium composition.

(c) The pressure on the system is decreased. Explain in terms of K_c what happens to the equilibrium composition.

2 The equilibrium position does not change with changes of pressure when there is the same number of gaseous moles on both sides.
Explain this statement, in terms of K_c.

By the end of this spread, you should be able to ...

∗ Describe an acid as a proton donor and a base as a proton acceptor.

∗ Understand that scientific knowledge is always evolving.

Acids and bases from AS chemistry

During your GCSE studies of acids and bases, you learned that the pH scale measures the strengths of acids and bases. You also learned about the reactions of acids with metals, carbonates and alkalis.

During AS chemistry, you revisited these reactions and found out how titrations can be used to determine the concentration of an unknown acid or base. (See *AS Chemistry* spreads 1.1.10–1.1.15.)

In A2 chemistry, you will learn about the importance of the H^+ ion in acid–base reactions, acid strength, pH, buffers and neutralisation.

The road to acids

Acids have been known about for hundreds of years, but knowledge was very sketchy until comparatively recently. Acids were known to be 'sour' and to change the colour of some vegetable dyes (indicators), but little else was known. This spread outlines how scientists were finally able to understand the reactions of acids. You will also see how acids link to bases and alkalis.

Figure 1 Laboratory of Joseph Priestley (1733–1804) showing various pieces of experimental apparatus for investigating gases. Priestley discovered ten gases including oxygen, hydrogen chloride, nitrous oxide and sulfur dioxide

Acids and oxygen

In the 1770s several chemists were independently carrying out experiments on the air and gases. Joseph Priestley, a British chemist, reported on his amazing discovery: 'This air is of exalted nature ... a candle burned in this air with an amazing strength of flame; and a bit of red-hot wood crackled and burned with a prodigious rapidity. But to complete the proof of the superior quality of this air, I introduced a mouse into it; and in a quantity in which, had it been common air, it would have died in about a quarter of an hour; it lived at two different times, a whole hour, and was taken out quite vigorous.'

Karl Scheele, a Swedish chemist, also discovered a gas that he referred to as 'fire air' because it supported combustion. In 1777 the Frenchman Antoine-Laurent de Lavoisier (see *AS Chemistry* spread 1.3.1) carried out further experiments based on the work of Priestley and Scheele. He concluded that combustion involves the combination of a substance with this mysterious gas. He also demonstrated the role of the gas in the rusting of metal, as well as its role in animal and plant respiration. Lavoisier tried to take credit for Priestley's discoveries and even claimed to have discovered the gas himself.

Some of Lavoisier's combustion experiments used non-metals such as phosphorus and sulfur, and these produced acidic gases. In 1778 Lavoisier mistakenly proposed that this new gas was the source of acidity. In 1779 he named this part of the gas 'oxygen' – after the Greek words *oxys* meaning 'acid' or 'sharp' and *geinomai* for 'engender' or 'give'. So 'oxygen' incorrectly means 'acid giver'. The name for this gas is unlikely to be changed despite its true meaning implying incorrect science.

Acids and hydrogen

In 1815 Humphrey Davy showed that some acidic substances, such as HCl, did not contain oxygen. In about 1832 the German chemist Justus Liebig carried out detailed studies of the chemical composition of organic acids. Liebig defined an acid as a hydrogen-containing substance in which the hydrogen could be replaced by a metal. Aspects of this definition still apply today and it was almost 50 years before Liebig's definition was updated.

Module 1
Rates, equilibrium and pH
The road to acids

Arrhenius's acid

In the late 1800s the Swedish chemist Svante Arrhenius proposed that some substances split up or dissociated in solution to form positive and negative ions. Arrhenius thought that acids and bases also dissociated when they formed solutions in water.

- An Arrhenius acid dissociates when dissolved in water to form hydrogen ions, H^+.
- An Arrhenius base dissociates when dissolved in water to form hydroxide ions, OH^-.

The Arrhenius definitions form a very good working model for acids and **alkalis**, but they are restricted to solutions in water and to bases that are soluble hydroxides.

Brønsted–Lowry acids and bases

In 1923 the Danish chemist Johannes Brønsted and the British chemist Thomas Lowry independently proposed definitions for acids and bases. This is usually referred to as the Brønsted–Lowry theory of acids and bases.

In this model, an acid–base reaction involves transfer of hydrogen ions, H^+, usually referred to as protons.

- A **Brønsted–Lowry acid** is any substance from which a proton can be removed.
- A **Brønsted–Lowry base** is any substance that can remove a proton from an acid.

As a result, many more substances could now be considered to be acids or bases than using the Arrhenius model. For example, the Brønsted–Lowry definition could be applied to include bases such as ammonia, NH_3, that are not soluble hydroxides, and also to ions. The Brønsted–Lowry model also allowed for solvents other than just water.

The Brønsted–Lowry definition is the one used at A level.

Figure 2 Svante Arrhenius (1859–1927) discovered that aqueous solutions of acids, bases and salts conduct electricity because charged ions are formed in the water – a finding that was at first scoffed at, but which later won him the 1903 Nobel Prize for Chemistry. In later years he investigated the effect of temperature on reaction rates (Arrhenius equation) and he discovered the greenhouse effect

Key definition

A **Brønsted–Lowry acid** is a proton, H^+, donor.

A **Brønsted–Lowry base** is a proton, H^+, acceptor.

An **alkali** is a base that dissolves in water forming OH^-(aq) ions.

Examiner tip

The Brønsted–Lowry theory of acids and bases is used in this A2 course. You must be comfortable with this idea of acids and bases in terms of proton transfer. It underpins all the work in this section.

STRETCH and CHALLENGE

For some solvents and chemicals, the Brønsted–Lowry definitions of acids and bases are inadequate. An even more general theory has been devised by the American chemist Gilbert Lewis. Lewis's theory is expressed in terms of electron-pair transfer, rather than proton transfer:

- a Lewis acid is an electron-pair acceptor
- a Lewis base is an electron-pair donor.

Lewis acids and bases can be atoms, ions or molecules, provided that they are capable or donating or accepting electron pairs.

Applications of this definition of acids and bases are beyond the scope of this A2 course. If you are interested, you can research the details of this theory on the Internet.

Figure 3 Johannes Brønsted proposed a new definition for acids and bases

Questions

1 The model of an acid proposed by Lavoisier correctly identifies some chemicals as being acids, but omits others. State the formulae of two acids that fit into Lavoisier's defintion and two acids that are missed.

2 What are the differences in the models of acids and bases proposed by Arrhenius, Brønsted and Lowry, and by Lewis?

By the end of this spread, you should be able to ...

✱ Illustrate the role of H⁺ in the reactions of acids with carbonates, bases, alkalis and metals.

The role of H⁺ ions in acid–base reactions

A hydrogen ion is simply a hydrogen atom that has lost its single electron. All that remains is a proton, so 'hydrogen ion', 'H⁺' and 'proton' all have the same meaning.

When an acid is added to water, the acid dissociates (splits up) releasing H⁺ ions (protons) into solution:

* hydrogen chloride, HCl

$$HCl(g) + aq \longrightarrow H^+(aq) + Cl^-(aq)$$

* sulfuric acid, H_2SO_4

$$H_2SO_4(l) + aq \longrightarrow H^+(aq) + HSO_4^-(aq)$$

In these equations, 'aq' is used to show an excess of water.

Mono-, di- and tri-basic acids

Depending on their formulae and bonding, different acids can release different numbers of protons.

In the reactions below, sometimes the release of protons is complete. This has been shown in the equations as $\longrightarrow$. If the release of protons is incomplete, this has been shown by use of the equilibrium sign, $\rightleftharpoons$, in the equations. This aspect of acid–base chemistry is discussed in detail in spread 2.1.16.

HCl is a monobasic acid because each molecule can release 1 proton:
* $HCl(aq) \longrightarrow H^+(aq) + Cl^-(aq)$ Cl^- = chloride

H_2SO_4 is a dibasic acid because each molecule can release 2 protons. This is done in two stages:
* $H_2SO_4(aq) \longrightarrow H^+(aq) + HSO_4^-(aq)$ HSO_4^- = hydrogensulfate
* $HSO_4^-(aq) \rightleftharpoons H^+(aq) + SO_4^{2-}(aq)$ SO_4^{2-} = sulfate

H_3PO_4 is a tribasic acid because each molecule can release 3 protons. This is done in three stages:
* $H_3PO_4(aq) \rightleftharpoons H^+(aq) + H_2PO_4^-(aq)$ $H_2PO_4^-$ = dihydrogenphosphate
* $H_2PO_4^-(aq) \rightleftharpoons H^+(aq) + HPO_4^{2-}(aq)$ HPO_4^{2-} = hydrogenphosphate
* $HPO_4^{2-}(aq) \rightleftharpoons H^+(aq) + PO_4^{3-}(aq)$ PO_4^{3-} = phosphate

Acid–base reactions

Aqueous acids take part in typical acid–base reactions with:
* carbonates
* bases
* alkalis.

In these reactions, the aqueous acid is **neutralised** and water is formed as one of the products.

In the reactions of acids that follow, ionic equations are used to show the important role of the H⁺ ion.
* Hydrochloric acid, HCl, nitric acid, HNO_3, and sulfuric acid, H_2SO_4, are used.
* The ionic equations include only H⁺(aq) from the acid.
* All three acids have the same ionic equation for each type of reaction.

Key definition

Neutralisation is a chemical reaction in which an acid and a base react together to produce a salt and water.

State symbols are important in ionic equations:
- ions are dissociated only in solution, so a solid ionic compound is shown undissociated
- we normally cancel only species that do not change – this includes physical states.

In all these reactions, it does not matter if the release of H^+ ions is incomplete (see above). As H^+ ions react, the acid releases more $H^+(aq)$ ions (remember le Chatelier's principle). The reaction will stop only when all the H^+ ions have been released. If the H^+ ions are all released from the start, the reaction just takes place more quickly.

Reactions with carbonates

Aqueous acids react with solid carbonates forming a salt, carbon dioxide and water:
- full equation: $2HCl(aq) + CaCO_3(s) \longrightarrow CaCl_2(aq) + CO_2(g) + H_2O(l)$
- ions: $2H^+(aq) + \cancel{2Cl^-(aq)} + CaCO_3(s) \longrightarrow Ca^{2+}(aq) + \cancel{2Cl^-(aq)} + CO_2(g) + H_2O(l)$
- ionic equation: $2H^+(aq) + CaCO_3(s) \longrightarrow Ca^{2+}(aq) + CO_2(g) + H_2O(l)$

If the carbonate is in solution, the final ionic equation simplifies because the carbonate is dissociated:
- full equation: $2HCl(aq) + Na_2CO_3(aq) \longrightarrow 2NaCl(aq) + CO_2(g) + H_2O(l)$
- ions: $2H^+(aq) + \cancel{2Cl^-(aq)} + \cancel{2Na^+(aq)} + CO_3^{2-}(aq) \longrightarrow \cancel{2Na^+(aq)} + \cancel{2Cl^-(aq)} + CO_2(g) + H_2O(l)$
- ionic equation: $2H^+(aq) + CO_3^{2-}(aq) \longrightarrow CO_2(g) + H_2O(l)$

Reactions with bases

Aqueous acids react with bases forming a salt and water:
- full equation: $2HNO_3(aq) + MgO(s) \longrightarrow Mg(NO_3)_2(aq) + H_2O(l)$
- all ions: $2H^+(aq) + \cancel{2NO_3^-(aq)} + MgO(s) \longrightarrow Mg^{2+}(aq) + \cancel{2NO_3^-(aq)} + H_2O(l)$
- ionic equation: $2H^+(aq) + MgO(s) \longrightarrow Mg^{2+}(aq) + H_2O(l)$

Reactions with alkalis

Aqueous acids react with alkalis forming a salt and water:
- full equation: $H_2SO_4(aq) + 2KOH(aq) \longrightarrow K_2SO_4(aq) + 2H_2O(l)$
- ions: $2H^+(aq) + \cancel{SO_4^{2-}(aq)} + \cancel{2K^+(aq)} + 2OH^-(aq) \longrightarrow \cancel{2K^+(aq)} + \cancel{SO_4^{2-}(aq)} + 2H_2O(l)$
- ionic equation: $2H^+(aq) + 2OH^-(aq) \longrightarrow 2H_2O(l)$
- This cancels to: $H^+(aq) + OH^-(aq) \longrightarrow H_2O(l)$

Redox reactions of acids with metals

The reaction of an acid with a metal is a redox reaction, so does not fit in with the acid–base model.

Aqueous acids react with metals forming a salt and hydrogen:
- full equation: $2HCl(aq) + Mg(s) \longrightarrow MgCl_2(aq) + H_2(g)$
- all ions: $2H^+(aq) + \cancel{2Cl^-(aq)} + Mg(s) \longrightarrow Mg^{2+}(aq) + \cancel{2Cl^-(aq)} + H_2(g)$
- ionic equation: $2H^+(aq) + Mg(s) \longrightarrow Mg^{2+}(aq) + H_2(g)$

Although most acids react with metals in this way, you have to be careful. Some acids, such as sulfuric and nitric acids, are powerful oxidising agents and other reactions may also take place, especially when the acids are concentrated. It is best to steer clear of H_2SO_4 and HNO_3 when giving examples of acid reactions with metals.

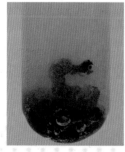

Figure 2 Zinc reacting with hydrochloric acid, an example of a metal–acid reaction. The bubbles are hydrogen gas and the salt formed is zinc chloride, $ZnCl_2$

$Zn(s) + 2HCl(aq) \rightarrow ZnCl_2(aq) + H_2(g)$
$Zn(s) + 2H^+(aq) \rightarrow Zn^{2+}(aq) + H_2(g)$

Figure 1 Reaction of solid sodium carbonate in ethanoic acid, CH_3COOH, an example of an acid–carbonate reaction. The bubbles are carbon dioxide gas and the salt formed is sodium ethanoate, CH_3COONa

$Na_2CO_3(s) + 2CH_3COOH(aq) \longrightarrow$
$\qquad 2CH_3COONa(aq) + CO_2(g) + H_2O$
$Na_2CO_3(s) + 2H^+(aq) \longrightarrow$
$\qquad 2Na^+(aq) + CO_2(g) + H_2O$

Questions

1 Write equations for dissociation of the following acids in water:
   (a) nitric acid, HNO_3;
   (b) chromic acid, H_2CrO_4.
2 Write full and ionic equations for the following acid–base reactions:
   (a) sulfuric acid and solid magnesium carbonate;
   (b) sulfuric acid and aqueous potassium carbonate;
   (c) hydrochloric acid and solid calcium oxide;
   (d) nitric acid and aqueous sodium hydroxide;
   (e) hydrochloric acid and aluminium.

By the end of this spread, you should be able to ...

✳ Describe and use the term *conjugate acid–base pairs*.

Acids as proton donors

A molecule of an acid must contain hydrogen that can be released as H^+. Although an acid is defined as a proton donor, it is important to realise that this does not just happen on its own. An acid will release a proton only if there is something (a base) that will accept it.

Hydronium ions, H_3O^+

The release of a proton from an acid often takes place when an acid is added to water. Water molecules accept protons in this process. For example, when hydrogen chloride gas is dissolved in water, the HCl molecules give up their protons to water molecules. This results in formation of a *hydronium* (or *oxonium*) *ion*, $H_3O^+(aq)$:

$$HCl(aq) + H_2O(l) \longrightarrow H_3O^+(aq) + Cl^-(aq)$$

Figure 1 shows how the hydronium ion forms in this reaction.

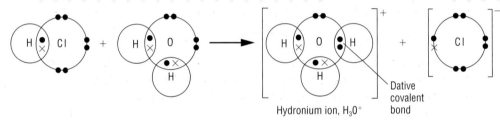

Figure 1 Formation of the hydronium ion from HCl and H_2O

The equation for this process is often shown in a simplified form, with $H^+(aq)$ being used rather than $H_3O^+(aq)$:

$$HCl(aq) \longrightarrow H^+(aq) + Cl^-(aq)$$

Using $H^+(aq)$ makes the equation easier to understand. But do remember that $H^+(aq)$ and $H_3O^+(aq)$ both represent H^+ ions in aqueous solution.

Acid–base pairs

An **acid–base pair** is a set of two species that transform into each other by gain or loss of a proton. Figure 2 identifies the acid–base pair for the dissociation of nitrous acid, HNO_2. It also shows how the acid and base in an acid–base pair are linked by H^+.

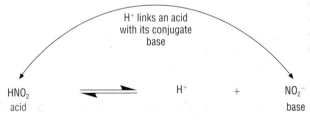

Figure 2 Conjugate acid–base pair

Key definition

An **acid–base pair** is a pair of two species that transform into each other by gain or loss of a proton.

Acid–base equilibria involve two acid–base pairs. The equilibrium below shows the dissociation of nitrous acid, HNO_2, in water:

$$HNO_2(aq) + H_2O(l) \rightleftharpoons H_3O^+(aq) + NO_2^-(aq)$$

In the forward direction:
- the acid HNO_2 releases a proton to form its conjugate base NO_2^-
- the base H_2O accepts the proton to form its conjugate acid H_3O^+.

In the reverse direction:
- the acid H_3O^+ releases a proton to form its conjugate base H_2O
- the base NO_2^- accepts the proton to form its conjugate acid HNO_2.

HNO_2 and NO_2^- differ by H^+ and make up one acid–base pair: acid 1 and base 1.

H_3O^+ and H_2O differ by H^+ and make up a second acid–base pair: acid 2 and base 2.

$$HNO_2(aq) + H_2O(l) \rightleftharpoons H_3O^+(aq) + NO_2^-(aq)$$
acid 1   base 1
         base 2   acid 2

How Alka-Seltzer works

If you look at the ingredients for Alka-Seltzer, you will find that it contains citric acid and sodium hydrogencarbonate, $NaHCO_3$. When you drop a tablet in water, the acid and sodium hydrogencarbonate react forming carbon dioxide – this produces the fizz.

But why is there no reaction until the tablet is added to water?

When the Alka-Seltzer tablet is added to water, the citric acid in the tablet reacts with water in an acid–base equilibrium to release the hydronium ion, $H_3O^+(aq)$. In the equation below, the formula of citric acid has been simplified as HA:

$$HA(aq) + H_2O(l) \rightleftharpoons H_3O^+(aq) + A^-(aq)$$

The sodium hydrogencarbonate in the tablet dissolves in the water releasing hydrogencarbonate ions, HCO_3^-:

$$NaHCO_3(s) + aq \longrightarrow Na^+(aq) + HCO_3^-(aq)$$

The hydronium ion, H_3O^+, can now react with hydrogencarbonate ions, HCO_3^-. Without water, there are no hydronium ions. Without hydronium ions, there is no reaction.

$$H_3O^+(aq) + HCO_3^-(aq) \longrightarrow 2H_2O(l) + CO_2(g)$$

Remember that the hydronium ion is usually simplified in equations as $H^+(aq)$:

$$H^+(aq) + HCO_3^-(aq) \longrightarrow H_2O(l) + CO_2(g).$$

Each effervescent tablet contains:
Aspirin 324 mg,
Sodium Hydrogen Carbonate 1625 mg,
Citric Acid 965 mg.
Sodium content 445 mg per tablet.

Alka-Seltzer dissolved in water becomes Sodium Acetylsalicylate, Sodium Citrate and Sodium Hydrogen Carbonate.

Figure 3 Alka-Seltzer tablets. The list of contents is taken from the label of a packet of tablets. Note that 'Hydrogen Carbonate' should be a single word: 'Hydrogencarbonate'!

Questions

1 Identify the acid–base pairs in the acid–base equilibria below:
   (a) $HIO_3 + H_2O \rightleftharpoons H_3O^+ + IO_3^-$
   (b) $CH_3COOH + H_2O \rightleftharpoons CH_3COO^- + H_3O^+$
   (c) $NH_3 + H_2O \rightleftharpoons NH_4^+ + OH^-$
2 Write an equation for the formation of the hydronium ion from hydrofluoric acid, HF, and water.

By the end of this spread, you should be able to ...

* Define pH as pH = −log[H⁺(aq)].
* Define [H⁺] = 10⁻ᵖᴴ.
* Convert between pH values and [H⁺(aq)].

The pH scale

The Danish chemist Søren Sørensen worked at the Carlsberg laboratories in Copenhagen. He carried out research work on amino acids, proteins and enzymes – hydrogen ion concentration plays a key role in enzyme reactions.

Aqueous solutions have concentrations of H⁺(aq) ions in the range:

* from about 10 mol dm⁻³ (10¹ mol dm⁻³)
* to about 0.000 000 000 000 001 mol dm⁻³ (10⁻¹⁵ mol dm⁻³).

This is an enormous range – difficult to understand and very cumbersome.

In 1909 Sørenson devised the pH scale as a more convenient way of measuring concentrations of H⁺(aq) ions. Sørensen's pH scale removes the awkward negative power for hydrogen ion concentrations. The pH scale was devised partly because it is easier to compare 2 with 13 than to compare 10^{-2} with 10^{-13}. Numerical values based on pH are used universally in modern science.

Figure 1 Søren Sørensen

Table 1 compares pH values with [H⁺(aq)] concentrations.

pH	−1	0	1	2	3	4	5	6	7	8	9	10	11	12	13	14	15
[H⁺(aq)]/ mol dm⁻³	10	1	10^{-1}	10^{-2}	10^{-3}	10^{-4}	10^{-5}	10^{-6}	10^{-7}	10^{-8}	10^{-9}	10^{-10}	10^{-11}	10^{-12}	10^{-13}	10^{-14}	10^{-15}

Table 1 pH and [H⁺(aq)]

Key definition

pH = −log[H⁺(aq)]

[H⁺(aq)] = 10⁻ᵖᴴ

Examiner tip

In exams, it is perfectly acceptable (and a lot easier) to define pH as pH = −log[H⁺(aq)] rather than trying to express this formula as a sentence.

If you compare the pH and [H⁺(aq)] values, you can see a pattern:

* the pH is the negative power of [H⁺(aq)]
* [H⁺(aq)] = 10⁻ᵖᴴ.

The pH scale is logarithmic – the difference between each successive whole number pH value is a factor of 10.

For example, the difference between pH 3 and pH 5, is a factor of 10 × 10 = 100.

* pH is equal to the negative logarithm to the base 10 of [H⁺(aq)]
* pH = −log[H⁺(aq)].

What does a pH value mean?

The relationship between pH and [H⁺(aq)] is sometimes called a 'see-saw' relationship – 'when one is up, the other is down'.

* A *low* pH value means a *large* [H⁺(aq)].
* A *high* pH value means a *small* [H⁺(aq)].
* A pH change of 1 changes [H⁺(aq)] by 10 times.
* An acid with a pH of 2 contains 1000 times the [H⁺(aq)] of an acid with a pH of 5.

STRETCH and CHALLENGE

In truth, no one really knows any more what pH stands for. Books often claim 'power of hydrogen' but there are many other suggestions. Carry out some Internet research and see how many you can find.

Converting between pH and [H⁺(aq)]

Table 2 shows examples of the relationship between pH and $[H^+(aq)]$ for which pH is not a whole number and $[H^+(aq)]$ is not an exact multiple of 10.

pH	4.52	5.79	10.63	13.41
[H⁺(aq)]/mol dm⁻³	$10^{-4.52}$	$10^{-5.79}$	$10^{-10.63}$	$10^{-13.41}$

Table 2

Unfortunately the $[H^+(aq)]$ values above are not shown conventionally in standard form. To get them into standard form, we must turn to a calculator for help. Simply input 10^{-pH}. Figure 2 shows the calculator key to use.

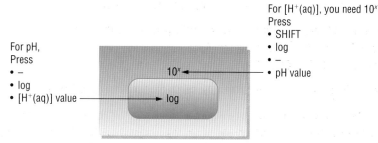

For pH,
Press
- –
- log
- [H⁺(aq)] value

For [H⁺(aq)], you need 10^x
Press
- SHIFT
- log
- –
- pH value

Figure 2 The calculator key for pH calculations

pH	4.52	5.79	10.63	13.41
[H⁺(aq)]/mol dm⁻³	3.02×10^{-5}	1.62×10^{-6}	2.34×10^{-11}	3.89×10^{-14}

- Practise this on your calculator.
- Then try to convert the $[H^+(aq)]$ values back to pH.
- Repeat until you are comfortable with converting between pH and $[H^+(aq)]$.

Hints for pH calculations

pH calculations are easy if you learn how to use your calculator! Different makes and versions of calculators may need different keys inputs, so *don't* rely on borrowing a calculator in lessons or for your exam or you may get confused. The example shown in Figure 2 may not work for all brands of calculator.

Finally, do decide if the calculator's answer looks sensible. A calculator is only as accurate as the person pressing the keys! How many times have you heard someone say 'My calculator won't give me the right answer'?

Questions

1 How many times more hydrogen ions are in an acidic solution of pH 2 than in one of pH 4?
2 Calculate the pH of solutions with the following $[H^+(aq)]$ values. Give your answers to two decimal places.
  (a) 3.33×10^{-3} mol dm⁻³;
  (b) 4.73×10^{-4} mol dm⁻³
  (c) 2.39×10^{-12} mol dm⁻³;
  (d) 2.30 mol dm⁻³.
3 Calculate $[H^+(aq)]$ of solutions with the following pH values. Give your answers to three significant figures.
  (a) pH 6.53;
  (b) pH 2.87;
  (c) pH 9.58;
  (d) pH –0.12.

By the end of this spread, you should be able to ...

* Explain qualitatively the differences between strong and weak acids.
* Explain that the acid dissociation constant, K_a, shows the extent of acid dissociation.
* Deduce expressions for K_a and pK_a for weak acids.

HCl	hydrochloric acid
HNO_3	nitric acid
H_2SO_4	sulfuric acid
HBr	hydrobromic acid
HI	hydriodic acid
$HClO_4$	chloric(VII) (perchloric) acid

Table 1 Strong acids

Key definitions

A **strong acid** is an acid that completely dissociates in solution.

A **weak acid** is an acid that partially dissociates in solution.

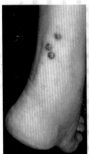 

Figure 1 Formic acid (methanoic acid, HCOOH) is a weak acid that occurs naturally in bee and ant stings. Formic acid makes up 55–60% of the body mass of a typical ant; its name comes from 'formica', the Latin word for ant. The acid was once prepared by distilling ants!

Key definition

The acid dissociation constant, K_a, of an acid HA is defined as

$$K_a = \frac{[H^+(aq)][A^-(aq)]}{[HA(aq)]}$$

Acid–base equilibria

In aqueous solution, acids dissociate and equilibrium is set up. The equilibrium below shows the dissociation of an acid, HA, in water:

$$HA(aq) + H_2O(l) \rightleftharpoons H_3O^+(aq) + A^-(aq)$$

or more simply:

$$HA(aq) \rightleftharpoons H^+(aq) + A^-(aq).$$

The *strength* of an acid HA is the extent of its dissociation into H^+ and A^- ions.

Strong acids

Some acids are *strong* – they are 100% dissociated in aqueous solution. There are comparatively few **strong acids**; the commonest are listed in Table 1. Almost every other acid is weak.

Weak acids

A **weak acid** only partially dissociates in aqueous solution. Many naturally occurring acids are weak.

The equilibrium that is set up when ethanoic acid, CH_3COOH, dissociates in water is shown below:

$$CH_3COOH(aq) \rightleftharpoons H^+(aq) + CH_3COO^-(aq)$$

* The equilibrium position lies well over to the left.
* There are only small concentrations of dissociated ions, $H^+(aq)$ and $CH_3COO^-(aq)$, compared with the concentration of undissociated $CH_3COOH(aq)$.

The acid dissociation constant, K_a

The actual extent of acid dissociation is measured by an equilibrium constant called the **acid dissociation constant, K_a**.

A weak acid HA has the following equilibrium in aqueous solution:

$$HA(aq) \rightleftharpoons H^+(aq) + A^-(aq)$$

The expression for the acid dissociation constant is:

$$K_a = \frac{[H^+(aq)][A^-(aq)]}{[HA(aq)]}$$

The units of K_a are always mol dm^{-3}. You can show this by cancelling the units in the K_a expression:

$$K_a = \frac{(\text{mol dm}^{-3})(\cancel{\text{mol dm}^{-3}})}{\cancel{(\text{mol dm}^{-3})}} = \text{mol dm}^{-3}$$

The terms *strong* and *weak* describe the extent of dissociation of an acid given by the K_a value:

* a large K_a value indicates a large extent of dissociation – the acid is strong
* a small K_a value indicates a small extent of dissociation – the acid is weak

Module 1
Rates, equilibrium and pH
Strong and weak acids

K_a and pK_a

As with pH and [H^+(aq)] (see spread 2.1.15), values of K_a can be made more manageable if expressed in a logarithmic form called pK_a.

K_a and pK_a conversions are just like those between pH and H^+:

- p$K_a = -\log_{10} K_a$
- $K_a = 10^{-pK_a}$
- A low value of K_a matches a high value of pK_a
- A high value of K_a matches a low value of pK_a
- The smaller the pK_a value, the stronger the acid.

Table 2 below compares K_a and pK_a values for some weak acids.

Key definition

p$K_a = -\log_{10} K_a$

$K_a = 10^{-pK_a}$

Acid	Relative strength	K_a/mol dm^{-3}	pK_a
Phosphoric acid H_3PO_4	Stronger acid	7.9×10^{-3}	$-\log(7.9 \times 10^{-3})$ = 2.10
Sulfurous acid H_2SO_3		1.5×10^{-3}	$-\log(1.5 \times 10^{-2})$ = 2.82
Methanoic acid HCOOH		1.6×10^{-4}	$-\log(1.6 \times 10^{-4})$ = 3.80
Ethanoic acid CH_3COOH	Weaker acid	1.7×10^{-5}	$-\log(1.7 \times 10^{-5})$ = 4.77

Table 2 K_a and pK_a values of weak acids

STRETCH and CHALLENGE

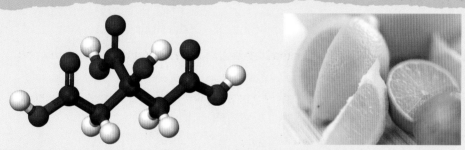

Figure 2 Juices of citrus fruits, such as lemon and lime, contain citric acid

Citric acid makes up as much as 8% of the dry mass of a citrus fruit.

Citric acid is a weak tribasic acid. It releases its three acidic protons in three steps, each with its own pK_a value:

$$C_3H_4OH(COOH)_3 \rightleftharpoons H^+(aq) + C_3H_4OH(COOH)_2(COO^-) \qquad pK_a = 3.15$$
$$C_3H_4OH(COOH)_2(COO^-) \rightleftharpoons H^+(aq) + C_3H_4OH(COOH)(COO^-)_2 \qquad pK_a = 4.77$$
$$C_3H_4OH(COOH)(COO^-)_2 \rightleftharpoons H^+(aq) + C_3H_4OH(COO^-)_3 \qquad pK_a = 6.40$$

Notice how the acid becomes weaker at each dissociation. This is typical of a multibasic acid.

Questions

1 For each of the following acid–base equilibria, write down the expression for K_a:
   (a) HCOOH(aq) $\rightleftharpoons$ H^+(aq) + HCOO$^-$(aq)
   (b) CH_3CH_2COOH(aq) $\rightleftharpoons$ H^+(aq) + $CH_3CH_2COO^-$(aq).
2 Calculate pK_a from the following K_a values. Give you answers to two decimal places.
   (a) $K_a = 2.3 \times 10^{-1}$ mol dm^{-3}
   (b) $K_a = 2.5 \times 10^{-3}$ mol dm^{-3}
   (c) $K_a = 4.8 \times 10^{-11}$ mol dm^{-3}.
3 Calculate K_a from the following pK_a values. Give your answers to three significant figures.
   (a) p$K_a = 2.90$;   (b) p$K_a = 7.20$;   (c) p$K_a = 10.60$.

By the end of this spread, you should be able to . . .

* Calculate pH from [H⁺(aq)] and [H⁺(aq)] from pH for strong and weak acids.
* Calculate [H⁺(aq)] from pH for strong and weak acids.
* Calculate K_a for a weak acid.

Calculating the pH of strong acids

A strong monobasic acid, HA, has virtually complete dissociation in water:

$$HA(aq) \longrightarrow H^+(aq) + A^-(aq)$$

This means that [H⁺(aq)] of a strong acid is equal to the concentration of the acid, [HA(aq)]:

$$[H^+(aq)] = [HA(aq)]$$

The pH can then be calculated using pH = –log[H⁺(aq)].

> **Calculating the pH of strong acids**
>
> For a *strong* acid HA, [H⁺(aq)] = [HA(aq)]

Worked example 1

A sample of hydrochloric acid, HCl, has a concentration of 1.22×10^{-3} mol dm⁻³. What is its pH?

HCl(aq) is a strong acid and completely dissociates:
$$HCl(aq) \longrightarrow H^+(aq) + Cl^-(aq)$$

$$\therefore [H^+(aq)] = [HCl(aq)] = 1.22 \times 10^{-3} \text{ mol dm}^{-3}$$
$$pH = -\log[H^+(aq)]$$
$$= -\log(1.22 \times 10^{-3})$$
$$= 2.91$$

Worked example 2

A sample of nitric acid, HNO₃, has a pH of 5.63. What is the concentration of the HNO₃(aq)?

$$[H^+(aq)] = 10^{-pH} = 10^{-5.63} = 2.34 \times 10^{-6} \text{ mol dm}^{-3}$$

HNO₃(aq) is a strong acid and completely dissociates:
$$HNO_3(aq) \longrightarrow H^+(aq) + NO_3^-(aq)$$

$$\therefore [HNO_3(aq)] = [H^+(aq)]$$
$$= 2.34 \times 10^{-6} \text{ mol dm}^{-3}$$

Calculating the pH of weak acids

In aqueous solution, a weak monobasic acid, HA, partially dissociates, setting up the equilibrium:

$$HA(aq) \rightleftharpoons H^+(aq) + A^-(aq)$$

$$K_a = \frac{[H^+(aq)][A^-(aq)]}{[HA(aq)]}$$

To work out the pH we need to know [H⁺(aq)]. This depends on:
* the concentration of the acid, [HA(aq)]
* the acid dissociation constant, K_a.

Unlike strong acids, [H⁺(aq)] is much less than [HA(aq)] because the extent of dissociation is small.

When HA molecules dissociate, [H⁺(aq)] and [A⁻(aq)] ions are formed in equal quantities.

$$\therefore [H^+(aq)] = [A^-(aq)] \ldots \text{ so } [H^+(aq)][A^-(aq)] = [H^+(aq)]^2.$$

Because relatively few HA molecules have dissociated, [HA(aq)] will have reduced slightly. The equilibrium concentration of HA(aq) will be [HA(aq)] – [H⁺(aq)].

The K_a expression now becomes:

$$K_a = \frac{[H^+(aq)]^2}{[HA(aq)] - [H^+(aq)]}$$

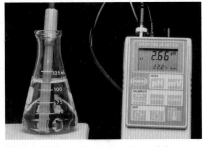

Figure 1 The pH of a solution of the weak acid methanoic acid is being measured using a pH meter. The concentration of the methanoic acid is 0.030 mol dm⁻³ and the pH reading can be seen to be 2.66

Making an approximation

Finally, we make an approximation to simplify the pH calculation. Only a very small proportion of HA dissociates. We can assume that the equilibrium concentration of HA will be very nearly the same as the concentration of undissociated HA(aq).

The K_a expression approximates to $K_a = \dfrac{[H^+(aq)]^2}{[HA(aq)]}$

Rearranging the equation, $[H^+(aq)]^2 = K_a[HA(aq)]$

$\therefore [H^+(aq)] = \sqrt{K_a \times [HA(aq)]}$

The pH can then be calculated using $pH = -\log[H^+(aq)]$.

Examiner tip

The key to calculating the pH of a weak acid, HA, is to find $[H^+(aq)]$.

$[H^+(aq)]$ depends on $[HA(aq)]$ and K_a.

For a *weak* acid HA,
$[H^+(aq)] = \sqrt{K_a \times [HA(aq)]}$

STRETCH and CHALLENGE

If you take A-level mathematics, you should be able to calculate the pH of a weak acid, HA, without making the approximation $[HA(aq)] \approx [HA(aq)] - [H^+(aq)]$.

For this you will need to use a quadratic equation. Try Worked example 3 without using the approximation and see what difference it makes to the result.

Hence, make a judgement about whether the approximation is valid.

When will the approximation be valid and when will it be invalid?

Worked example 3

The concentration of a sample of nitrous acid, HNO_2, is 0.055 mol dm^{-3}. $K_a = 4.70 \times 10^{-4}$ mol dm^{-3} at 25 °C. Calculate the pH.

First we find $[H^+(aq)]$. HNO_2 is a weak acid and partially dissociates:

$$HNO_2(aq) \rightleftharpoons H^+(aq) + NO_2^-(aq)$$

$$K_a = \frac{[H^+(aq)][NO_2^-(aq)]}{[HNO_2(aq)]} \approx \frac{[H^+aq]^2}{[HNO_2(aq)]}$$

$$\therefore [H^+(aq)]^2 = K_a \times [HNO_2(aq)]$$

$$\therefore [H^+(aq)] = \sqrt{K_a \times [HNO_2(aq)]} = \sqrt{(4.70 \times 10^{-4} \times 0.055)} = 5.08 \times 10^{-3} \text{ mol dm}^{-3}$$

Now we calculate pH:

$$pH = -\log[H^+(aq)] = -\log(5.08 \times 10^{-3}) = 2.29$$

Examiner tip

When working out the square root, use brackets around the expression being square-rooted. If you don't do this, you will only take the square root of the first value.

Calculating K_a for weak acids

To determine K_a for a weak acid, we need to measure the pH of a solution of the weak acid using a pH meter (see Figure 1). We also need the concentration of the weak acid.

Worked example 4

A sample of 0.030 mol dm^{-3} methanoic acid, HCOOH, has a pH of 2.66. Calculate K_a.

First we find $[H^+(aq)]$ from the pH:

$$[H^+(aq)] = 10^{-pH} = 10^{-2.66} = 2.19 \times 10^{-3} \text{ mol dm}^{-3}$$

Now we can calculate K_a:

$$K_a = \frac{[H^+(aq)][HCOO^-(aq)]}{[HCOOH(aq)]} \approx \frac{[H^+(aq)]^2}{[HCOOH(aq)]}$$

$$\therefore K_a = \frac{(2.19 \times 10^{-3})^2}{0.030} = 1.6 \times 10^{-4} \text{ mol dm}^{-3}$$

Examiner tip

How accurately should you show pH values?

A typical pH meter, such as that in Figure 1, shows pH values to 2 decimal places. The whole number before the decimal place is the logarithmic way for showing powers of 10. In terms of significant figures, pH starts after the decimal place. So a pH of 2.66 is to 2 significant figures only! So don't 'round' pH values to 1 decimal place as this is only 1 significant figure.

As a general rule, show pH values to 2 decimal places.

Questions

1 Find the pH of the following solutions of weak acids:
   (a) 0.65 mol dm^{-3} CH_3COOH ($K_a = 1.7 \times 10^{-5}$ mol dm^{-3});
   (b) 4.4×10^{-2} mol dm^{-3} HClO ($K_a = 3.7 \times 10^{-8}$ mol dm^{-3}).
2 Find the values of K_a and pK_a for the following weak acids:
   (a) 0.13 mol dm^{-3} solution with a pH of 3.52;
   (b) 7.8×10^{-2} mol dm^{-3} solution with a pH of 5.19.

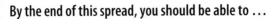
By the end of this spread, you should be able to ...

* State and use the expression for the ionic product of water, K_w.
* Understand the importance of K_w in controlling the concentrations of $H^+(aq)$ and $OH^-(aq)$ in aqueous solutions.

The ionisation of water and K_w

Pure water ionises into $H^+(aq)$ and $OH^-(aq)$ ions but the extent of ionisation is tiny – only 1 molecule dissociates out of about every 500 000 000 molecules of H_2O.

The equation below shows the ionisation of water. The position of equilibrium lies well to the left.

$$H_2O(l) \underset{\text{equilibrium}}{\rightleftharpoons} H^+(aq) + OH^-(aq)$$

At equilibrium, $K_c = \dfrac{[H^+(aq)][OH^-(aq)]}{[H_2O(l)]}$

$\therefore K_c \times [H_2O(l)] = [H^+(aq)][OH^-(aq)]$

$[H_2O(l)]$ is equal to the concentration of water in mol dm^{-3}. This is a constant value, equal to $\dfrac{1000}{18.0} = 55.6$ mol dm^{-3}

So, K_c and $[H_2O(l)]$ are both constants – we combine these together to give a new constant called the **ionic product of water**. 'Ionic product' gets its name from 'the ions in water multiplied together'.

$$K_w = \underbrace{K_c \times [H_2O(l)]}_{\text{constant, } K_w} = \underbrace{[H^+(aq)][OH^-(aq)]}_{\text{ionic product}}$$

At 25 °C, the measured pH of water is 7 and $[H^+(aq)] = 10^{-7}$ mol dm^{-3}.
* When pure water, H_2O, ionises as in the equation above, $[H^+(aq)] = [OH^-(aq)]$.
* This means that the concentration of $OH^-(aq)$ is also equal to 10^{-7} mol dm^{-3}.

$$K_w = [H^+(aq)][OH^-(aq)] = 10^{-7} \times 10^{-7}$$
$$= 10^{-14} \text{ mol}^2 \text{ dm}^{-6}$$
$$\therefore \text{ at } 25\,°C, K_w = 1.00 \times 10^{-14} \text{ mol}^2 \text{ dm}^{-6}.$$

The significance of K_w

K_w controls the balance between $[H^+(aq)]$ and $[OH^-(aq)]$ in all aqueous solutions.

At 25 °C, the pH value of 7 is the neutral point at which $H^+(aq)$ and $OH^-(aq)$ concentrations are the same, equal to 10^{-7} mol dm^{-3}. This applies to water and also to neutral solutions.

All aqueous solutions contain $H^+(aq)$ and $OH^-(aq)$ ions:
* in water and neutral solutions, $[H^+(aq)] = [OH^-(aq)]$
* in acidic solutions, $[H^+(aq)] > [OH^-(aq)]$
* in alkaline solutions, $[H^+(aq)] < [OH^-(aq)]$

The relative concentrations of $H^+(aq)$ and $OH^-(aq)$ ions in solution are determined by K_w.

At 25 °C, $K_w = [H^+(aq)] \times [OH^-(aq)]$ must always be equal to 1.00×10^{-14} mol^2 dm^{-6}

Figure 1 Drop of water hitting a shallow dish of water in a photograph taken with a high-speed flash with a duration of 1/20 000th of a second

Key definition

The **ionic product of water**, K_w, is defined as

$K_w = [H^+(aq)][OH^-(aq)]$

At 25 °C, $K_w = 1.00 \times 10^{-14}$ mol^2 dm^{-6}.

Examiner tip

Provided you know $[H^+(aq)]$, you can always find $[OH^-(aq)]$ using
$K_w = [H^+(aq)][OH^-(aq)]$
$= 1.00 \times 10^{-14}$ mol^2 dm^{-6}

This value is provided in exams on the Data sheet.

Module 1
Rates, equilibrium and pH
The ionisation of water

The link between [H⁺(aq)] and [OH⁻(aq)]

We can easily find the concentrations of $H^+(aq)$ and $OH^-(aq)$ in solutions with different pH values.

- We need to use $[H^+(aq)][OH^-(aq)] = 1.00 \times 10^{-14}$ mol^2 dm^{-6}.
- Table 2 shows how $[H^+(aq)]$ and $[OH^-(aq)]$ are related at different pH values at 25 °C.
- Note that the indices of $[H^+(aq)]$ and $[OH^-(aq)]$ always add up to -14.

pH	−1	0	1	2	3	4	5	6	7	8	9	10	11	12	13	14	15
[H⁺(aq)]/ mol dm⁻³	10^1	10^0 (1)	10^{-1}	10^{-2}	10^{-3}	10^{-4}	10^{-5}	10^{-6}	10^{-7}	10^{-8}	10^{-9}	10^{-10}	10^{-11}	10^{-12}	10^{-13}	10^{-14}	10^{-15}
[OH⁻(aq)]/ mol dm⁻³	10^{-15}	10^{-14}	10^{-13}	10^{-12}	10^{-11}	10^{-10}	10^{-9}	10^{-8}	10^{-7}	10^{-6}	10^{-5}	10^{-4}	10^{-3}	10^{-2}	10^{-1}	10^0 (1)	10^1

Table 2 At 25 °C, $[H^+(aq)]$ and $[OH^-(aq)]$ are linked by K_w:
$K_w = [H^+(aq)][OH^-(aq)] = 1.00 \times 10^{-14}$ mol^2 dm^{-6}.

How science works

Natural rain water is acidic ...

As rain falls through the atmosphere, some carbon dioxide gas dissolves in it.

The result is a weak acid called carbonic acid, H_2CO_3:

$$CO_2(g) \longrightarrow CO_2(aq) \longrightarrow H_2CO_3(aq)$$

Carbonic acid partially dissociates and the concentration of $H^+(aq)$ ions increases:

$$H_2CO_3(aq) \rightleftharpoons H^+(aq) + HCO_3^-(aq)$$

- Because $[H^+(aq)]$ has increased, $[OH^-(aq)]$ must fall until $[H^+(aq)][OH^-(aq)] = 1.00 \times 10^{-14}$ mol^2 dm^{-6} (at 25 °C).

Figure 2 Rain water

The overall effect is that the pH decreases. Typical rain water has a pH value of about 5.6.

... and mineral water is alkaline

Mineral Analysis Typical Values mg/litre			
Calcium	35	Chloride	39
Magnesium	19	Sulphate	35
Potassium	1	Nitrate (as NO₃)	8
Sodium	15	Dry residue at 180°C	228
Bicarbonate	123	pH	8.0

Figure 3 Bottles of mineral water. The values are taken from the label of a bottle of mineral water. Mineral water from different regions will contain different amount of minerals. Note that the contents are really all ions. Do you know all their formulae and their charges? Note also the alkaline pH

Mineral water contains dissolved ions, formed as water seeps through natural rocks. Many mineral waters are collected from limestone areas. Limestone is mainly calcium carbonate, $CaCO_3$. The calcium carbonate reacts with $H^+(aq)$ ions present in natural water:

$$CaCO_3(s) + 2H^+(aq) \longrightarrow Ca^{2+}(aq) + 2HCO_3^-(aq)$$

- Because $[H^+(aq)]$ has decreased, $[OH^-(aq)]$ must increase until $[H^+(aq)][OH^-(aq)] = 1.00 \times 10^{-14}$ mol^2 dm^{-6} (at 25 °C).

The overall effect is that the pH increases. The pH of mineral water is typically in the range 7.0–8.0. Mineral water gets its name from the 'dissolved mineral rocks' that it contains. Mineral water from limestone rocks contains mainly $Ca^{2+}(aq)$ and $HCO_3^-(aq)$ ions. HCO_3^- is 'hydrogencarbonate', shown as the alternative common name 'bicarbonate' in Figure 3.

STRETCH and CHALLENGE

At 25 °C, the pH of water is 7 and $[H^+(aq)] = 10^{-7}$ mol dm^{-3}. As temperature increases, K_w also increases. In response, more H_2O molecules dissociate, increasing the concentrations of $H^+(aq)$ and $OH^-(aq)$. With a higher concentration of $H^+(aq)$ ions, the pH of neutrality decreases. At body temperature, 37 °C, K_w has increased to 2.40×10^{-14} mol^2 dm^{-6} and the pH of neutral water is 6.81 at this temperature!

The water is still 'neutral' because $[OH^-(aq)]$ will also have increased to 2.40×10^{-14} mol^2 dm^{-6}. An aqueous solution is always 'neutral' if $[H^+(aq)] = [OH^-(aq)]$. So it is important to always quote the temperature for K_w and pH.

Questions

1 Define the term *ionic product of water*.

2 Some aqueous solutions have the following concentrations of $H^+(aq)$. In each solution, what is the concentration of $OH^-(aq)$?
   (a) $[H^+(aq)] = 10^{-6}$ mol dm^{-3};
   (b) $[H^+(aq)] = 10^{-2}$ mol dm^{-3};
   (c) $[H^+(aq)] = 10^{-11}$ mol dm^{-3}.

3 Some aqueous solutions have the following concentrations of $OH^-(aq)$. In each solution, what is the concentration of $H^+(aq)$?
   (a) $[OH^-(aq)] = 10^{-13}$ mol dm^{-3};
   (b) $[OH^-(aq)] = 10^{-4}$ mol dm^{-3};
   (c) $[OH^-(aq)] = 10^{-10}$ mol dm^{-3}.

By the end of this spread, you should be able to ...

✳ Calculate pH from [H⁺(aq)] for strong bases, using K_w.

✳ Calculate [H⁺(aq)] from pH for strong bases, using K_w.

Bases

Dictionaries define a base as a substance capable of reacting with an acid to form a salt and water or, more broadly, as a substance that can accept protons. An alkali is a soluble base that releases hydroxide ions, OH⁻, when dissolved in water.

But where did the strange name 'base' come from? The concept of a base was first introduced in 1754 by the French chemist Guillaume François Rouelle. Most acids known at the time were volatile liquids. These liquids could be turned into solid salts by the addition of certain substances. Rouelle named these substances 'bases' because he thought that they provided a concrete 'base' for the salt. We now know that a base is a substance that can accept protons – but the name 'base' has stuck.

Examiner tip

Remember that an alkali is a soluble base that releases hydroxide ions, OH⁻.

Base strength

The *strength* of a base is a measure of its dissociation in solution to generate OH⁻ ions.

Sodium hydroxide, NaOH, is an example of a strong base. In aqueous solution, NaOH dissociates completely – in the equation below, the complete dissociation is shown as ⟶:

$$NaOH(aq) + aq \longrightarrow Na^+(aq) + OH^-(aq)$$

Ammonia, NH_3, is a weak base. In aqueous solution, NH_3 dissociates only partially and an equilibrium is set up. The equilibrium position lies well to the left-hand side:

$$\overset{\text{equilibrium}}{\underset{}{\longleftarrow}}$$
$$NH_3(aq) + H_2O(l) \rightleftharpoons NH_4^+(aq) + OH^-(aq)$$

Strong bases

Bases that dissociate in water to release hydroxide ions are called alkalis. Strong bases are alkalis – they are 100% dissociated in aqueous solution. Strong bases tend be hydroxides of the metals in Groups 1 and 2 in the Periodic Table.

The strong bases most commonly met are NaOH, KOH and $Ca(OH)_2$.

Figure 1 Making soap. In this historical artwork of men working in a soap factory, soap is being made by melting a fat and an alkali together. After the mixture is bought to the boil, the fat reacts with the alkali, producing soap and glycerol. Typically, the sodium hydroxide used has a concentration of about 8 mol dm⁻³ and a pH of 14.9. At this concentration the alkali is extremely corrosive and harmful to organic material (such as skin!)

Calculating the pH of strong bases

To work out the pH of a strong base we need to know $[H^+(aq)]$ and this depends on:
- the concentration of the base
- the ionic product of water, $K_w = 1.00 \times 10^{-14}$ mol^2 dm^{-6}.

A strong monobasic alkali, e.g. NaOH, is completely dissociated in aqueous solution. This means that $[OH^-(aq)]$ of a strong base is equal to the concentration of the base.
- $NaOH(aq) \longrightarrow Na^+(aq) + OH^-(aq)$
- So $[OH^-(aq)] = [NaOH(aq)]$

We can find $[H^+(aq)]$ from K_w and $[OH^-(aq)]$:
- $K_w = [H^+(aq)][OH^-(aq)]$
- $\therefore [H^+(aq)] = \dfrac{K_w}{[OH^-(aq)]}$

The pH can then be calculated using $pH = -\log[H^+(aq)]$.

You can also work out the pH of a buffer using the pOH value. This method is shown below. Use the method that you have been taught – both methods give the same answer.

Worked example

A solution of KOH has a concentration of 0.050 mol dm^{-3}. What is the pH?

KOH(aq) is a strong base, so there will be complete dissociation.

$\therefore [OH^-(aq)] = [KOH(aq)] = 0.050$ mol dm^{-3}

Method 1: Using K_w
First we find $[H^+(aq)]$ from $[OH^-(aq)]$:

$K_w = [H^+(aq)][OH^-(aq)]$
$\quad = 1.00 \times 10^{-14}$ mol^2

$\therefore [H^+(aq)] = \dfrac{K_w}{[OH^-(aq)]} = \dfrac{1.00 \times 10^{-14}}{0.050}$

$\quad = 2.0 \times 10^{-13}$ mol dm^{-3}

Now we calculate pH:

$pH = -\log[H^+(aq)] = -\log(2.0 \times 10^{-13})$
$\quad = \textbf{12.70}$

Method 2: Using pOH
pOH is defined as: $pOH = -\log[OH^-(aq)]$

$pOH = -\log[OH^-(aq)] = -\log(0.050)$
$\quad = \textbf{1.30}$

To get the pH, subtract 1.30 from 14:

$pH = 14 - 1.30 = \textbf{12.70}$

Returning to K_w

All aqueous solutions contain both $H^+(aq)$ and $OH^-(aq)$ ions. In the previous spread, we looked at the importance of K_w in controlling the balance between $[H^+(aq)]$ and $[OH^-(aq)]$ in all aqueous solutions.

This relationship is critical:
- at 25°C, $K_w = [H^+(aq)][OH^-(aq)] = 1.00 \times 10^{-14}$ mol^2 dm^{-6}

Using K_w:
- we can find the concentration of $OH^-(aq)$ if we know the concentration of $H^+(aq)$
- we can find the concentration of $H^+(aq)$ if we know the concentration of $OH^-(aq)$.

K_w rules!

STRETCH and CHALLENGE

How the pOH method works
We know that

$[H^+(aq)][OH^-(aq)] = 10^{-14}$ mol^2 dm^{-6}

As logarithms:

$\log[H^+(aq)] + \log[OH^-(aq)] = -14$

or:

$-\log[H^+(aq)] + -\log[OH^-(aq)] = 14$

$\therefore pH + pOH = 14$

Rearranging: $pH = 14 - pOH$
In this example pOH = **1.30**,
so pH = 14 − 1.30 = **12.70**

Questions

1 Find the $[H^+(aq)]$ and pH of the following alkalis at 25°C:
   (a) 2.0×10^{-3} mol dm^{-3} OH$^-$(aq);
   (b) 5.7×10^{-1} mol dm^{-3} OH$^-$(aq).
2 Find the pH of the following solutions at 25°C:
   (a) 0.0050 mol dm^{-3} KOH(aq);
   (b) 3.56×10^{-2} mol dm^{-3} NaOH(aq).
3 Find the concentration, in mol dm^{-3}, of OH$^-$(aq) at 25°C:
   (a) pH = 12.43;
   (b) pH = 13.82.

By the end of this spread, you should be able to . . .

* Describe what is meant by a *buffer solution*.
* State that a buffer solution can be made from a weak acid and a salt of the weak acid.
* Explain the role of the conjugate acid–base pair in an acid buffer solution.

Buffer solutions

A **buffer solution** 'resists' changes in pH during the addition of small amounts of acid or alkali. A buffer solution cannot prevent the pH from changing completely, but pH changes are minimised, at least for as long as some of the buffer solution remains.

Buffer solutions are very important in the control of pH in living systems.

A buffer solution is a mixture of:
* a weak acid, HA
* its conjugate base, A⁻.

It can be made from a weak acid and a salt of the weak acid – for example, ethanoic acid, CH_3COOH/sodium ethanoate, CH_3COONa.

Alternatively, a weak acid such as methanoic acid, $HCOOH$, can be partially neutralised by an aqueous alkali, such as $NaOH(aq)$, to give a solution containing a mixture of the salt and the excess of weak acid.

In the CH_3COOH/CH_3COONa buffer system:
* the weak acid, CH_3COOH, dissociates *partially*:

$$CH_3COOH(aq) \rightleftharpoons H^+(aq) + CH_3COO^-(aq)$$

* the salt dissociates *completely*, generating the conjugate base, CH_3COO^-:

$$CH_3COO^-Na^+(aq) \longrightarrow CH_3COO^-(aq) + Na^+(aq)$$

The equilibrium mixture formed contains a high concentration of the undissociated weak acid, CH_3COOH, and its conjugate base, CH_3COO^-. The high concentration of the conjugate base pushes the equilibrium to the left, so the concentration of $H^+(aq)$ ions is very small.

The resulting buffer solution contains large 'reservoirs' of the weak acid and its conjugate base. You can see this in Figure 1.

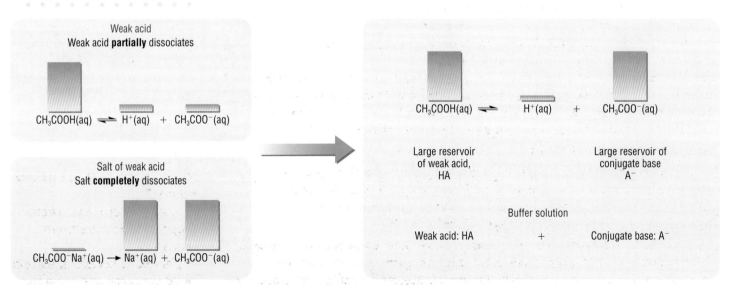

Figure 1 Making a buffer solution from a weak acid and a salt of the weak acid

How does a buffer act?

In a buffer solution, the weak acid, HA, and the conjugate base, A^-, are both responsible for controlling pH.

The buffer solution minimises pH changes by using the equilibrium:

$$HA(aq) \rightleftharpoons H^+(aq) + A^-(aq)$$

The overall principle is simple:

- the weak acid, HA, removes added alkali
- the conjugate base, A^-, removes added acid.

On addition of an acid, $H^+(aq)$, to a buffer solution:

- $[H^+(aq)]$ is increased
- the conjugate base, $A^-(aq)$, reacts with $H^+(aq)$ ions
- the equilibrium shifts to the left, removing most of the added $H^+(aq)$ ions.

$$\xleftarrow{\text{equilibrium}}$$
$$HA(aq) \rightleftharpoons H^+(aq) + A^-(aq)$$

On addition of an alkali, $OH^-(aq)$, to a buffer mixture:

- $[OH^-(aq)]$ is increased
- the small concentration of $H^+(aq)$ ions react with the $OH^-(aq)$ ions

  $$H^+(aq) + OH^-(aq) \longrightarrow H_2O(l)$$

- HA dissociates, shifting the equilibrium to the right to restore most of the $H^+(aq)$ ions that have reacted

$$\xrightarrow{\text{equilibrium}}$$
$$HA(aq) \rightleftharpoons H^+(aq) + A^-(aq)$$

$$\underset{\text{Added alkali}\ \longrightarrow}{\overset{\longleftarrow\ \text{Added acid}}{HA(aq) \rightleftharpoons H^+(aq) + A^-(aq)}}$$

Figure 2 Shifting the buffer equilibrium

Examiner tip

The weak acid, HA, removes most of any added alkali. The conjugate base, A^-, removes most of any added acid.

Added acid and added alkali shift the buffer equilibrium in opposite directions.

Make sure that you are able to explain the role of each component in a buffer system. Also remember that the buffer cannot remove *all* of any added acid or alkali, but pH changes are minimised.

The mode of action of a buffer solution is often tested in exams.

Buffers in cosmetics

For effective treatment of the skin, it is important that the pH of cosmetics is controlled closely. Products are sold in a buffered form with different pH ranges for different uses. Many skin-care products and shampoos are buffered to the natural pH 5.5 of normal healthy skin. This is important because the normal pH value of your skin provides a protective acid mantle. This helps to protect the skin from environmental damage and dryness.

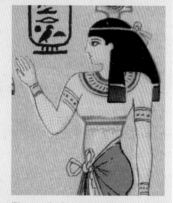

Figure 3 Cleopatra

There are many skin-care products containing alpha-hydroxy acids, AHAs, on the market. AHAs are not new to skin care – Cleopatra was famous for her beautiful skin, and her secret was to soak in a bath of raw fresh milk. Chemists now know why milk worked such wonders for her skin – milk contains lactic acid, which is an AHA. Cosmetic scientists are now using AHAs in skin creams to combat the appearance of ageing. These 'anti-ageing creams' are buffered at pH values lower than 5.5. The low pH irritates the skin, causing it to swell – hiding wrinkles as it does so. More concentrated solutions of AHAs have lower pH values and are used to remove dead skin and hair.

Does your skin cream contain AHAs?

AHA ingredients may be listed as any of the following: glycolic acid, lactic acid, malic acid, citric acid, glycolic acid + ammonium glycolate, alpha-hydroxyethanoic acid + ammonium alpha-hydroxyethanoate, alpha-hydroxyoctanoic acid, alpha-hydroxycaprylic acid, hydroxycaprylic acid, mixed fruit acid, tri-alpha hydroxy fruit acids.

It is very difficult to obtain raw fresh milk nowadays. Many countries have banned its sale on health grounds as raw milk is unpasteurised. Its ban has caused some controversy as some people believe that raw milk actually has health benefits. Cleopatra did not have this problem in her day. She may not have been as beautiful if she could not have obtained her 'bathwater'!

Questions

1 What is meant by the term *buffer solution*?

2 A buffer solution is made from propanoic acid and sodium propanoate.
Explain, using equations, the role of the conjugate acid–base pair in this buffer solution.

By the end of this spread, you should be able to . . .

✳ Calculate the pH of a buffer solution from the K_a value of a weak acid and the equilibrium concentrations of the conjugate acid–base pair.

✳ Explain the role of the carbonic acid–hydrogencarbonate ion system as a buffer in the control of blood pH.

Calculations involving buffer solutions

The pH of a buffer solution depends on:
- the acid dissociation constant, K_a, of the buffer system
- the concentration ratio of the weak acid and its conjugate base.

There are two methods commonly used to work out the pH of a buffer solution. Use the method that you have been taught.

Method 1

For a buffer consisting of a weak acid, HA, and its conjugate base, A⁻

$$K_a = \frac{[\text{H}^+(aq)][\text{A}^-(aq)]}{[\text{HA}(aq)]}$$

$$\therefore [\text{H}^+(aq)] = K_a \times \frac{[\text{HA}(aq)]}{[\text{A}^-(aq)]}$$

We need to check on the concentrations of HA(aq) and A⁻(aq). Only a very small proportion of HA dissociates, so we can assume that $[\text{HA}(aq)]_{\text{equilibrium}} = [\text{HA}(aq)]_{\text{undissociated}}$. See spread 2.1.17 for more details.

The salt of the weak acid is ionic and dissociates completely in aqueous solution, for example:

$$\text{CH}_3\text{COO}^-\text{Na}^+(aq) \longrightarrow \text{CH}_3\text{COO}^-(aq) + \text{Na}^+(aq)$$

$$\therefore \text{CH}_3\text{COO}^-(aq) = \text{CH}_3\text{COO}^-\text{Na}^+(aq)$$

Once [H⁺(aq)] is known, the pH can be calculated using pH = –log[H⁺(aq)].

Method 2

You can also work out the pH of a buffer solution using the Henderson–Hasselbalch relationship:

$$\text{pH} = \text{p}K_a + \log\frac{[\text{A}^-(aq)]}{[\text{HA}(aq)]}$$

where $\text{p}K_a = -\log K_a$.

Worked example

The worked example below shows each method for working out the pH of buffer solutions.

Calculate the pH at 25 °C of a buffer solution containing 0.050 mol dm⁻³ CH₃COOH(aq) and 0.10 mol dm⁻³ CH₃COO⁻Na⁺(aq). For CH₃COOH, $K_a = 1.7 \times 10^{-5}$ mol dm⁻³ at 25 °C.

Method 1

Using the first method above:

$$[\text{H}^+(aq)] = K_a \times \frac{[\text{HCOOH}(aq)]}{[\text{HCOO}^-(aq)]}$$

$$= 1.7 \times 10^{-5} \times \frac{0.050}{0.10}$$

$$= 8.5 \times 10^{-6} \text{ mol dm}^{-3}$$

$$\therefore \text{pH} = -\log[\text{H}^+(aq)]$$

$$= -\log(8.5 \times 10^{-6})$$

$$= 5.07$$

Method 2

Using the Henderson–Hasselbalch relationship:

$$\text{pH} = \text{p}K_a + \log\frac{[\text{HCOO}^-(aq)]}{[\text{HCOOH}(aq)]}$$

$$= -\log K_a + \log\frac{[\text{HCOO}^-(aq)]}{[\text{HCOOH}(aq)]}$$

$$= 4.77 + 0.30$$

$$= 5.07$$

The carbonic acid–hydrogencarbonate buffer system

Healthy human blood plasma needs to have a pH between 7.35 and 7.45. If the pH falls below 7.35, a condition called acidosis is produced. If the pH rises above 7.45, then the condition is called alkalosis.

The pH of blood is controlled by a mixture of buffers. The carbonic acid–hydrogencarbonate ion buffer is the most important buffer system in the blood:
- carbonic acid, H_2CO_3, acts as the weak acid
- hydrogencarbonate, HCO_3^-, acts as the conjugate base.

$$H_2CO_3(aq) \rightleftharpoons H^+(aq) + HCO_3^-(aq)$$
weak acid $\qquad$ conjugate base

Any increase in $H^+(aq)$ ions in the blood is removed by the conjugate base, $HCO_3^-(aq)$:
- the equilibrium shifts to the left, removing most of the $H^+(aq)$ ions

equilibrium
$$\xleftarrow{\hspace{3cm}}$$
$$H_2CO_3(aq) \rightleftharpoons H^+(aq) + HCO_3^-(aq)$$

Any increase in $OH^-(aq)$ ions in the blood is removed by the weak acid $H_2CO_3(aq)$:
- the small concentration of $H^+(aq)$ ions reacts with the $OH^-(aq)$ ions:

$$H^+(aq) + OH^-(aq) \longrightarrow H_2O(l)$$

- H_2CO_3 dissociates, shifting the equilibrium to the right to restore most of the $H^+(aq)$ ions:

equilibrium
$$\xrightarrow{\hspace{3cm}}$$
$$H_2CO_3(aq) \rightleftharpoons H^+(aq) + HCO_3^-(aq)$$

The shifts in the buffer equilibrium are shown in Figure 2.

The acid dissociation constant, K_a, for this equilibrium is 4.3×10^{-7} mol dm^{-3}.

Most materials released into the blood are acidic and the hydrogencarbonate ions effectively remove these by being converted into H_2CO_3. The carbonic acid is converted into aqueous carbon dioxide through the action of an enzyme. In the lungs, the dissolved carbon dioxide is converted into carbon dioxide gas, which is then exhaled.

$$CO_2(aq) \rightleftharpoons CO_2(g)$$

The amount of $CO_2(aq)$ in the blood can be controlled by simply changing the rate of breathing – heavy breathing removes more $CO_2(g)$; breathing less fast removes less $CO_2(g)$.

The acid–base balance of your blood is vital to your survival. If the pH of your blood drops below 7.2 or rises above 7.6, then you are in deep trouble!

Figure 1 Blood sample – healthy blood has a pH of 7.40

Added acid
$$\xleftarrow{\hspace{2cm}}$$
$$H_2CO_3(aq) \rightleftharpoons H^+(aq) + HCO_3^-(aq)$$
Added alkali
$$\xrightarrow{\hspace{2cm}}$$

Figure 2 Carbonic acid–hydrogencarbonate ion equilibrium

Worked example

This example shows that the blood's buffer system contains about 20 times more hydrogencarbonate ions than carbonic acid.

The pH of healthy blood = 7.40
$$\therefore [H^+(aq)] = 10^{-pH} = 10^{-7.40}$$
$$= 3.98 \times 10^{-8} \text{ mol dm}^{-3}$$
$$[H^+(aq)] = K_a \times \frac{[H_2CO_3(aq)]}{[HCO_3^-(aq)]}$$
$$\therefore \frac{[HCO_3^-(aq)]}{[H_2CO_3(aq)]} = \frac{K_a}{[H^+(aq)]}$$
$$= \frac{4.3 \times 10^{-7}}{3.98 \times 10^{-8}}$$
$$= \frac{10.8}{1}$$

When there are equal concentrations of a weak acid and its conjugate base, the acid : conjugate base concentration ratio is 1 : 1.
$$\therefore [H^+(aq)] = K_a \times \frac{1}{1} = K_a$$
$$\therefore pH = -\log[H^+(aq)] = pK_a$$
(see spread 2.1.16).

For ethanoic acid at 25 °C, $K_a = 1.7 \times 10^{-5}$ mol dm^{-3}.
With a 1 : 1 concentration ratio of $CH_3COOH : CH_3COO^-$,
$$pH = pK_a = -\log(1.7 \times 10^{-5}) = 4.77$$

Using this buffer system:
- the middle pH of the range of buffering capacity is 4.77
- the pH can by fine-tuned by using different ratios of the two buffer components.

Using weak acid : conjugate base ratios from 10 : 1 to 1 : 10, the pH range is equal to $pK_a \pm 1$:
- a 10 : 1 ratio of $CH_3COOH(aq) : CH_3COO^-(aq)$ gives a pH of 3.77
- a 10 : 1 ratio of $CH_3COO^-(aq) : CH_3COOH(aq)$ gives a pH of 5.77.

Questions

1 Calculate the pH of a buffer solution containing 0.15 mol dm^{-3} methanoic acid, HCOOH(aq), and 0.065 mol dm^{-3} sodium methanoate, HCOONa(aq), at 25 °C. For HCOOH, $K_a = 1.6 \times 10^{-4}$ mol dm^{-3} at 25 °C.

2 If extra acid is produced in the blood, how do buffers prevent the pH from falling?

By the end of this spread, you should be able to . . .

* Interpret and sketch acid–base titration pH curves for strong and weak acids and bases.
* Explain the choice of suitable indicators for acid–base titrations.

Titrations for AS chemistry

During AS chemistry, you carried out acid–base titrations using indicators.

In a titration, you are measuring the volume of one solution that reacts *exactly* with a known volume of the other solution. This is called the **equivalence point** of the titration. At this point, the solution in the conical flask has exactly reacted with the solution in the burette.

This spread considers pH changes from acid to base. Titrations can be carried out from acid to base or from base to acid – the same principles apply to both routes.

Key features of titration curves

Acid–base titration pH curves can be plotted. The pH is measured successively after the addition of small volumes of the solution from the burette – the pH is measured using a pH meter. Alternatively, the pH can be data logged continuously and a titration curve plotted by a computer.

For a titration in which the acid is in the conical flask and the base in the burette, the following changes in pH will take place.

* When the base is first added, the pH increases very slightly – the acid is in great excess.
* Within 1–2 cm^3 of the equivalence point, the pH starts to increase more quickly. There is now only a small excess of acid present. Eventually, there is a very sharp increase in pH brought about by a very small addition of base, typically just one drop. This shows on the titration curve as a vertical section. The equivalence point is at the centre of this vertical section.
* As further base is added, there is little additional change in pH – the base is now in great excess.

Figure 1 shows these changes for the titration of a strong acid by a strong alkali.

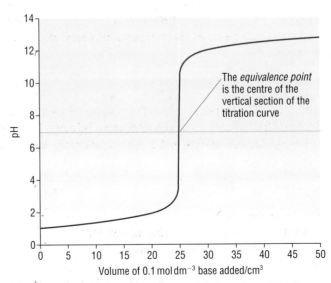

The *equivalence point* is the centre of the vertical section of the titration curve

Figure 1 Titration curve for a strong acid–strong base titration, showing the vertical section

Choosing the indicator

An acid–base indicator is a weak acid, often represented as HIn. An indicator has one colour in its acid form (HIn) and a different colour in its conjugate base form (In⁻).

For methyl orange, HIn is red and In⁻ is yellow:

$$\boxed{HIn} \rightleftharpoons H^+ + \boxed{In^-}$$

When there are equal amounts of the weak acid and the conjugate base present:
- $[HIn] = [In^-]$
- the indicator is at its **end point**.

The colour of the end point for methyl orange is orange, at which [HIn] (red) and [In⁻] (yellow) are equal.

Most indicators change colour over a range of about two pH units. The pH ranges for the indicators methyl orange, bromothymol blue and phenolphthalein are shown in Figure 2. The end point is typically at the middle of this range.

Indicators and titration curves

For a titration, an indicator is chosen so that the pH value of the end point is as close as possible to the pH value of the titration's equivalence point. In practice, a suitable indicator changes colour within the pH range in the vertical section of the titration curve. This often coincides with just a single drop of the base.

To produce the titration curves in Figure 3, different combinations of strong and weak acids were used.

The pH ranges for phenolphthalein and methyl orange have been shown.

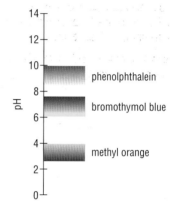

Figure 2 pH colour ranges for some common indicators

strong acid–strong base
- phenolphthalein suitable
- methyl orange suitable

strong acid–weak base
- phenolphthalein unsuitable
- methyl orange suitable

weak acid–strong base
- phenolphthalein suitable
- methyl orange unsuitable

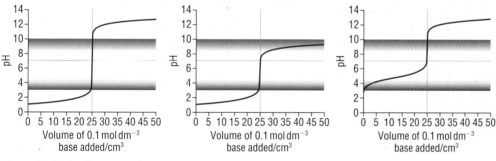

Figure 3 pH titration curves for different combinations of acids and bases with phenolphthalein and methyl orange indicator ranges

Titration of weak acid and weak alkali

As the base is added, the pH changes slowly through the equivalence point and there is no vertical section to the titration curve. An indicator would change colour gradually over a few cm³. There are no indicators really suitable for a titration of a weak acid with a weak base.

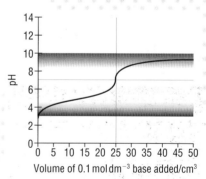

Figure 4 Weak acid–weak base titration curve with phenolphthalein and methyl orange indicator ranges

Questions

1 What is meant by the terms:
   (a) equivalence point;   (b) end point?
2 The pH ranges for two indicators are shown below:
   bromocresol green pH 3.8–5.4; thymolphthalein pH 9.3–10.5
   Use the titration curves in Figure 3 to identify whether these indicators are suitable for the titration of each combination of acid and base.

By the end of this spread, you should be able to ...

* Define and use the term *enthalpy change of neutralisation*.
* Calculate enthalpy changes from appropriate experimental results.

Standard enthalpy change of neutralisation

When an aqueous acid is neutralised by an aqueous base, a salt and water are formed.

For example, hydrochloric acid is neutralised by sodium hydroxide to form sodium chloride and water:

$$HCl(aq) + NaOH(aq) \longrightarrow NaCl(aq) + H_2O(l) \qquad \Delta H^\ominus_{neut} = -57.9 \text{ kJ mol}^{-1}$$
$$\text{acid} \qquad \text{base} \longrightarrow \qquad 1 \text{ mol}$$

The ionic equation is much simpler:
* all ions: $\quad H^+(aq) + \cancel{Cl^-(aq)} + \cancel{Na^+(aq)} + OH^-(aq) \longrightarrow \cancel{Na^+(aq)} + \cancel{Cl^-(aq)} + H_2O(l)$
* ionic equation: $H^+(aq) + OH^-(aq) \longrightarrow H_2O(l)$

$Na^+(aq)$ and $Cl^-(aq)$ ions are *spectator ions* and do not take part in the reaction.

The enthalpy change for this reaction is called the **enthalpy change of neutralisation**, $\Delta H^\ominus_{neut}$, and refers to the neutralisation of an aqueous acid by an aqueous base to form one mole of water.

Determination of enthalpy change of neutralisation

The enthalpy change of neutralisation can be determined easily by experiment. The simplest experiment uses a plastic coffee cup to hold the reactants as shown in Figure 1.

Key definition

The **standard enthalpy change of neutralisation** $\Delta H^\ominus_{neut}$ is the energy change that accompanies the neutralisation of an aqueous acid by an aqueous base to form one mole of $H_2O(l)$, under standard conditions.

Figure 1 A simple coffee cup calorimeter

Worked example

25.0 cm³ of dilute nitric acid, of concentration 2.00 mol dm⁻³, is added to 25.0 cm³ of aqueous potassium hydroxide of concentration 2.00 mol dm⁻³. The temperature increases from 22.0 °C to 35.5 °C.

Calculate the enthalpy change of neutralisation.
* specific heat capacity of solution, $c = 4.18 \text{ J g}^{-1} \text{ K}^{-1}$
* density of solution = 1.00 g cm⁻³

Step 1: Work out the energy change
Total volume of solution = 25.0 + 25.0 = 50.0 cm³
50.0 cm³ of solution has a mass of 50.0 g
Temperature change, $\Delta T = (35.5 - 22.0)°C$
$$= +13.5 °C$$
Heat *gained* by surroundings, $Q = m \times c \times \Delta T = 50 \times 4.18 \times 13.5 \text{ J}$
$$= +2821.5 \text{ J}$$
This heat has been released by the chemical system.
∴ heat *lost* by chemical system = −2821.5 J

Step 2: Work out the amount, in mol, that reacted
Amount of HNO_3 that reacted $= \dfrac{2.00 \times 25.0}{1000}$
$$= 0.0500 \text{ mol}$$
Amount of KOH that reacted $= \dfrac{2.00 \times 25.0}{1000}$
$$= 0.0500 \text{ mol}$$

Step 3: Scale the quantities to match the molar quantities needed to form one mole of H_2O.

$$HNO_3(aq) + KOH(aq) \longrightarrow KNO_3(aq) + H_2O(l)$$
$$\text{1 mol} \qquad \text{1 mol} \qquad \text{1 mol} \qquad \text{1 mol}$$

For reaction of 0.0500 mol HNO_3 with 0.0500 mol KOH, $\Delta H = -2821.5$ J

For 1 mol of HNO_3 and 1 mol KOH, $\Delta H = \dfrac{-2821.5}{0.0500}$

$$= -564\,30 \text{ J}$$
$$= -56.4 \text{ kJ}$$

Step 4: Finally, work out the enthalpy change for neutralisation to form 1 mole of H_2O, in kJ mol^{-1}.

$$HNO_3(aq) + KOH(aq) \longrightarrow KNO_3(aq) + H_2O(l)$$

all ions: $\qquad H^+(aq) + \cancel{NO_3^-(aq)} + \cancel{K^+(aq)} + OH^-(aq) \longrightarrow \cancel{K^+(aq)} + \cancel{NO_3^-(aq)} + H_2O(l)$

ionic equation: $H^+(aq) + OH^-(aq) \longrightarrow H_2O(l) \quad \Delta H = -56.4$ kJ mol^{-1}

$\therefore \Delta H_{neut} = -56.4$ kJ mol^{-1}

Comparison of enthalpy changes of neutralisation of different acids

The table shows values for the enthalpy changes of neutralisation of different acids. Each result was obtained using the same aqueous base, NaOH(aq).

Acid	Reaction	$\Delta H^{\ominus}_{neut}$/kJ mol^{-1}
Hydrochloric acid	$HCl(aq) + NaOH(aq) \longrightarrow NaCl(aq) + H_2O(l)$	-57.9
Nitric acid	$HNO_3(aq) + NaOH(aq) \longrightarrow NaNO_3(aq) + H_2O(l)$	-57.6
Hydrobromic acid	$HBr(aq) + NaOH(aq) \longrightarrow NaBr(aq) + H_2O(l)$	-57.6
Ethanoic acid	$CH_3COOH(aq) + NaOH(aq) \longrightarrow CH_3COONa(aq) + H_2O(l)$	-57.1

The first three acids – HCl(aq), HNO_3(aq) and HBr(aq) – are all strong acids.

In aqueous solution, strong acids and bases are completely dissociated. Consequently the first three equations all simplify down to the same ionic equation and the $\Delta H^{\ominus}_{neut}$ values are virtually the same:

- HCl(aq): $\qquad H^+(aq) + \cancel{Cl^-(aq)} + \cancel{Na^+(aq)} + OH^-(aq) \longrightarrow \cancel{Na^+(aq)} + \cancel{Cl^-(aq)} + H_2O(l)$
- HNO_3(aq): $\qquad H^+(aq) + \cancel{NO_3^-(aq)} + \cancel{Na^+(aq)} + OH^-(aq) \longrightarrow \cancel{Na^+(aq)} + \cancel{NO_3^-(aq)} + H_2O(l)$
- HBr(aq): $\qquad H^+(aq) + \cancel{Br^-(aq)} + \cancel{Na^+(aq)} + OH^-(aq) \longrightarrow \cancel{Na^+(aq)} + \cancel{Br^-(aq)} + H_2O(l)$
- ionic equation: $H^+(aq) + OH^-(aq) \longrightarrow H_2O(l)$

Ethanoic acid is a weak acid, only partially dissociated. Although the same neutralisation reaction takes place, most of the ethanoic acid molecules must first dissociate by breaking the O–H bond. This requires energy, and consequently the enthalpy change recorded is slightly less exothermic at -57.1 kJ mol^{-1}.

- bond breaking: $CH_3COOH(aq) \longrightarrow CH_3COO^-(aq) + H^+(aq) \qquad$ endothermic
- ions $\qquad H^+(aq) + \cancel{CH_3COO^-(aq)} + \cancel{Na^+(aq)} + OH^-(aq) \longrightarrow$
$$\cancel{Na^+(aq)} + \cancel{CH_3COO^-(aq)} + H_2O(l)$$
- ionic equation: $H^+(aq) + OH^-(aq) \longrightarrow H_2O(l)$

Questions

1 75.0 cm^3 of dilute hydrochloric acid of concentration 4.00 mol dm^{-3} is added to 75.0 cm^3 of aqueous sodium hydroxide of concentration 4.00 mol dm^{-3}. The temperature increases from 21.0 °C to 48.5 °C. The specific heat capacity, c, of solution is 4.18 J g^{-1} K^{-1}. Calculate the enthalpy change of neutralisation.

2 (a) Why do we get the same value for the enthalpy change of neutralisation for different strong acids?

(b) Why do weak acids give a less exothermic value for the enthalpy change of neutralisation using strong acids?

Strong and weak acids

Acid is a proton donor
Base is a proton acceptor
A strong acid completely dissociates:

$$HCl \longrightarrow H^+ + Cl^-$$

A weak acid partially dissociates:

$$HCOOH \rightleftharpoons H^+ + HCOO^-$$

Extent of dissociation = 'strength' measured by acid dissociation constant, K_a:

$$K_a = \frac{[H^+][HCOO^-]}{[HCOOH]} \qquad pK_a = -\log K_a$$

The stronger the acid, the **greater** the value of K_a and the **smaller** the value of pK_a

Rates

For reaction: $X + Y + Z \longrightarrow$ products
rate equation determined experimentally as $rate = k[X]^2[Y]$

- 2nd order with respect to X
- 1st order with respect to Y
- Zero order with respect to Z

Concentration–time graphs
1st order has a constant half-life

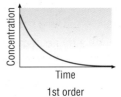

1st order

Rate–concentration graphs

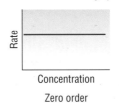
Zero order

1st order

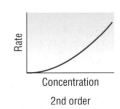
2nd order

Acid–base pairs

Acid–base pairs differ by H^+

Acid 1   Base 1
$$HA + H_2O \rightleftharpoons H_3O^+ + A^-$$
Base 2   Acid 2

Equilibrium constant, K_c

$$N_2(g) + 3H_2(g) \rightleftharpoons 2NH_3(g)$$
reactants   products

$$K_c = \frac{[NH_3]^2}{[N_2][H_2]^3}$$

products over reactants

K_c only changes with temperature
Depends on sign of ΔH for forward reaction:
ΔH +ve (endo), K_c increases with increasing T
ΔH −ve (exo), K_c decreases with increasing T

pH calculations

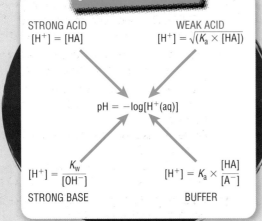

STRONG ACID
$$[H^+] = [HA]$$

WEAK ACID
$$[H^+] = \sqrt{(K_a \times [HA])}$$

$$pH = -\log[H^+(aq)]$$

$$[H^+] = \frac{K_w}{[OH^-]}$$
STRONG BASE

$$[H^+] = K_a \times \frac{[HA]}{[A^-]}$$
BUFFER

Buffers

Buffer minimises pH changes
A weak acid, HA, and its conjugate base, A^-

The equilibrium shifts in response to addition of small amounts of acid and alkali

addition of alkali
$$HA(aq) \rightleftharpoons H^+(aq) + A^-(aq)$$
addition of acid

Practice questions

1 In a reaction between **W**, **X** and **Y**:
when the concentration of **W** is tripled, the rate increases by 9 times;
when the concentration of **X** is halved, the rate is halved;
when the concentration of **X** is doubled and **Y** is doubled, the rate doubles.

(a) How is the rate of a reaction measured from a concentration–time graph?

(b) What are the units of rate of reaction?

(c) Deduce the orders with respect to **W**, **X** and **Y**.

(d) Write the rate equation for the reaction.

(e) How will the rate change if the concentrations of **W**, **X** and **Y** are all increased by a factor of 4?

2 **R**, **S** and **T** react together. Four experiments are carried out to investigate the kinetics of this reaction. The results are shown in the table below.

Experiment	Initial [R]/ mol dm^{-3}	Initial [S]/ mol dm^{-3}	Initial [T]/ mol dm^{-3}	Initial rate/ mol dm^{-3} s^{-1}
1	0.00100	0.00200	0.00300	4.56×10^{-8}
2	0.00100	0.00400	0.00300	9.12×10^{-8}
3	0.00050	0.00400	0.00300	2.28×10^{-8}
4	0.00050	0.01200	0.00600	6.84×10^{-8}

(a) Deduce the order of the reaction with respect to **R**, **S** and **T**. Show your reasoning,

(b) Write the rate equation.

(c) Calculate the rate constant, k, for the reaction.

3 **A** and **B** react together as in the overall equation:

$$2\textbf{A} + \textbf{B} \longrightarrow 2\textbf{C}$$

The rate equation for this reaction is rate = $k[\textbf{A}][\textbf{B}]$.

(a) Explain what is meant by *rate-determining step*.

(b) What does the rate equation above tell you about the rate-determining step?

(c) Suggest a possible two-step mechanism for this reaction.

4 For each of the following equilibria: **(i)** write the expression for K_c and state the units; **(ii)** state what happens to the value of K_c when the temperature is increased.

(a) $CH_4(g) + H_2O(g) \rightarrow CO(g) + 3H_2(g)$ $\quad \Delta H = +210$ kJ mol^{-1}

(b) $H_2(g) + I_2(g) \rightarrow 2HI(g)$ $\quad \Delta H = -10$ kJ mol^{-1}

(c) $2SO_2(g) + O_2(g) \rightarrow 2SO_3(g)$ $\quad \Delta H = -197$ kJ mol^{-1}

5 For the reaction $H_2(g) + I_2(g) \rightarrow 2HI(g)$, 0.200 mol H_2 and 0.400 mol I_2 are placed in a sealed container of volume 4.0 dm^3. At equilibrium, 0.140 mol H_2 remains.

(a) Determine the equilibrium amounts of I_2 and HI present.

(b) Calculate the value of K_c under these conditions.

6 **(a)** Identify the acid–base pairs in the acid–base equilibria:

(i) $CH_3COOH + H_2O \rightarrow CH_3COO^- + H_3O^+$

(ii) $CH_3COOH + HCl \rightarrow CH_3COOH_2^+ + Cl^-$

(b) Write full and ionic equations for the following acid–base reactions:

(i) hydrochloric acid and solid calcium carbonate;

(ii) nitric acid and magnesium oxide;

(iii) ethanoic acid and aqueous calcium hydroxide.

7 **(a)** Calculate the pH values of solutions of the following weak acids:

(i) 7.38×10^{-2} mol dm^{-3} methanoic acid, HCOOH ($K_a = 1.60 \times 10^{-4}$ mol dm^{-3})

(ii) 6.31×10^{-4} mol dm^{-3} chloric(l) acid, HClO ($K_a = 3.70 \times 10^{-8}$ mol dm^{-3})

(b) Find the values of K_a and pK_a for the following weak acids:

(i) a weak acid that has a pH of 3.54 at a concentration of 0.267 mol dm^{-3} solution;

(ii) a weak acid that has a pH of 5.31 at a concentration of 5.35×10^{-3} mol dm^{-3} solution.

8 **(a)** Calculate the pH of the following strong alkalis at 25°C:

(i) 0.125 mol dm^{-3} NaOH(aq)

(ii) 6.91×10^{-4} mol dm^{-3} KOH(aq)

(iii) 2.35×10^{-3} mol dm^{-3} Ca(OH)$_2$.

(b) Calculate the concentrations, in mol dm^{-3}, of the following solutions of strong bases at 25°C:

(i) KOH with a pH of 11.34;

(ii) NaOH with a pH of 13.72;

(iii) KOH with a pH of 14.55.

9 Buffer solutions can be made by mixing together ethanoic acid, CH_3COOH, and sodium ethanoate, CH_3COONa. For CH_3COOH, $K_a = 1.7 \times 10^{-5}$ mol dm^{-3} at 25°C.

(a) Explain how buffer solutions made from CH_3COOH and CH_3COONa minimise pH changes.

(b) Two buffer solutions, **A** and **B**, were made using CH_3COOH and CH_3COONa. Calculate the pH of each buffer solution.
Buffer **A** has the same concentrations of CH_3COOH and CH_3COONa.
Buffer **B** has the concentrations 0.750 mol dm^{-3} CH_3COOH and 0.250 mol dm^{-3} CH_3COONa.

(c) What ratio of [CH_3COOH] : [CH_3COONa] would be needed to make a solution that buffers at a pH of 5.47?

1 Hydrogen peroxide, H_2O_2, is a colourless liquid, widely used as an oxidising agent, antiseptic, and bleach for hair and cloth.

Hydrogen peroxide reacts with iodide ions, I^-, in the presence of acid, H^+, forming iodine, I_2.

(a) Suggest a balanced equation for the overall reaction between $H_2O_2(aq)$, $I^-(aq)$ and $H^+(aq)$ to form aqueous iodine. [2]

(b) Three experiments were carried out using different initial concentrations of $H_2O_2(aq)$, $I^-(aq)$ and $H^+(aq)$. The initial rate of formation of I_2 was measured for each experiment.

The experimental results are shown below.

Experiment	[H_2O_2(aq)] /mol dm^{-3}	[I^-(aq)] /mol dm^{-3}	[H^+(aq)] /mol dm^{-3}	Rate /mol dm^{-3} s^{-1}
1	0.010	0.010	0.005	1.15×10^{-6}
2	0.010	0.020	0.005	4.60×10^{-6}
3	0.010	0.020	0.010	4.60×10^{-6}

(i) Showing all your reasoning, determine the reaction orders for I^- and for H^+. [4]

(ii) This reaction is first order with respect to H_2O_2. Use this information and your answers to (i) to complete the rate equation for this reaction. [2]

(iii) Calculate the rate constant k for this reaction. State the units for k. [3]

(c) This reaction was shown to be first order with respect to H_2O_2 by plotting a rate–concentration graph.

Using the axes below, sketch a graph to show how the rate of this reaction changes with increasing H_2O_2 concentration. [2]

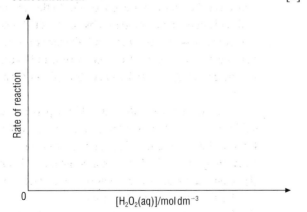

(d) Hydrogen peroxide readily decomposes to give water and oxygen.

Hydrogen peroxide is sold by volume strength. For example, 20-volume H_2O_2 yields 20 volumes of oxygen gas for each volume of aqueous H_2O_2 solution.

(i) Construct an equation for the decomposition of hydrogen peroxide. [1]

(ii) Determine the concentration, in mol dm^{-3}, of 20-volume hydrogen peroxide. Show all your working clearly. [3]

[Total: 17]

(OCR 2816/1 Jun04)

2 Propanone reacts with iodine in the presence of dilute hydrochloric acid.

A student carried out an investigation into the kinetics of this reaction.

He measured how the concentration of propanone changes with time. He also investigated how different concentrations of iodine and hydrochloric acid affect the initial rate of the reaction.

The graph and results are shown below.

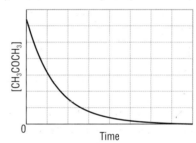

[CH_3COCH_3] /mol dm^{-3}	[I_2] /mol dm^{-3}	[H^+] /mol dm^{-3}	Initial rate /mol dm^{-3} s^{-1}
1.5×10^{-3}	0.0300	0.0200	2.1×10^{-9}
1.5×10^{-3}	0.0300	0.0400	4.2×10^{-9}
1.5×10^{-3}	0.0600	0.0400	4.2×10^{-9}

The overall equation for the reaction is given below.

$$CH_3COCH_3 + I_2 \rightarrow CH_3COCH_2I + HI$$

This is a multi-step reaction.

- What conclusions can be drawn about the kinetics of this reaction from the student's experiment? Justify your reasoning.
- Calculate the rate constant for this reaction.
- Suggest the equations for a possible two-step mechanism for this reaction. Label the rate-determining step and explain your reasoning.

[Total: 14]

(OCR 2816/1 Jan06)

3 The formation of ethyl ethanoate and water from ethanoic acid and ethanol is a reversible reaction, which can be allowed to reach equilibrium. The equilibrium is shown below.

$$CH_3COOH + C_2H_5OH \rightleftharpoons CH_3COOC_2H_5 + H_2O$$

(a) Write the expression for K_c for this equilibrium system. [2]

(b) A student mixed together 6.0 mol ethanoic acid and 12.5 mol ethanol. A small amount of hydrochloric acid was also added to catalyse the reaction. He left the mixture for two days to reach equilibrium, after which time 1.0 mol ethanoic acid remained.

(i) Complete the table below to show the equilibrium composition of the equilibrium mixture.

Component	CH₃COOH	C₂H₅OH	CH₃COOC₂H₅	H₂O
Initial amount/mol	6.0	12.5	0.0	0.0
Equilibrium amount/mol				

[2]

(ii) Calculate K_c to two significant figures. State the units, if any. The total volume of the equilibrium mixture is 1.0 dm³. [2]

(c) The student was concerned that the mixture may **not** have reached equilibrium.
What could he do to be sure that equilibrium had been reached? [2]

(d) The student added more ethanol to the mixture.
(i) State, giving a reason, what would happen to the composition of the equilibrium mixture. [2]
(ii) What happens to the value of K_c? [1]

(e) The student added more of the acid catalyst to the mixture.
State, giving a reason, what would happen to the composition of the equilibrium mixture. [2]

(f) The student repeated the experiment at a higher temperature and found that the value of K_c decreased.
(i) State, giving a reason, what would happen to the composition of the equilibrium mixture. [2]
(ii) What additional information does this information give you about the reaction? [1]

[Total: 16]
(OCR 2816/1 Jan04)

4 The preparation of hydrogen iodide, HI, from hydrogen and iodine gases is a reversible reaction that reaches equilibrium at constant temperature.

$$H_2(g) + I_2(g) \rightleftharpoons 2HI(g)$$

(a) Write the expression for K_c for this equilibrium system. [1]

(b) A student mixed together 0.30 mol $H_2(g)$ with 0.20 mol $I_2(g)$ and the mixture was allowed to reach equilibrium. At equilibrium, 0.14 mol $H_2(g)$ was present.
(i) Complete the table below to show the amount of each component in the equilibrium mixture.

Component	H₂(g)	I₂(g)	HI(g)
Initial amount /mol	0.30	0.20	0
Equilibrium amount /mol			

[2]

(ii) Calculate K_c to an appropriate number of significant figures. State the units, if any. [3]

(c) The student compressed the equilibrium mixture so that its volume was reduced. The temperature was kept constant.
Comment on the value of K_c **and** the composition of the equilibrium mixture under these new conditions. [2]

(d) The student repeated the experiment at a higher temperature and found that less HI was present at equilibrium.
Explain what additional information this gives you about the reaction. [2]

(e) Hydriodic acid, HI(aq), is a strong acid that is an aqueous solution of hydrogen iodide. In the laboratory, hydriodic acid can be prepared by the method below.
A mixture of 480 g of iodine and 600 cm³ of water was put into a flask. The mixture was stirred and hydrogen sulfide gas, $H_2S(g)$, was bubbled through for several hours.
The mixture became yellow as sulfur separated out. The sulfur was filtered off and the solution was purified by fractional distillation. A fraction was collected containing 440 g of HI in a total volume of 750 cm³.
(i) Construct a balanced equation, with state symbols, for the preparation of hydriodic acid. [2]
(ii) Determine the percentage yield of hydriodic acid. [3]
(iii) Calculate the pH of the hydriodic acid fraction. [2]

[Total: 17]
(OCR 2816/1 Jan06)

5 Carbonic acid, H_2CO_3, is a weak Brønsted–Lowry acid formed when carbon dioxide dissolves in water. Blood contains several buffer solutions, and healthy blood is buffered to a pH of 7.40. The most important buffer solution in blood is a mixture of carbonic acid and hydrogencarbonate ions, HCO_3^-.

The equilibrium in the carbonic acid/hydrogencarbonate buffer system is shown below.

$H_2CO_3(aq) \rightleftharpoons H^+(aq) + HCO_3^-(aq)$ $K_a = 4.17 \times 10^{-7}$ mol dm^{-3}

(a) Carbonic acid is a weak Brønsted–Lowry acid.
What is meant by the following terms:

   (i) a Brønsted–Lowry acid   [1]

   (ii) a weak acid   [1]

   (iii) pH   [1]

   (iv) a buffer solution?   [1]

(b) In this question, one mark is available for the quality of written communication.
Explain how the carbonic acid/hydrogencarbonate buffer works. Use equations to help your answer.   [4]
Quality of Written Communication   [1]

(c) Calculate the ratio of $HCO_3^-:H_2CO_3$ in healthy blood with a pH of 7.40.   [5]

[Total: 14]
(OCR 2816/1 Jan04)

6 Methanoic acid, HCOOH, is a weak organic acid that occurs naturally in ants and stinging nettles.

(a) Use an equation for the dissociation of methanoic acid to show what is meant by a *weak acid*.   [1]

(b) A 1.50×10^{-2} mol dm^{-3} solution of HCOOH has $[H^+] = 1.55 \times 10^{-3}$ mol dm^{-3}.

   (i) Calculate the pH of this solution and give one reason why the pH scale is a more convenient measurement for measuring acid concentrations than $[H^+]$.   [2]

   (ii) Write the expression for K_a for methanoic acid. [1]

   (iii) Calculate the values of K_a and pK_a for methanoic acid.   [3]

   (iv) Estimate the percentage of HCOOH molecules that have dissociated in this aqueous solution of methanoic acid.   [1]

(c) A student titrated the 1.50×10^{-2} mol dm^{-3} methanoic acid with aqueous sodium hydroxide.
A 25.00 cm^3 sample of the HCOOH(aq) was placed in a conical flask and the NaOH(aq) was added from a burette until the pH no longer changed.

   (i) Write a balanced equation for the reaction between HCOOH(aq) and NaOH(aq).   [1]

   (ii) The pH curve for this titration is shown below.

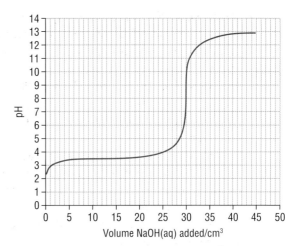

Calculate the concentration, in mol dm^{-3}, of the aqueous sodium hydroxide.   [3]

   (iii) Calculate the pH of the aqueous sodium hydroxide.   [2]

   (iv) The pH ranges in which the colour changes for three acid–base indicators are shown below.

Indicator	pH range
Metacresol purple	7.4–9.0
2,4,6-trinitrotoluene	11.5–13.0
Ethyl orange	3.4–4.8

Explain which of the three indicators is suitable for this titration.   [2]

[Total: 16]
(OCR 2816/1 Jan06)

7 The K_a values for three acids are shown in the table below.

Acid		K_a/mol dm^{-3}
Ethanoic acid	CH_3COOH	1.70×10^{-5}
Phenol	C_6H_5OH	1.28×10^{-10}
Sulfurous acid	H_2SO_3	1.50×10^{-2}

(a) What information is provided by K_a values?   [1]

(b) When sulfurous acid and ethanoic acid are mixed together, an acid–base reaction takes place.
$H_2SO_3(aq) + CH_3COOH(aq) \rightleftharpoons$
$\qquad\qquad\qquad HSO_3^-(aq) + CH_3COOH_2^+(aq)$

   (i) Copy the equation and
label one **conjugate acid–base pair** as acid 1 and base 1,
label the other **conjugate acid–base pair** as acid 2 and base 2.   [2]

   (ii) Predict and explain the acid–base reaction that would take place if ethanoic acid were mixed with phenol. Include an equation in your answer.   [2]

(c) The pH value of 0.0450 mol dm^{-3} hydrochloric acid is different from that of 0.0450 mol dm^{-3} ethanoic acid. Calculate the pH values of these two acids. Show all your working. [5]

(d) An excess of magnesium was added to 100 cm^3 of 0.0450 mol dm^{-3} hydrochloric acid. The same mass of magnesium was added to 100 cm^3 of 0.0450 mol dm^{-3} ethanoic acid.
Both reactions produced 54 cm^3 of hydrogen gas, measured at room temperature and pressure, but the reaction with ethanoic acid took much longer to produce this gas volume.
Explain why the reactions produced the same volume of a gas but at different rates.
Use equations in your answer. [4]

[Total: 14]
(OCR 2816/1 Jun06)

8 In sewage plants, biological activity can be reduced by increasing the pH of the water. This is achieved by adding small amounts of solid calcium hydroxide, Ca(OH)$_2$, to the sewage water.
In all parts of this question, assume that measurements have been made at 25 °C.

(a) The pH of aqueous solutions is determined by K_w. K_w has a value of 1.0×10^{-14} mol^2 dm^{-6} at 25 °C.
   (i) What name is given to K_w? [1]
   (ii) Write the expression for K_w. [1]

(b) A chemist checked the concentration of aqueous calcium hydroxide, Ca(OH)$_2$, in the sewage water by titration with 5.00×10^{-3} mol dm^{-3} hydrochloric acid.
Ca(OH)$_2$(aq) + 2HCl(aq) → CaCl$_2$(aq) + 2H$_2$O(l)
The chemist titrated 25.0 cm^3 of the sewage water with 21.35 cm^3 of HCl to reach the endpoint of the titration. Calculate the concentration, in mol dm^{-3}, of the calcium hydroxide in the sewage water. [3]

(c) The chemist analysed a sample of water from another part of the sewage works and he found that the calcium hydroxide concentration was 2.7×10^{-3} mol dm^{-3}.
When solid calcium hydroxide dissolves in water, its ions completely dissociate.
Ca(OH)$_2$(s) → Ca^{2+}(aq) + 2OH$^-$(aq)
Calculate the pH of this sample. [3]

(d) After further treatment, the water could be used for drinking. In the drinking water produced, the OH$^-$ concentration was 100 times greater than the H$^+$ concentration.
What was the pH of this drinking water? [1]

[Total: 9]
(OCR 2816/1 Jan07)

9 This question looks at two acids:
methanoic acid, HCOOH, a weak organic acid;
nitric acid, HNO$_3$, a strong acid which can also act as a powerful oxidising agent.

(a) Methanoic acid is a weak Brønsted–Lowry acid. Explain what is meant by a *weak Brønsted–Lowry acid*. [2]

(b) Calculate the pH of a 0.025 mol dm^{-3} solution of methanoic acid. Show your working.
For HCOOH, $K_a = 1.58 \times 10^{-4}$ mol dm^{-3}. [3]

(c) Methanoic acid is a component of a buffer solution used in shampoos. The buffer solution can be made by mixing methanoic acid with another chemical.
   (i) Explain what is meant by a *buffer solution*. [1]
   (ii) Suggest a chemical that could be added to methanoic acid to prepare a buffer solution. Explain your answer. [2]
   (iii) What factors determines the pH of this buffer solution? [2]

(d) Nitric acid, HNO$_3$, is sold by a chemical supplier as a 65% solution, by mass. As supplied, each cubic decimetre of this nitric acid has a mass of 1400 g. Calculate the pH of this solution. [3]

(e) When dilute, nitric acid behaves as a typical acid. Write an equation for the reaction of nitric acid with limestone. [2]

(f) When nitric acid is added to methanoic acid, the acid–base equilibrium below is set up.
HNO$_3$ + HCOOH ⇌ NO$_3^-$ + HCOOH$_2^+$
Use this equilibrium to explain what the meant by the term *conjugate acid–base pairs*. [3]

(g) Concentrated nitric acid is a powerful oxidising agent. Concentrated nitric acid oxidises sulfur to sulfuric acid, nitrogen dioxide and another product.
   (i) Suggest a balanced equation for this reaction. [1]
   (ii) Deduce the change in oxidation number of nitrogen in this reaction. [1]

[Total: 20]
(OCR 2816/1 Jun07)

Module 2
Energy

Introduction

Some chemical and physical changes happen by themselves. If ice is taken out of the refrigerator, it melts; when sugar is added to hot tea it dissolves; and when iron is left in the atmosphere, it rusts. These changes are spontaneous – they just happen!

The strength of ionic bonds, like the ones in this NaCl lattice, is reflected in their lattice enthalpy, but unfortunately it is impossible to measure lattice enthalpy in the laboratory environment. The study of other energy changes and the use of energy cycles enable these values to be calculated and to be used to compare the stability of ionic materials.

As supplies of natural gas and oil run out and the cost of running our cars increases, scientists are attempting to convert chemical energy into electrical energy using fuel cells. This is electrochemistry.

In this module, you will study important energy topics and you will learn how to calculate energy changes and electrode potentials.

Test yourself

1. Define the term *enthalpy change of formation*.
2. Write the equation for the lattice enthalpy of KCl.
3. What does the symbol $E^{\ominus}$ represent?
4. How does a fuel cell work?
5. What is meant by the term *redox*?

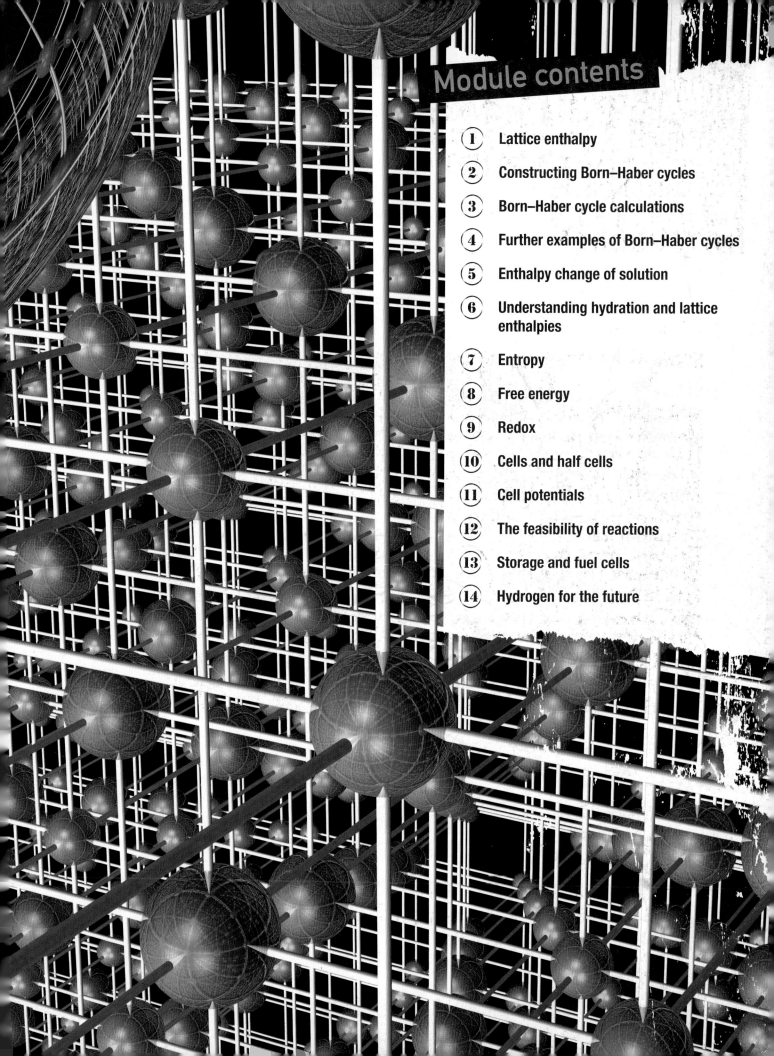

By the end of this spread, you should be able to . . .

✱ **Explain and use the term *lattice enthalpy*.**

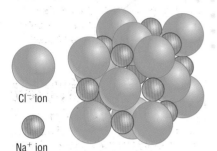

Cl⁻ ion

Na⁺ ion

Figure 1 Ionic lattice of sodium chloride – lattice enthalpy indicates the strength of attraction between the oppositely charged ions

Lattice enthalpy

From your study of bonding in AS chemistry, you learned that:
- ionic bonds are electrostatic forces of attraction between oppositely charged ions
- ionic bonds are strong.

Consequently the melting point of an ionic compound is very high. The strength of ionic bonding is directly related to the **lattice enthalpy**, $\Delta H^{\ominus}_{LE}$, of the ionic compound.

For the **lattice enthalpy**, $\Delta H^{\ominus}_{LE}$, one mole of a solid ionic lattice is formed from its gaseous ions under standard conditions.

$\Delta H^{\ominus}_{LE}$ for potassium chloride is:

$$K^+(g) + Cl^-(g) \longrightarrow KCl(s) \qquad \Delta H = -711 \text{ kJ mol}^{-1}.$$

Features of lattice enthalpy

There are some key points to learn about lattice enthalpy.
- Lattice enthalpy is an exothermic change. Lattice enthalpies have a negative sign because energy is given out when ionic bonds are being formed from gaseous ions.
- Lattice enthalpy indicates the strength of an ionic lattice and is a measure of the ionic bond strength. A large negative value for lattice enthalpy shows that there are strong electrostatic forces of attraction between the oppositely charged ions in the lattice.
- A covalent substance does not have a lattice enthalpy because there are no ions in its structure.
- Lattice enthalpies cannot be measured directly because it is impossible to form one mole of an ionic lattice from gaseous ions.

It is possible to calculate lattice enthalpies using other known energy changes by applying Hess' law. **Hess' law** was introduced in *AS Chemistry* spread 2.3.8.

An energy diagram known as a Born–Haber cycle is used to represent these energy changes. Before constructing a Born–Haber cycle, it is important to know some key definitions and how to write balanced equations, including state symbols, to represent the changes taking place.

Key enthalpy changes

The standard enthalpy change of formation, $\Delta H_f^{\ominus}$

For the **standard enthalpy change of formation**, $\Delta H_f^{\ominus}$, one mole of a compound is formed from its constituent elements in their standard states.

This is usually an exothermic process for an ionic compound.

$\Delta H_f^{\ominus}$ for potassium chloride is:

$$K(s) + \tfrac{1}{2}Cl_2(g) \longrightarrow KCl(s) \qquad \Delta H_f^{\ominus} = -437 \text{ kJ mol}^{-1}$$

Key definitions

Lattice enthalpy is the enthalpy change that accompanies the formation of one mole of an ionic compound from its gaseous ions under standard conditions.

Hess' law states that if a reaction can take place by more than one route and the initial and final conditions are the same, the total enthalpy change is the same for each route.

Key definition

The **standard enthalpy change of formation** is the enthalpy change that takes place when one mole of a compound is formed from its constituent elements in their standard states under standard conditions.

The standard enthalpy change of atomisation, $\Delta H_{at}^{\ominus}$

For the **standard enthalpy change of atomisation**, $\Delta H_{at}^{\ominus}$, one mole of gaseous atoms is formed from its element in its standard state.

$\Delta H_{at}^{\ominus}$ is always an endothermic process because bonds have to be broken.

For potassium metallic bonds are broken:

$$K(s) \longrightarrow K(g) \qquad \Delta H_{at}^{\ominus} = +89 \text{ kJ mol}^{-1}$$

For chlorine covalent bonds are broken:

$$\tfrac{1}{2}Cl_2(g) \longrightarrow Cl(g) \qquad \Delta H_{at}^{\ominus} = +121 \text{ kJ mol}^{-1}$$

<aside>
Key definition

The **enthalpy change of atomisation** is the enthalpy change that takes place when one mole of gaseous atoms forms from the element in its standard state.
</aside>

The first ionisation energy, $\Delta H_{I1}^{\ominus}$

For the **first ionisation energy**, $\Delta H_{I1}^{\ominus}$, one mole of gaseous 1+ ions is formed from gaseous atoms.

$\Delta H_{I1}^{\ominus}$ is an endothermic process because the electron being lost has to overcome attraction from the nucleus in order to leave the atom.

$\Delta H_{I1}^{\ominus}$ for potassium is:

$$K(g) \longrightarrow K^+(g) + e^- \qquad \Delta H_{I1}^{\ominus} = +419 \text{ kJ mol}^{-1}$$

<aside>
Examiner tip

When writing an atomisation equation remember that only one mole of atoms is formed. Students frequently get this equation wrong in exams!

Learn the definitions below.
</aside>

The second ionisation energy, $\Delta H_{I2}^{\ominus}$

For the **second ionisation energy**, $\Delta H_{I2}^{\ominus}$, one mole of gaseous 2+ ions are formed from one mole of gaseous 1+ ions.

As with the first ionisation energy, this is an endothermic process because the electron being lost has to overcome attraction from the nucleus.

$\Delta H_{I2}^{\ominus}$ for calcium is:

$$Ca^+(g) \longrightarrow Ca^{2+}(g) + e^- \qquad \Delta H_{I2}^{\ominus} = +1145 \text{ kJ mol}^{-1}$$

<aside>
Key definitions

The **first ionisation energy** is the enthalpy change accompanying the removal of one electron from each atom in one mole of gaseous atoms to form one mole of gaseous 1+ ions.

The **second ionisation energy** is the enthalpy change accompanying the removal of one electron from each ion in one mole of gaseous 1+ ions to form one mole of gaseous 2+ ions.

The **first electron affinity** is the enthalpy change accompanying the addition of one electron to each atom in one mole of gaseous atoms to form one mole of gaseous 1− ions.

The **second electron affinity** is the enthalpy change accompanying the addition of one electron to each ion in one mole of gaseous 1− ions to form one mole of gaseous 2− ions.
</aside>

The first electron affinity, $\Delta H_{EA1}^{\ominus}$

For the **first electron affinity**, $\Delta H_{EA1}^{\ominus}$, one mole of gaseous 1− ions is formed from gaseous atoms.

This is an exothermic process because the electron is attracted into the outer shell of an atom by the nucleus.

$\Delta H_{EA1}^{\ominus}$ for chlorine is:

$$Cl(g) + e^- \longrightarrow Cl^-(g) \qquad \Delta H_{EA1}^{\ominus} = -346 \text{ kJ mol}^{-1}$$

The second electron affinity, $\Delta H_{EA2}^{\ominus}$

For the **second electron affinity**, $\Delta H_{EA2}^{\ominus}$, one mole of gaseous 2− ions is formed from gaseous 1− ions.

This is an endothermic process because the electron is repelled by the 1− ion. This repulsion has to be overcome.

$\Delta H_{EA2}^{\ominus}$ for oxygen is:

$$O^-(g) + e^- \longrightarrow O^{2-}(g) \qquad \Delta H_{EA2}^{\ominus} = +798 \text{ kJ mol}^{-1}$$

<aside>
Examiner tip

In these changes, the state symbols are critical. It is important to write all state symbols in your equations.
</aside>

Questions

1 Define the term *lattice enthalpy*.
2 Why can lattice energies not be determined directly?
3 Write equations to represent the following enthalpy changes:
  (a) the first ionisation energy of lithium;
  (b) the enthalpy change of formation of calcium chloride;
  (c) the lattice enthalpy of calcium chloride.

By the end of this spread, you should be able to . . .

✳ **use the lattice enthalpy of a simple ionic solid and relevant energy terms to construct Born–Haber cycles.**

Born–Haber cycles

A Born–Haber cycle is similar to a Hess' law energy cycle. It allows the calculation of an enthalpy change which cannot be measured directly, providing that all other relevant energy changes are known.

* Start your cycle with the elements in their standard states at zero energy. The elements always sit on the zero line – sometimes called the datum line.
* The ionic solid can form from its elements in their standard states. This is the enthalpy change of formation of the ionic solid, which is an exothermic process ($\Delta H = -ve$).

In the Born–Haber cycle:

* all ΔH values pointing upwards are endothermic values
* all ΔH values pointing downwards are exothermic values.

The first step in drawing a cycle is to place the ionic solid on the line immediately below the elements, as shown in Figure 1.

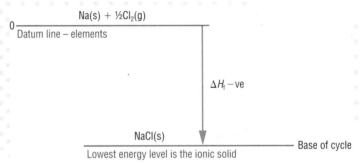

Figure 1 First step in constructing the Born–Haber cycle for sodium chloride

The ionic solid can also be made from its ions in the gas state (lattice enthalpy). This step is very exothermic and the gaseous ions will have much more energy than the ionic solid. Figure 2 shows this step added to the cycle.

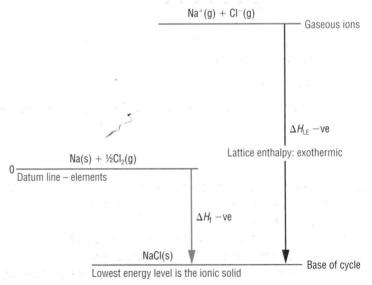

Figure 2 Adding the second step to the Born–Haber cycle

The cycle has to be completed by converting the elements into gaseous ions in a number of steps using equations and provided data. The following steps are needed to complete the cycle:

- atomisation of sodium: $Na(s) \longrightarrow Na(g)$ endothermic
- atomisation of chlorine: $\frac{1}{2}Cl_2(g) \longrightarrow Cl(g)$ endothermic
- ionisation of sodium: $Na(g) \longrightarrow Na^+(g) + e^-$ endothermic
- electron affinity of chlorine: $Cl(g) + e^- \longrightarrow Cl^-(g)$ exothermic

The completed Born–Haber cycle for sodium chloride is shown in Figure 3.

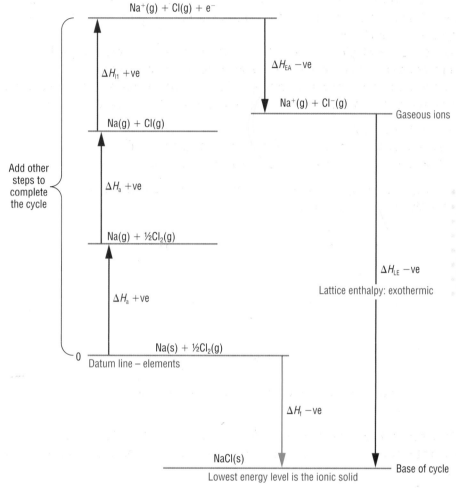

Figure 3 Completed Born–Haber cycle for sodium chloride

Born–Haber cycles, such as the one shown above, can be used to calculate an unknown enthalpy change. We can apply Hess' law to the cycle such that:

ΔH_f = the sum of all the other enthalpy changes

or   sum of anti-clockwise enthalpy changes = sum of clockwise enthalpy changes
          (arrows in green)   (arrows in purple)

Questions

1 Write the equations for making the ionic solid $MgBr_2$:
   (a) from its constituent elements under standard conditions;
   (b) from its constituent gaseous ions.
   Name the enthalpy changes represented by your two equations.
2 Why is the atomisation of chlorine a positive enthalpy change?
3 Why do you have to use a Born–Haber cycle to determine the lattice enthalpy of an ionic solid?

③ Born–Haber cycle calculations

By the end of this spread, you should be able to ...

* Construct Born–Haber cycles.
* Carry out related calculations.

Worked example 1

Construct a Born–Haber cycle from the following data, and calculate the lattice enthalpy for caesium chloride.

Enthalpy change of formation of CsCl = -433 kJ mol^{-1} Cs(s) + ½Cl$_2$(g) $\longrightarrow$ CsCl(s)
Enthalpy change of atomisation of Cs = $+79$ kJ mol^{-1} Cs(s) $\longrightarrow$ Cs(g)
First ionisation energy of Cs   = $+376$ kJ mol^{-1} Cs(g) $\longrightarrow$ Cs$^+$(g) + e$^-$
Enthalpy change of atomisation of Cl = $+121$ kJ mol^{-1} ½Cl$_2$(g) $\longrightarrow$ Cl(g)
First electron affinity of Cl   = -346 kJ mol^{-1} Cl(g) + e$^-$ $\longrightarrow$ Cl$^-$(g)

Before starting to draw your cycle, you should construct the equations alongside the data as shown above.

The question asks you to calculate the lattice enthalpy for Cs$^+$(g) + Cl$^-$(g) $\longrightarrow$ CsCl(s)

The cycle is drawn in steps:
* place the elements at zero energy and add the enthalpy change of formation
* add the lattice enthalpy
* complete the cycle and insert the enthalpy values.

The complete Born–Haber cycle for caesium chloride is shown below.

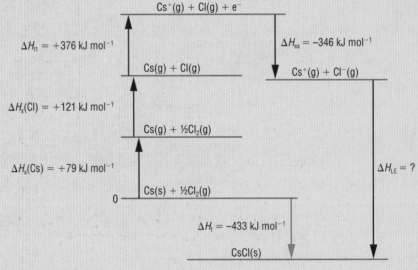

Figure 1 Born–Haber cycle for caesium chloride

Calculating the lattice enthalpy is the final stage. Using Hess' law:

Sum of clockwise enthalpies = sum of anti-clockwise enthalpies
$$\Delta H_f = \Delta H_a(\text{Cs}) + \Delta H_a(\text{Cl}) + \Delta H_{I1} + \Delta H_{EA} + \Delta H_{LE}$$
$$-433 = +79 + 121 + 376 + (-346) + \Delta H_{LE}$$

Rearranging:

$$\Delta H_{LE} = -433 - (79 + 121 + 376 - 346)$$

∴ lattice enthalpy of CsCl = -663 kJ mol^{-1}

Worked example 2

Construct a Born–Haber cycle from the following data, and calculate the lattice enthalpy for sodium oxide.

Enthalpy change of formation
of sodium oxide $= -414$ kJ mol^{-1} $2Na(s) + \frac{1}{2}O_2(g) \longrightarrow Na_2O(s)$
Enthalpy change of atomisation
of sodium $= +108$ kJ mol^{-1} $Na(s) \longrightarrow Na(g)$
Enthalpy change of atomisation
of oxygen $= +249$ kJ mol^{-1} $\frac{1}{2}O_2(g) \longrightarrow O(g)$
First ionisation energy of sodium $= +496$ kJ mol^{-1} $Na(g) \longrightarrow Na^+(g) + e^-$
First electron affinity of oxygen $= -141$ kJ mol^{-1} $O(g) + e^- \longrightarrow O^-(g)$
Second electron affinity of oxygen $= +790$ kJ mol^{-1} $O^-(g) + e^- \longrightarrow O^{2-}(g)$

The question requires the calculation of the lattice enthalpy for
$2Na^+(g) + O^{2-}(g) \longrightarrow Na_2O(s)$

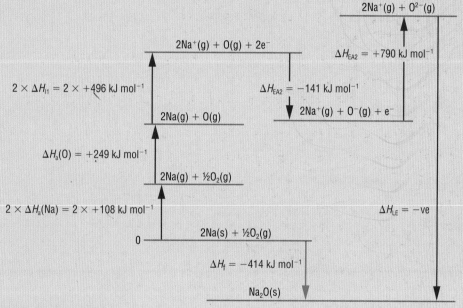

Figure 2 Born–Haber cycle for sodium oxide

Calculating the lattice enthalpy is the final stage. Using Hess' law:

Sum of clockwise = sum of anti-clockwise
  enthalpies   enthalpies

$$\Delta H_f = 2 \times \Delta H_a(Na) + \Delta H_a(O) + 2 \times \Delta H_{I1} + \Delta H_{EA1} + \Delta H_{EA2} + \Delta H_{LE}$$
$$-414 = (+108 \times 2) + 249 + (+496 \times 2) + (-141) + 790 + \Delta H_{LE}$$

Rearranging:

$$\Delta H_{LE} = -414 - ((+108 \times 2) + 249 + (+496 \times 2) - 141 + 790)$$

$\therefore$ lattice enthalpy of $Na_2O = -2520$ kJ mol^{-1}

Question

1 Calculate the lattice enthalpy for rubidium chloride given the following data:
ΔH(formation) of RbCl $= -435$ kJ mol^{-1}; $\quad \Delta H$(atomisation) of Rb $= +81$ kJ mol^{-1};
First ionisation energy of Rb $= +403$ kJ mol^{-1}; $\quad \Delta H$(atomisation) of Cl $= +121$ kJ mol^{-1};
First electron affinity of Cl $= -346$ kJ mol^{-1}

Further examples of Born–Haber cycles

By the end of this spread, you should be able to ...

* Construct Born–Haber cycles.
* Carry out related calculations.

There are a number of different Born–Haber cycles that may be set on exam papers and in this spread we look at some of the more complicated examples.

Worked example 1

The following data is provided for use in the Born–Haber cycle for calcium chloride:

$$Ca(s) \longrightarrow Ca(g) \qquad \Delta H = +178 \text{ kJ mol}^{-1}$$
$$Ca(g) \longrightarrow Ca^+(g) + e^- \qquad \Delta H = +590 \text{ kJ mol}^{-1}$$
$$Ca^+(g) \longrightarrow Ca^{2+}(g) + e^- \qquad \Delta H = +1145 \text{ kJ mol}^{-1}$$
$$\tfrac{1}{2}Cl_2(g) \longrightarrow Cl(g) \qquad \Delta H = +121 \text{ kJ mol}^{-1}$$
$$Ca(s) + Cl_2(g) \longrightarrow CaCl_2(s) \qquad \Delta H = -795 \text{ kJ mol}^{-1}$$
$$Ca^{2+}(g) + 2Cl^-(g) \longrightarrow CaCl_2(s) \qquad \Delta H = -2258 \text{ kJ mol}^{-1}$$

Draw a Born–Haber cycle for calcium chloride, and determine the electron affinity of chlorine.

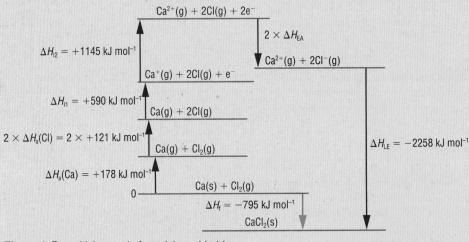

Figure 1 Born–Haber cycle for calcium chloride

Calculating the electron affinity of chlorine is the final stage. The equation for electron affinity is

$$Cl(g) + e^- \longrightarrow Cl^-(g)$$

Using Hess' law:

Sum of clockwise = sum of anti-clockwise
enthalpies   enthalpies

$$\Delta H_f = \Delta H_a(Ca) + 2 \times \Delta H_a(Cl) + \Delta H_{I1} + \Delta H_{I2} + 2 \times \Delta H_{EA} + \Delta H_{LE}$$
$$-795 = +178 + 242 + 590 + 1145 + 2 \times \Delta H_{EA} + (-2258)$$

Rearranging:

$$-795 - (178 + 242 + 590 + 1145 - 2258) = 2 \times \Delta H_{EA}$$

$$\therefore 2 \times \Delta H_{EA} = -692 \text{ kJ mol}^{-1}$$

$$\therefore \Delta H_{EA} = \frac{-692}{2} = -346 \text{ kJ mol}^{-1}$$

Worked example 2

Using the enthalpy data below, draw a Born–Haber cycle, and hence calculate a value for the lattice enthalpy of copper(II) oxide.

Process	$\Delta H^{\ominus}$/kJ mol^{-1}
Atomisation of copper	+339
First ionisation of copper	+745
Second ionisation of copper	+1960
Atomisation of oxygen	+249
First electron affinity of oxygen	−141
Second electron affinity of oxygen	+790
Formation of copper(II) oxide	−155

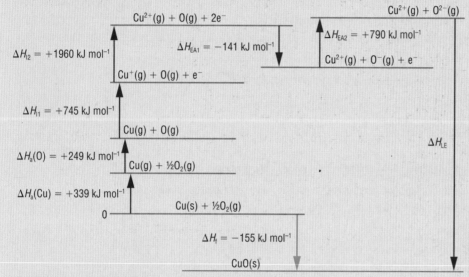

Figure 2 Born–Haber cycle for copper(II) oxide

Calculating the lattice enthalpy is the final stage. Using Hess' law:

Sum of clockwise = sum of anti-clockwise
enthalpies   enthalpies

$$\Delta H_f = \Delta H_a(Cu) + \Delta H_a(O) + \Delta H_{I1} + \Delta H_{I2} + \Delta H_{EA1} + \Delta H_{EA2} + \Delta H_{LE}$$
$$-155 = +339 + 249 + 745 + 1960 - 141 + 790 + \Delta H_{LE}$$

Rearranging:

$$\Delta H_{LE} = -155 - (339 + 249 + 745 + 1960 - 141 + 790)$$

∴ lattice enthalpy of copper(II) oxide = −4097 kJ mol^{-1}

Examiner tip

In this example you have to remember to represent the two electron affinities correctly, because the first is exothermic and the second is endothermic.

Question

1 (a) Draw a labelled Born–Haber cycle for the formation of solid MgCl$_2$ showing the species present at each stage.

(b) (i) What is meant by the term *electron affinity*?

(ii) Use the data below to find the value of the electron affinity of chlorine:

enthalpy change of formation of MgCl$_2$	−642 kJ mol^{-1}
lattice enthalpy of MgCl$_2$	−2493 kJ mol^{-1}
first ionisation energy of Mg	+736 kJ mol^{-1}
second ionisation energy of Mg	+1450 kJ mol^{-1}
enthalpy change of atomisation of Cl	+121 kJ mol^{-1}
enthalpy change of atomisation of Mg	+150 kJ mol^{-1}

By the end of this spread, you should be able to ...

* Use the enthalpy change of solution of a simple ionic solid and relevant enthalpy change terms to construct Born–Haber cycles.
* Carry out related calculations.

Making a solution

When calcium chloride dissolves in water, the temperature of the solution increases as heat is generated – this is an exothermic reaction. However, when ammonium nitrate dissolves in water the temperature of the water falls – the reaction is endothermic. These reactions are simple to perform and can be carried out very quickly in the laboratory, but the process of dissolving may not be quite as simple as you think!

The standard enthalpy change of solution, $\Delta H_s^{\ominus}$

For the **standard enthalpy change of solution**, $\Delta H_s^{\ominus}$, one mole of a compound is dissolved in water under standard conditions.

The process of dissolving can be exothermic or endothermic.

$\Delta H_s^{\ominus}$ for potassium chloride is the enthalpy change for:

$$KCl(s) + aq \longrightarrow K^+(aq) + Cl^-(aq)$$

What happens when a solid dissolves?

When an ionic solid dissolves in water, two processes take place:
* breakdown of the ionic lattice into gaseous ions
* hydration of the ions.

The resulting enthalpy change is known as the enthalpy change of solution and its value depends on the balance between these two processes.

Breakdown of the ionic lattice

The ionic solid must be broken down. This involves breaking down the crystal lattice and separating the ions. Overcoming the attractive forces between the oppositely charged ions requires energy. The process is the exact opposite of that producing the lattice enthalpy:
* Enthalpy change involved in breaking ionic lattice = $-\Delta H_{LE}$
* So the energy involved has the same magnitude as lattice enthalpy but the sign is opposite.

Lattice enthalpy of potassium chloride:

$$K^+(g) + Cl^-(g) \longrightarrow KCl(s) \qquad \Delta H_{LE} = -711 \text{ kJ mol}^{-1}$$

Enthalpy change for breakdown of ionic lattice:

$$KCl(s) \longrightarrow K^+(g) + Cl^-(g) \qquad \Delta H = +711 \text{ kJ mol}^{-1}$$

Hydration

The second stage of the dissolving process involves the hydration of the gaseous ions. In this process the gaseous ions bond with the water molecules. The positively charged ions will be attracted to the slightly negative oxygen atoms in water, and the negatively charged ions will be attracted to the slightly positive hydrogen atoms in water. This can be seen in Figure 2.

Figure 1 Determining the enthalpy change of solution of potassium chloride

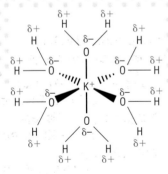

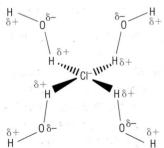

Figure 2 Ionic lattice dissolves in polar solvents – each ion is surrounded by water molecules to become 'hydrated'

Examiner tip

Lattice energy is an exothermic process and the sign is −ve. The process involved in breaking down the ionic lattice is endothermic and the sign is +ve.

The standard enthalpy change of hydration, $\Delta H^{\ominus}_{hyd}$

Energy is released when the ions form bonds with water molecules. This energy is known as the enthalpy change of hydration – different ions have different enthalpy changes of hydration.

For the **standard enthalpy change of hydration**, $\Delta H^{\ominus}_{hyd}$, one mole of aqueous ions are formed from their gaseous ions under standard conditions.

Hydration is an exothermic process.

$\Delta H^{\ominus}_{hyd}$ for potassium ions is:

$K^+(g) + aq \longrightarrow K^+(aq)$ $\Delta H^{\ominus}_{hyd} = -322$ kJ mol^{-1}

$\Delta H^{\ominus}_{hyd}$ for chloride ions is:

$Cl^-(g) + aq \longrightarrow Cl^-(aq)$ $\Delta H^{\ominus}_{hyd} = -363$ kJ mol^{-1}

Key definition

The **standard enthalpy change of hydration** is the enthalpy change that takes place when one mole of isolated gaseous ions is dissolved in water forming one mole of aqueous ions under standard conditions.

Calculating a lattice enthalpy from enthalpy changes of solution and hydration

The lattice enthalpy of an ionic solid can be calculated using
- the enthalpy changes of hydration of the constituent gaseous ions and
- the enthalpy change of solution of the ionic solid.

A Born–Haber cycle can be used to show the energy changes. The example below shows how the lattice enthalpy of potassium chloride, KCl, can be calculated.

The gaseous ions, $K^+(g)$ and $Cl^-(g)$, are placed at the top of the cycle. There are two routes that connect the gaseous ions to the ionic solid:

Route 1 Lattice enthalpy followed by enthalpy change of solution
- $K^+(g) + Cl^-(g) \longrightarrow KCl(s)$ ΔH_{LE}
- $KCl(s) + aq \longrightarrow K^+(aq) + Cl^-(aq)$ $\Delta H^{\ominus}_s = +26$ kJ mol^{-1}

Route 2 Hydration of $K^+(g)$ and $Cl^-(g)$ ions
- $K^+(g) + aq \longrightarrow K^+(aq)$ $\Delta H_{hyd} = -322$ kJ mol^{-1}
- $Cl^-(g) + aq \longrightarrow Cl^-(aq)$ $\Delta H_{hyd} = -363$ kJ mol^{-1}

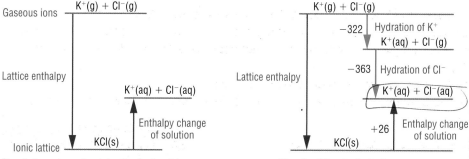

Step 1: the process involving the ionic solid   **Step 2:** adding the hydration processes

Figure 3 Born–Haber cycle for calculating the lattice enthalpy of KCl(s)

Examiner tip

The enthalpy change of solution can be +ve or –ve.

If you have to calculate ΔH_s, just use the same method. You still get the correct answer whether your arrow goes up or down.

These routes can be put together to give a Born–Haber cycle as shown below.

All that remains is to apply Hess' law to the cycle to calculate the lattice enthalpy:

Sum of clockwise enthalpy changes = sum of anti-clockwise enthalpy changes

$$\Delta H_{hyd}(K^+) + \Delta H_{hyd}(Cl^-) = \Delta H_{LE}(KCl) + \Delta H_s(KCl)$$
$$-322 + (-363) = \Delta H_{LE}(KCl) + 26$$

Rearranging:

$-322 - 363 - 26 = \Delta H_{LE}(KCl)$

$\therefore \Delta H_{sLE}(KCl) = -711$ kJ mol^{-1}

Question

1 You are provided with the following enthalpy changes:

$Na^+(g) + F^-(g) \longrightarrow NaF(s)$
$\Delta H = -918$ kJ mol^{-1}

$Na^+(g) + aq \longrightarrow Na^+(aq)$
$\Delta H = -390$ kJ mol^{-1}

$Na^+F^-(s) + aq \longrightarrow$
$Na^+(aq) + F^-(aq)$
$\Delta H = +71$ kJ mol^{-1}

(a) Name these three enthalpy changes.

(b) Write an equation, including state symbols, for the change that accompanies the enthalpy change of hydration of F^- ions.

(c) Calculate the enthalpy change of hydration of F^- ions.

By the end of this spread, you should be able to ...

✳ **Explain, in qualitative terms, the effect of ionic charge and ionic radius on lattice enthalpy and enthalpy change of hydration.**

Lattice enthalpies

A large exothermic value for a lattice enthalpy means that there is a large electrostatic force of attraction between the oppositely charged ions and that the ionic bonds are strong. There are two factors that govern the size of lattice enthalpy:

- ionic size
- ionic charge.

Ionic size

Small ions can pack together closely in a lattice and attract each other strongly. Larger ions are further apart in their lattice and the forces of attraction between them are weaker.

As the ionic radius increases:

- the attraction between the ions decreases
- the lattice enthalpy becomes less negative (less exothermic).

Lattice enthalpies /kJ mol^{-1}

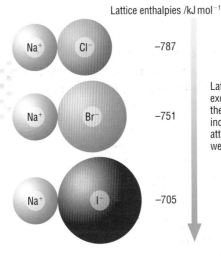

−787

−751

−705

Lattice enthalpy becomes less exothermic and *less negative* as the size of the negative ions increases. This indicates weaker attraction between ions and hence weaker ionic bonding

Figure 1 Trend in lattice enthalpies of the sodium halides

Ionic charge

The compounds with the most negative lattice enthalpies are those which have small, highly charged ions.

When two highly charged ions of opposite charge are present in the lattice they attract each other strongly. A very exothermic reaction will take place when a lattice is formed from two highly charged ions.

Charge increases and produces a greater attraction between these positive ions and negative ions.
The ionic radius decreases resulting in the ions in the lattice being closer together producing more attraction.
Lattice enthalpies become more exothermic – *more negative*

Figure 2 Lattice enthalpy becomes more exothermic as the ionic charge increases

Worked example 1

(a) Describe how, and explain why, the lattice enthalpy of magnesium oxide differs from that of barium oxide.

The lattice enthalpy of magnesium oxide is more exothermic than that of barium oxide. Mg^{2+} ions have a smaller ionic radius than Ba^{2+} ions, meaning that the ions in MgO are closer together than the ions in BaO. Therefore there is stronger attraction between Mg^{2+} and O^{2-} ions in magnesium oxide than between Ba^{2+} and O^{2-} ions in barium oxide.

(b) Give one reason why magnesium oxide is a good material to use for making the lining of furnaces.

Magnesium oxide has a highly negative lattice enthalpy and this means that there is a strong attraction between its ions. Compounds with highly exothermic lattice enthalpies have high melting points and so will be stable inside the working furnace at high temperatures.

Worked example 2

State and explain which compound has the most exothermic lattice enthalpy: $MgCl_2$, $MgBr_2$ or MgI_2.

Magnesium chloride has the most exothermic lattice enthalpy because the chloride ion is smaller than both bromide ions and iodide ions. This means the Mg^{2+} and Cl^- ions in the $MgCl_2$ lattice can pack closer together and exert a greater attraction on each other than the ions in $MgBr_2$ or MgI_2.

Examiner tip

It is important when answering questions such as this that you state that *ions* affect the magnitude of the lattice enthalpy. Sloppy wording such as 'chlorine' and 'bromine' instead of chloride and bromide must be avoided at all costs.

Enthalpy change of hydration

The standard enthalpy change of hydration is the enthalpy change that takes place when one mole of isolated gaseous ions dissolve in water forming one mole of aqueous ions under standard conditions.

As with lattice enthalpies, the factors that affect the size of hydration enthalpy are ionic size and ionic charge.

Ionic size

As the ionic radius become smaller, the value of the enthalpy change of hydration becomes more negative. Hydration depends on the ability of an ion to attract and bond with water molecules. Small ions exert more attraction on water molecules and more energy is released. This trend can be seen in Figure 3.

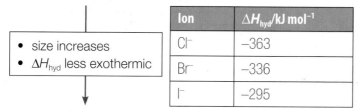

Ion	ΔH_{hyd}/kJ mol^{-1}
Cl$^-$	−363
Br$^-$	−336
I$^-$	−295

- size increases
- ΔH_{hyd} less exothermic

Figure 3 Enthalpy changes of hydration of halide ions

Ionic charge

As the charge on an ion increases it has a greater attraction for water molecules, and hence the hydration enthalpy is more negative. Across Period 3, the ions Na$^+$, Mg^{2+} and Al^{3+} progressively decrease in size (see Figure 2). For these three ions, the effect of the increasing charge and decreasing size on the hydration enthalpy can be seen in Figure 4.

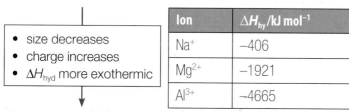

Ion	ΔH_{hy}/kJ mol^{-1}
Na$^+$	−406
Mg^{2+}	−1921
Al^{3+}	−4665

- size decreases
- charge increases
- ΔH_{hyd} more exothermic

Figure 4 Enthalpy changes of hydration of some common cations

Questions

1 The lattice enthalpy of sodium chloride is more exothermic than the lattice enthalpy of caesium chloride. State and explain the relative strengths of the ionic bonding in these two chlorides.

2 Lattice enthalpy is used to compare the strengths of ionic bonds.
   (a) Define the term *lattice enthalpy*.
   (b) Describe and explain the effect of ionic charge and ionic radius on the magnitude of lattice enthalpy.

3 Describe how, and explain why, the enthalpy change of hydration of sodium ions differs from that of rubidium ions.

By the end of this spread, you should be able to ...

* Explain that entropy is a measure of the disorder of a system, and that a system becomes energetically more stable when it becomes more disordered.
* Explain: the difference in entropy of a solid and a gas; the change when a solid lattice dissolves; the change in a reaction in which there is a change in the number of gaseous molecules.
* Calculate the entropy change for a reaction given the entropies of reactants and products.

What is entropy?

In AS chemistry, you met enthalpy as a measure of the heat content of a chemical system. There is another important thermodynamic aspect of all chemical reactions – it is called **entropy**.

Entropy, S, is sometimes described as 'a measure of disorder' or the 'amount of randomness' in a system.

All substances possess some degree of disorder because particles are always in constant motion. Thus, S is always a positive number.

As with enthalpy changes, chemists are interested in entropy changes, ΔS, during chemical reactions:

Because entropy is related to energy, a better description for entropy might well be 'the dispersal of energy'. There is a general tendency towards greater entropy.

Entropy

One of the ideas involved in the concept of entropy is that nature tends to move from order to disorder in isolated systems. For example, gas molecules spread out over time to fill a space, increasing their entropy over time.

Key definition

Entropy, S, is the quantitative measure of the degree of disorder in a system.

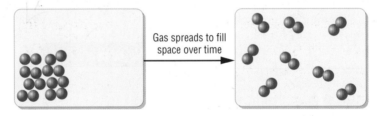

Gas spreads to fill space over time

Pile of bricks dropped from a lorry
Which arrangement is the more probable?

Disorder is more probable than order

Figure 1 Increasing entropy

What can entropy tell us?

Entropy helps us to explain things that happen naturally:

* a gas spreading through a room
* heat from a fire spreading through a room
* ice melting in a warm room
* salt dissolving in water.

In all these examples, energy is changing from being localised (*concentrated*) to becoming more spread out (*diluted*). Entropy increases whenever particles become more disordered.

Chemists are usually concerned with the 'entropy of a chemical system'. This is the distribution of energy within the chemicals themselves.
- At 0 K, perfect crystals have zero entropy.

The entropy of a pure substance increases with increasing temperature.
- Entropy increases during the changes in state that give more randomness:

$$\text{solid} \longrightarrow \text{liquid} \longrightarrow \text{gas}$$

- When water changes state from liquid to gas, its entropy rises from +70 to +189 J K^{-1} mol^{-1}.

- Entropy increases when a solid lattice dissolves – for example:

$$CuSO_4 \cdot 5H_2O(s) + aq \longrightarrow Cu^{2+}(aq) + SO_4^{2-}(aq) + 5H_2O(l)$$

- Entropy increases in a reaction in which there is an increase in the number of gaseous molecules – i.e. when a gas is evolved. For example:

$$Mg(s) + 2HCl(aq) \longrightarrow MgCl_2(aq) + H_2(g)$$
$$MgCO_3(s) \longrightarrow MgO(s) + CO_2(g)$$

Calculating entropy changes

The standard entropy, $S^{\ominus}$, of a substance is the entropy content of one mole of the substance under standard conditions. Standard entropies have units of J K^{-1} mol^{-1}.

The **entropy change of a reaction** can be calculated using the standard entropies of the reactants and products:

$$\Delta S = \Sigma S^{\ominus}_{(products)} - \Sigma S^{\ominus}_{(reactants)} \qquad \text{(where } \Sigma = \text{'sum of')}$$

Entropy is a measure of randomness. Systems that are more chaotic have a higher entropy value.
- If a change makes a system more random, ΔS is positive.
- If a change makes a system more ordered, ΔS is negative.

Figure 2 Crystals of $CuSO_4 \cdot 5H_2O(s)$ and a solution of $CuSO_4(aq)$ with some undissolved crystals. The blue crystals are soluble and dissolving them produces a blue solution – copper and sulfate ions are spreading out and entropy is increasing

Figure 3 Magnesium reacting with hydrochloric acid, forming hydrogen – entropy is increasing because the number of moles of gas is increasing

Worked example

This equation shows the formation of ammonia from nitrogen and hydrogen:

$$N_2(g) + 3H_2(g) \longrightarrow 2NH_3(g)$$

Use the data in the table to calculate ΔS for this reaction.

	$N_2(g)$	$H_2(g)$	$NH_3(g)$
$S^{\ominus}$/J K^{-1} mol^{-1}	+192	+131	+193

$$\begin{aligned} \Delta S &= \Sigma S^{\ominus}_{(products)} - \Sigma S^{\ominus}_{(reactants)} \\ &= (2 \times 193) - [(+192) + (3 \times 131)] \\ &= -199 \text{ J K}^{-1} \text{ mol}^{-1} \end{aligned}$$

This decrease in entropy is to be expected because a more ordered system with fewer gaseous molecules is produced: 2 moles as opposed to 4 moles.

Questions

1 What is the sign of ΔS for the following reactions?
   (a) $NaCl(s) \longrightarrow Na^+(aq) + Cl^-(aq)$ (b) $2SO_2(g) + O_2(g) \longrightarrow 2SO_3(g)$
   (c) $C_8H_{18}(l) + 12\frac{1}{2}O_2(g) \longrightarrow 8CO_2(g) + 9H_2O(l)$
2 Use the data in the table below to calculate ΔS for these reactions:
   (a) $2NO(g) + O_2(g) \longrightarrow N_2O_4(g)$ (b) $C_6H_6(l) + 7\frac{1}{2}O_2(g) \longrightarrow 6CO_2(g) + 3H_2O(l)$

	$NO(g)$	$O_2(g)$	$N_2O_4(g)$	$C_6H_6(l)$	$CO_2(g)$	$H_2O(l)$
$S^{\ominus}$/J K^{-1} mol^{-1}	+211	+205	+304	+173	+214	+70

By the end of this spread, you should be able to ...

✱ Explain that the tendency of a process to take place depends on temperature, T, the entropy change in the system, ΔS, and the enthalpy change, ΔH, with the surroundings.
✱ Explain that the balance between entropy and enthalpy change is the free energy change, ΔG.
✱ State and use the relationship $\Delta G = \Delta H - T\Delta S$.
✱ Explain how endothermic reactions are able to take place spontaneously.

Spontaneous processes

A spontaneous process proceeds on its own. Spontaneous processes lead to lower energy and increased stability.

• Many exothermic reactions take place spontaneously at room temperature. The enthalpy content of the chemical system decreases during the reaction and excess energy is released to the surroundings – this increases stability.
• Some endothermic reactions also take place spontaneously at room temperature. The enthalpy content of the chemical system increases during the reaction with energy being taken in from the surroundings. This must also increase stability. It seems surprising that some endothermic reactions can take place spontaneously – the reason lies with entropy.

A process is spontaneous if a chemical system becomes more stable and its overall energy decreases. The overall energy decrease results from contributions from *both* enthalpy *and* entropy.

Energy from entropy

The entropy contribution towards the overall energy depends on temperature.

• Energy derived from entropy = $T\Delta S$, where T is measured in Kelvin.
• As the temperature increases, the energy derived from entropy becomes more significant.

Free energy and feasibility

Whether a process is spontaneous or not depends on three factors:

• the temperature, T, in Kelvin
• the entropy change in the system, ΔS
• the enthalpy change, ΔH, with the surroundings.

The relationship between these three factors is expressed by a term known as the **free energy change**, ΔG:

$$\Delta G = \Delta H - T\Delta S$$

Free energy is sometimes called the 'Gibbs free energy' after the American physicist Josiah Willard Gibbs.

The feasibility of a reaction depends on the balance between enthalpy and entropy. Spontaneous processes happen when ΔG is negative.

For a reaction to be feasible, there must be a decrease in the overall energy derived from both the enthalpy change and the entropy of the system.

$$\therefore \Delta G < 0$$

Using the expresion $\Delta G = \Delta H - T\Delta S$, Table 1 shows how ΔG depends on ΔH and ΔS at different temperatures.

Key definition

The **free energy change** ΔG is the balance between enthalpy, entropy and temperature for a process: $\Delta G = \Delta H - T\Delta S$. A process can take place spontaneously when $\Delta G < 0$.

Figure 1 Josiah Willard Gibbs (1839–1903), US mathematician and theoretical physicist. Between 1873 and 1876, Gibbs developed the theory of thermodynamics in a long series of papers. His name is also associated with Gibbs free energy, which determines the conditions under which a chemical reaction will occur

ΔH	ΔS	ΔG	Feasibility
–ve	+ve	always –ve	reaction feasible
+ve	–ve	always +ve	reaction never feasible
–ve	–ve	–ve at low temperatures	feasible at low temperatures
+ve	+ve	–ve at high temperatures	feasible at high temperatures

Table 1

Generally you will find that most exothermic reactions are spontaneous, even if entropy decreases (the system becomes more ordered), because enthalpy contributes more to ΔG than does entropy. The exceptions are reactions occurring at high temperatures in which entropy decreases.

How do endothermic reactions take place?

For an endothermic reaction to take place spontaneously:

- ΔS must be positive
- the temperature must be high enough so that $T\Delta S > \Delta H$.

Worked example

The thermal decomposition of zinc carbonate, $ZnCO_3$, is represented by

$$ZnCO_3(s) \longrightarrow ZnO(s) + CO_2(g) \qquad \Delta H = +71 \text{ kJ mol}^{-1}$$

Use the data in the table to calculate the minimum temperature, in °C, for thermal decomposition to take place.

	$ZnCO_3(s)$	$ZnO(s)$	$CO_2(g)$
$S^{\ominus}/\text{J K}^{-1} \text{mol}^{-1}$	+82	+44	+214

Step 1: Calculate ΔS in units of $\text{kJ K}^{-1} \text{mol}^{-1}$

$\Delta S = \Sigma S^{\ominus}_{(products)} - \Sigma S^{\ominus}_{(reactants)}$

$= 44 + 214 - 82$

$= 176 \text{ J K}^{-1} \text{mol}^{-1}$

$= 0.176 \text{ kJ K}^{-1} \text{mol}^{-1}$

Step 2: Calculate T when $\Delta G = 0$

$\Delta G = \Delta H - T\Delta S$;

When $\Delta G = 0$, $\Delta H - T\Delta S = 0$

$\therefore T = \dfrac{\Delta H}{\Delta S} = \dfrac{71}{0.176}$

$= 403 \text{ K}$

$= 403 - 273 = 130\,°C$

Questions

1 The combustion of ethanol is represented by:

$$C_2H_5OH(l) + 3O_2(g) \longrightarrow 2CO_2(g) + 3H_2O(l) \qquad \Delta H = -786 \text{ kJ mol}^{-1};$$
$$\Delta S = +68 \text{ J K}^{-1} \text{mol}^{-1}$$

Calculate ΔG for this reaction at 25 °C.

2 The equation for the thermal decomposition of ammonium chloride, NH_4Cl is:

$$NH_4Cl(s) \longrightarrow NH_3(g) + HCl(g) \qquad \Delta H = +176 \text{ kJ mol}^{-1}$$

Calculate the minimum temperature, in °C, for thermal decomposition to take place spontaneously.

	$NH_4Cl(s)$	$NH_3(g)$	$HCl(g)$
$S^{\ominus}/\text{J K}^{-1} \text{mol}^{-1}$	95	+192	+187

By the end of this spread, you should be able to ...

* Explain the terms redox, oxidation number, half-reaction, oxidising agent and reducing agent for simple redox reactions.
* Construct redox equations using relevant half-equations or oxidation numbers.
* Interpret and make predictions for reactions involving electron transfer.

Species	Oxidation number
Uncombined element	0
Combined oxygen	−2
Combined hydrogen	+1
Simple ion	Ionic charge
Combined fluorine	−1

Table 1 Oxidation number rules

Key definitions

Oxidation is loss of electrons, or an increase in oxidation number.

Reduction is gain of electrons, or a decrease in oxidation number.

During the study of redox in AS chemistry (see *AS Chemistry* spreads 1.1.14 and 1.1.15), you learned:
* that oxidation and reduction involve electron transfer
* the rules for assigning oxidation states
* that oxidation and reduction can be expressed in terms of changes in oxidation number.

The most important oxidation number rules are listed in Table 1 as a reminder.

Redox

In a **redox reaction**, both reduction and oxidation take place.

Oxidation is
* the loss of electrons
* an increase in oxidation number.

Reduction is
* the gain of electrons
* a decrease in oxidation number.

The substance that is reduced is an **oxidising agent** – it takes electrons from the substance that is oxidised.

The substance that is oxidised is a **reducing agent** – it gives electrons to the substance that is reduced.
* Reduction must always be accompanied by oxidation.

Constructing redox equations using relevant half-equations

Redox reactions take place when electrons are transferred. We can construct an overall equation for a redox reaction if we know the oxidation and reduction half-reactions.
* In the oxidation half-reaction, electrons are lost.
* In the reduction half-reaction, electrons are gained.

To construct a balanced equation, we must follow a series of steps. These are illustrated in the worked example below.

Worked example 1

Iron reacts with aqueous copper(II) ions to form iron(III) ions and copper metal.

Step 1: Identify the oxidation and reduction half-equations
Equations for the separate oxidation and reduction reactions are shown lined up.

$$Fe \longrightarrow Fe^{3+} + 3e^- \qquad \text{oxidation (loss of electrons)}$$
$$Cu^{2+} + 2e^- \longrightarrow Cu \qquad \text{reduction (gain of electrons)}$$

Step 2: Balance the electrons
The half-equations are scaled so that the number of electrons is the same in each.

Here, we need to multiply the oxidation half-equation by 2 and the reduction half-equation by 3:

$$2Fe \longrightarrow 2Fe^{3+} + 6e^- \qquad \text{oxidation}$$
$$3Cu^{2+} + 6e^- \longrightarrow 3Cu \qquad \text{reduction}$$

Step 3: Add the two half equations together and cancel the electrons.

$$3Cu^{2+} + 2Fe + \cancel{6e^-} \longrightarrow 2Fe^{3+} + 3Cu + \cancel{6e^-}$$

So the balanced equation is:

$$3Cu^{2+} + 2Fe \longrightarrow 2Fe^{3+} + 3Cu$$

Constructing redox equations using oxidation numbers

Oxidation and reduction can be described in terms of oxidation numbers.

You can easily identify the oxidation and reduction processes using oxidation numbers. An oxidation number is assigned to each atom in the redox reaction. You can then easily see any *changes* in oxidation number. Oxidation numbers make it easy to account for electrons and are a great way of balancing complex redox reactions.

Worked example 2

Hydrogen iodide, HI, is oxidised to iodine, I_2, by concentrated sulfuric acid, H_2SO_4, which is reduced to hydrogen sulfide, H_2S.

Step 1: Identify the reactants and products from the information given
The reactants and products are written down to give an incomplete draft equation:

$$HI + H_2SO_4 \longrightarrow H_2S + I_2$$

Don't worry that there is no 'O' on the right-hand side of the equation – we will sort that out later.

Step 2: Identify the oxidation number changes
Write oxidation numbers underneath the atoms that change oxidation number. Balance any atom that changes oxidation number.
Identify the total changes in oxidation numbers.

$$2HI + H_2SO_4 \longrightarrow H_2S + I_2$$

$$+6 \xrightarrow{\quad -8 \quad} -2$$

$$2 \times -1 \xrightarrow[+2]{\quad} 0 \times 2$$

Step 3: Balance the oxidation numbers changes
The equation must be balanced so that:

total increase in oxidation number = total decrease in oxidation number.

Here, we need to multiply $2HI/I_2$ by 4 to balance the total oxidation number change for S:

$$8HI + H_2SO_4 \longrightarrow H_2S + 4I_2$$

$$+6 \xrightarrow{\quad -8 \quad} -2$$

$$8 \times -1 \xrightarrow[+8]{\quad} 4 \times 0 \times 2$$

Step 4: Check to see if anything else needs to be balanced
Here there are 8H and 4O extra on the left-hand side. We can easily complete the equation by adding $4H_2O$ to the right-hand side.

This gives the final equation:

$$8HI + H_2SO_4 \longrightarrow H_2S + 4I_2 + 4H_2O$$

Hints for completing the redox equations

Sometimes you will find that you are short of 'H' on one side. In general, you may have to add H^+, OH^- or H_2O.

- If the reaction has been carried out in acid conditions, then simply add the number of H^+ ions that that you need.
- For alkaline conditions, you can add OH^-.
- If 'H' and 'O' need to be added, think 'H_2O' (as in Step 4 in the worked example above).

Future use of redox equations

You will need to be able to combine half-equations in the next section on electrochemical cells (spreads 2.2.10–2.2.14). You will also need to combine half-equations and use oxidation numbers when you study redox titrations (spreads 2.3.9–2.3.11).

So this section is extremely important as it underpins much future work.

Examiner tip

If one species decreases its oxidation number (reduction), then another species must increase its oxidation number (oxidation).

The total increase in oxidation number equals the total decrease in oxidation number.

Questions

1 Construct redox equations from the following half-equations:
  (a) $Al \longrightarrow Al^{3+} + 3e^-$
      $Cu^{2+} + 2e^- \longrightarrow Cu$;
  (b) $2Br^- \longrightarrow Br_2 + 2e^-$
      $MnO_4^- + 8H^+ + 5e^- \longrightarrow Mn^{2+} + 4H_2O$.

2 Use oxidation states to construct redox equations for the following reactions:
  (a) Hydrogen bromide, HBr, is oxidised to bromine, Br_2, by concentrated sulfuric acid, H_2SO_4, which is reduced to sulfur dioxide, SO_2.
  (b) Hydrogen sulfide, H_2S, is oxidised to sulfur, S, by nitric acid, HNO_3, which is reduced to nitrogen monoxide, NO.

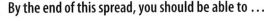

By the end of this spread, you should be able to . . .

* Describe simple half cells made from
    (i)   metals or non-metals in contact with their ions in aqueous solution;
    (ii)   ions of the same element in different oxidation states.
* Describe how half cells can be combined to make an electrochemical cell.

Electricity from chemical reactions

In redox reactions, electrons are transferred. An electrochemical cell controls the electron transfer to produce electrical energy. The controlled transfer of electrons is the basis of all cells and batteries.

Cells and half cells

A half cell comprises an element in two oxidation states. The simplest half cell has a metal placed in an aqueous solution of its ions.

For example, a copper half cell comprises a solution containing $Cu^{2+}(aq)$ ions (oxidation state +2) into which is placed a strip or rod of copper metal (oxidation state 0).

An equilibrium exists at the surface of the copper between these oxidation states of copper:

$$Cu^{2+}(aq) + 2e^- \rightleftharpoons Cu(s)$$

* The forward reaction involves electron gain and is reduction.
* The reverse reaction involves electron loss and is oxidation.

By convention, the equilibrium is written with the electrons on the left-hand side. The **electrode potential** of the half cell indicates its tendency to lose or gain electrons in this equilibrium.

Figure 1 Copper half cell – copper metal is placed in an aqueous solution containing $Cu^{2+}(aq)$ ions. An equilibrium is set up at the surface of the solid Cu with aqueous Cu^{2+} ions:

$$Cu^{2+}(aq) + 2e^- \rightleftharpoons Cu(s)$$

Cells from metal/metal ion half cells

A simple electrochemical cell can be made by connecting together two half cells with different electrode potentials:

* one half cell releases electrons
* the other half cell gains electrons.

The difference in electrode potential is measured with a voltmeter.

A copper–zinc cell is shown in Figure 2.

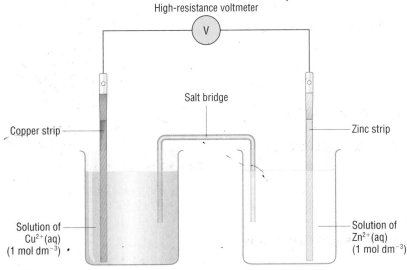

Figure 2 Copper–zinc electrochemical cell

The two half cells are joined using a wire and a salt bridge.

* The wire connects the two metals, allowing electrons to be transferred between the two half cells. In this cell, the wire connects the copper electrode in the copper half cell to the zinc electrode in the zinc half cell. The voltmeter has a high resistance and is used to minimise the current that flows. If a small bulb were to replace the voltmeter, the bulb would light up.
* The salt bridge connects the two solutions, allowing ions to be transferred between the half cells. A simple salt bridge can be made out of a strip of filter paper soaked in an aqueous solution of an ionic compound that does not react with either of the half-cell solutions. Usually aqueous $KNO_3(aq)$ or $NH_4NO_3(aq)$ is used.

The redox equilibrium in each half cell is shown below:

$$Zn^{2+}(aq) + 2e^- \rightleftharpoons Zn\,(s)$$

$$Cu^{2+}(aq) + 2e^- \rightleftharpoons Cu(s)$$

The Zn^{2+}/Zn equilibrium releases electrons more readily than the Cu^{2+}/Cu equilibrium.
- The Zn^{2+}/Zn equilibrium releases electrons into the wire, making zinc the negative electrode.
- Electrons flow along the wire to the Cu electrode of the Cu^{2+}/Cu half cell.
- The Zn^{2+}/Zn equilibrium loses electrons and moves to the left:

$$\overleftarrow{Zn^{2+} + 2e^-} \rightleftharpoons Zn$$

- The Cu^{2+}/Cu equilibrium gains the electrons and moves to the right:

$$\overrightarrow{Cu^{2+}(aq) + 2e^-} \rightleftharpoons Cu(s)$$

The reading on the voltmeter measures the potential difference of the cell – this measures the difference between the electrode potentials of the half cells.

Non-metal/non-metal ion half cells
A half cell can be made from a non-metal and its aqueous ions. For example, a hydrogen half cell comprises hydrogen gas, H_2, in contact with $H^+(aq)$ ions. The redox equilibrium is:

$$2H^+(aq) + 2e^- \rightleftharpoons H_2(g)$$

There is a problem in using this type of half cell when constructing a cell – there is no electrode to connect wire to. To overcome this, a platinum electrode is placed in the solution so that it is in contact with both $H_2(g)$ and $H^+(aq)$ ions. The platinum is inert and does not react at all – its sole purpose is to allow the transfer of electrons into and out of the half cell via a connecting wire. The surface of the platinum electrode is coated with *platinum black*, a spongy coating in which electrons can be transferred between the non-metal and its ions.

A standard hydrogen half cell is used as the reference for the measurement of voltages in electrochemical cells. A standard hydrogen half cell comprises:
- hydrochloric acid, $HCl(aq)$, of concentration 1 mol dm^{-3}, as the source of $H^+(aq)$
- hydrogen gas, $H_2(g)$, at 100 kPa (1 atmosphere) pressure
- an inert platinum electrode to allow electrons to pass into or out of the half cell via a connecting wire.

The use of a standard hydrogen half cell is discussed in the next spread.

Metal ion/metal ion half cells
This type of half cell contains ions of the same element in different oxidation states. For example, a half cell can contain $Fe^{3+}(aq)$ and $Fe^{2+}(aq)$ ions:

$$Fe^{2+}(aq) + e^- \rightleftharpoons Fe^{3+}(aq)$$

A standard $Fe^{2+}(aq)/Fe^{3+}(aq)$ half cell is made up of:
- a solution containing $Fe^{2+}(aq)$ and $Fe^{3+}(aq)$ ions with the same concentrations ('equimolar')
- an inert platinum electrode to allow electrons to pass into or out of the half cell via a connecting wire.

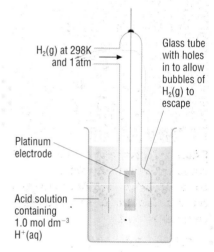

$H_2(g)$ at 298K and 1 atm

Glass tube with holes in to allow bubbles of $H_2(g)$ to escape

Platinum electrode

Acid solution containing 1.0 mol dm^{-3} $H^+(aq)$

Figure 3 Standard hydrogen half cell

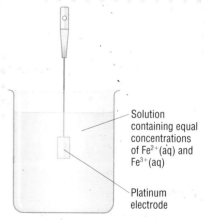

Solution containing equal concentrations of $Fe^{2+}(aq)$ and $Fe^{3+}(aq)$

Platinum electrode

Figure 4 Fe^{3+}/Fe^{2+} half cell

Question
1 (a) What does a standard hydrogen half cell consist of?
  (b) When two different half cells are connected, how does the electric current flow:
   (i) between the two electrodes;
   (ii) between the two solutions?

By the end of this spread, you should be able to . . .

* Define the term *standard electrode (redox) potential, $E^\ominus$*.
* Describe how to measure standard electrode potentials using a standard hydrogen half cell.
* Calculate a standard cell potential by combining two standard electrode potentials.

Standard electrode potentials

In spread 2.2.10 we discussed how two half cells can be connected together to make a cell.

* A standard hydrogen half cell is used as the reference standard for the measurement of cell e.m.f.
* An electromotive force, e.m.f., is the voltage produced by a cell when no current flows.

We can compare the tendency of different half cells to release or accept electrons by measuring the e.m.f. when the half cells are combined separately with a hydrogen half cell. To make this comparison meaningful, we use standard half cells and standard conditions.
* The resulting e.m.f. is called the **standard electrode potential** of a half cell.
* The standard electrode potential of a hydrogen half cell is defined as exactly 0 V.

Measuring standard electrode potentials

A standard electrode potential is measured by connecting a standard half cell to a standard hydrogen half cell. Figure 1 shows how a hydrogen half cell can be used to measure the standard electrode potential of a Zn^{2+}/Zn half cell. The reading on the voltmeter gives the standard electrode potential of the zinc half cell.

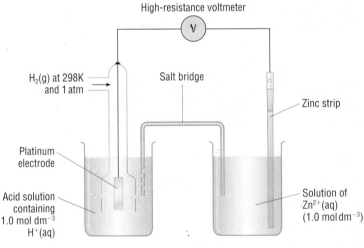

Figure 1 Measuring the standard electrode potential of a Zn/Zn^{2+} half cell

The measured standard electrode potential of the Zn^{2+}/Zn half cell is –0.76 V.

Standard cell potentials and cell reactions

In the previous spread, we saw that a cell can be constructed by connecting together two half cells.

The **standard cell potential** of a cell is the e.m.f. between the two half cells making up the cell under standard conditions.
* The standard cell potential of any cell is the difference between the standard electrode potentials of each half cell.

The **cell reaction** is the overall chemical reaction taking place in the cell.
* The cell reaction is the sum of the reduction and oxidation half reactions taking place in each half cell.

Worked example 1: the standard cell potential of a silver–copper cell

A silver–copper cell is made by connecting together two half cells:
- an Ag/Ag$^+$ half cell
- a Cu/Cu^{2+} half cell.

Step 1: Identify the two relevant redox equilibria and the sign of each electrode

The more negative of the two systems is the negative terminal of the cell.

Ag$^+$(aq) + e$^-$ $\rightleftharpoons$ Ag (s)	$E^{\ominus}$ = +0.80 V	positive terminal
Cu^{2+}(aq) + 2e$^-$ $\rightleftharpoons$ Cu(s)	$E^{\ominus}$ = +0.34 V	negative terminal

Step 2: Calculate the standard cell potential
The standard cell potential, $E^{\ominus}_{cell}$, is simply the difference between the standard electrode potentials of the half cells.

Subtract $E^{\ominus}$ of the negative terminal from $E^{\ominus}$ of the positive terminal:

$$E^{\ominus}_{cell} = E^{\ominus}(\text{positive terminal}) - E^{\ominus}(\text{negative terminal})$$

$$\therefore E^{\ominus}_{cell} = +0.80 - (+0.34)$$
$$= 0.46 \text{ V}$$

Worked example 2: the cell reaction in a silver–copper cell

Step 1: Work out the direction of electron flow in the redox equilibria
- From Worked example 1, we know the Cu^{2+}/Cu equilibrium has the more negative $E^{\ominus}$ value. This shows that the Cu^{2+}/Cu redox equilibrium has a greater tendency to lose electrons than the Ag$^+$/Ag half cell.
- The Cu^{2+}/Cu equilibrium loses electrons and moves to the left:

$$\overset{\longleftarrow}{Cu^{2+} + 2e^- \rightleftharpoons Cu}$$

The Cu^{2+}/Cu equilibrium releases electrons into the wire. Electrons flow along the wire to the Ag electrode of the Ag$^+$/Ag half cell.
- The Ag$^+$/Ag equilibrium gains the electrons and moves to the right:

$$\overset{\longrightarrow}{Ag^+(aq) + e^- \rightleftharpoons Ag(s)}$$

Step 2: Combine the half-equations to give the cell reaction
The equations taking place at each electrode are:

Cu(s)	$\longrightarrow$ Cu^{2+} (aq) + 2e$^-$	oxidation
Ag$^+$(aq) + e$^-$	$\longrightarrow$ Ag(s)	reduction

The Ag$^+$/Ag half-equation must be multiplied by 2 to balance the electrons:

Cu(s) $\longrightarrow$ Cu^{2+}(aq) + 2e$^-$
2Ag$^+$(aq) + 2e$^-$ $\longrightarrow$ 2Ag(s)

The two half-equations are then added together and electrons are cancelled:

Cu(s) + 2Ag$^+$(aq) + $\cancel{2e^-}$ $\longrightarrow$ 2Ag(s) + Cu^{2+}(aq) + $\cancel{2e^-}$

This gives the equation for the overall cell reaction:

Cu(s) + 2Ag$^+$(aq) $\longrightarrow$ 2Ag(s) + Cu^{2+}(aq)

Question

1 You are provided with the following standard electrode potentials:

½Cl$_2$(g) + e$^-$ $\rightleftharpoons$ Cl$^-$(aq)
       $E^{\ominus}$ = +1.36 V
Cu^{2+}(aq) + 2e$^-$ $\rightleftharpoons$ Cu(s)
       $E^{\ominus}$ = +0.34 V
Fe^{2+}(aq) + 2e$^-$ $\rightleftharpoons$ Fe(s)
       $E^{\ominus}$ = −0.44 V
Cr^{3+}(aq) + 3e$^-$ $\rightleftharpoons$ Cr(s)
       $E^{\ominus}$ = −0.74 V

(a) Calculate the standard cell potential for the following cells, prepared under standard conditions:
 (i) Cl$_2$/Cl$^-$ and Cu^{2+}/Cu
 (ii) Fe^{2+}/Fe and Cr^{3+}/Cr
 (iii) Cr^{3+}/Cr and Cl$_2$/Cl$^-$
(b) For each cell in (a), determine the overall cell reaction.

By the end of this spread, you should be able to . . .

* Predict, using standard cell potentials, the feasibility of reactions.
* Consider the limitations of predictions made using standard cell potentials, in terms of kinetics and concentration.

Using standard electrode potentials to predict redox reactions

Standard electrode potentials can be used to predict the feasibility of a redox reaction.

Worked example

Two redox equilibria are shown below:

$$Cu^{2+}(aq) + 2e^- \rightleftharpoons Cu(s) \qquad E^\ominus = +0.34 \text{ V}$$
$$NO_3^-(g) + 2H^+(aq) + e^- \rightleftharpoons NO_2(g) + H_2O(l) \qquad E^\ominus = +0.80 \text{ V}$$

We can predict the redox reaction that may take place between the species present in these redox equilibria by treating the redox systems as if they are half cells.

Step 1: Work out the direction of electron flow in the redox equilibria
* The Cu^{2+}/Cu equilibrium has the more negative (less positive) $E^\ominus$ value – this shows that the Cu^{2+}/Cu redox equilibrium supplies electrons and moves to the left.
* The NO_3^-/NO_2 equilibrium has the more positive (less negative) $E^\ominus$ value – this shows that the NO_3^-/NO_2 redox equilibrium accepts electrons and moves to the right.

$$\overleftarrow{Cu^{2+}(aq) + 2e^- \rightleftharpoons \mathbf{Cu(s)}} \qquad E^\ominus = +0.34 \text{ V} \qquad \textit{less positive}$$
$$\overrightarrow{\mathbf{NO_3^-(aq)} + \mathbf{2H^+(aq)} + e^- \rightleftharpoons NO_2(g) + H_2O(l)} \quad E^\ominus = +0.80 \text{ V} \qquad \textit{more positive}$$

We can therefore predict that **Cu** will react with $\mathbf{NO_3^-}$ and $\mathbf{H^+}$.

Step 2: Write down the half-equations for the oxidation and reduction
The half-equations taking place are:

$$Cu(s) \longrightarrow Cu^{2+}(aq) + 2e^- \qquad \text{oxidation}$$
$$NO_3^-(g) + 2H^+(aq) + e^- \longrightarrow NO_2(g) + H_2O(l) \qquad \text{reduction}$$

Step 3: Check the balance of electrons
The second half-equation must be multiplied by 2 to balance the electrons:

$$Cu(s) \longrightarrow Cu^{2+}(aq) + 2e^-$$
$$2NO_3^-(g) + 4H^+(aq) + 2e^- \longrightarrow 2NO_2(g) + 2H_2O(l)$$

Step 4: Construct the overall equation
The two half-equations are added together and the electrons cancelled:

$$Cu(s) + 2NO_3^-(g) + 4H^+(aq) + \cancel{2e^-} \longrightarrow Cu^{2+}(aq) + 2NO_2(g) + 2H_2O(l) + \cancel{2e^-}$$

This gives the equation for the overall reaction:

$$Cu(s) + 2NO_3^-(g) + 4H^+(aq) \longrightarrow Cu^{2+}(aq) + 2NO_2(g) + 2H_2O(l)$$

Figure 1 Redox reaction of concentrated nitric acid with copper to produce nitrogen dioxide gas, NO_2

Examiner tip

Notice that a reaction takes place between reactants taken from different sides of the two relevant half-equations.

The equilibrium with the more negative $E^\ominus$ value goes to the left.

The electrochemical series

When comparing two redox equilibria, it is conventional to show the electrons on the left-hand side. Shown like this, we can predict that a reaction may take place between:
* the species on the right-hand side of the redox equilibrium with the more negative $E^\ominus$ and
* the species on the left-hand side of the redox equilibrium with the more positive $E^\ominus$.

The list below shows several redox equilibria arranged with the most negative $E^\ominus$ value at the top. This is often called the **electrochemical series**.

The horizontal arrows show the preferred direction of reaction.

$$\begin{array}{ll}
\overleftarrow{Fe^{2+}(aq) + 2e^-} \rightleftharpoons Fe(s) & E^\ominus = -0.44 \text{ V} \\
2H^+(aq) + 2e^- \rightleftharpoons H_2(g) & E^\ominus = \quad 0 \text{ V} \\
Cu^{2+}(aq) + 2e^- \rightleftharpoons Cu(s) & E^\ominus = +0.34 \text{ V} \\
Br_2(g) + 2e^- \rightleftharpoons 2Br^-(aq) & E^\ominus = +1.07 \text{ V} \\
\underrightarrow{Cl_2(g) + 2e^- \rightleftharpoons 2Cl^-(aq)} & E^\ominus = +1.36 \text{ V}
\end{array}$$

Of those in the list, the redox equilibrium $Fe^{2+} + 2e^- \rightleftharpoons Fe$
- has the most negative $E^\ominus$ value
- has the greatest tendency to release electrons and shift the equilibrium to the left.

We can predict that Fe(s) would react with every species below and on the left in the list:
- H^+, Cu^{2+}, Br_2 and Cl_2

The redox equilibrium $Cu^{2+}(aq) + 2e^- \rightleftharpoons Cu(s)$
- has a more negative $E^\ominus$ value than the two redox equilibria below it in the list

We can predict that Cu(s) would react with every species below and on the left in the list:
- Br_2 and Cl_2

Limitations of predictions using standard electrode potentials

Electrode potentials and concentration

Non-standard conditions alter the value of an electrode potential. The half-equation for the copper half cell is:

$$Cu^{2+}(aq) + 2e^- \rightleftharpoons Cu(s)$$

From le Chatelier's principle, on increasing the concentration of $Cu^{2+}(aq)$:
- the equilibrium opposes the change by moving to the right
- electrons are removed from the equilibrium
- the electrode potential becomes less negative, or more positive.

A change in electrode potential resulting from concentration changes means that predictions made on the basis of the *standard* value may not be valid.

Will a reaction actually take place?

Remember that these are equilibrium processes in aqueous conditions.
- Predictions can be made about the equilibrium position but not about the reaction rate, which may be extremely slow because of a high activation energy.
- The actual conditions used for a reaction may be different from the standard conditions used to measure $E^\ominus$ values. This will affect the value of the electrode potential (see above).
- Standard electrode potentials apply to aqueous equilibria – many reactions take place under very different conditions.

As a general working rule:
- the larger the difference between $E^\ominus$ values, the more likely it is that a reaction will take place
- if the difference between $E^\ominus$ values is less than 0.4 V, then a reaction is unlikely to take place.

Question

1 Use the standard electrode potentials below to answer the questions that follow.

$$\begin{array}{ll}
Fe^{3+}(aq) + e^- \rightleftharpoons Fe^{2+}(aq) & E^\ominus = +0.77 \text{ V} \\
Br_2(aq) + 2e^- \rightleftharpoons 2Br^-(aq) & E^\ominus = +1.07 \text{ V} \\
Ni^{2+}(aq) + 2e^- \rightleftharpoons Ni(s) & E^\ominus = -0.25 \text{ V} \\
O_2(g) + 4H^+(aq) + 4e^- \rightleftharpoons 2H_2O(l) & E^\ominus = +1.23 \text{ V}
\end{array}$$

   (a) Predict the reactions that could take place under standard conditions.
   (b) Write an overall equation for each reaction in (a).

STRETCH and CHALLENGE

In the worked example, we know that 1 mol dm⁻³ nitric acid was used because the electrode potential used was the standard value. Predict the effect of using concentrated acid on the electrode potential given in the worked example.

Explain why the redox reaction between copper and nitric acid is *more* likely to take place with concentrated acid than using 1 mol dm⁻³ nitric acid.

By the end of this spread, you should be able to . . .

∗ **Apply the principles of electrode potentials to modern storage cells.**
∗ **Explain that a fuel cell uses the energy from the reaction of a fuel with oxygen to create a voltage.**
∗ **Explain the changes that take place at each electrode in a hydrogen–oxygen fuel cell.**

Electrochemical cells

Electrochemical cells are used widely in everyday life as a source of electrical energy. All cells work on the same principle – having two redox equilibria with different electrode potentials. For example, a simple cell can be set up based on zinc and copper (see also spread 2.2.10).

The redox equilibria in this cell are shown below together with their standard electrode potentials. The equilibrium with the more negative electrode potential is the negative terminal of the cell.

$$Zn^{2+}(aq) + 2e^- \rightleftharpoons Zn(s) \qquad E^\ominus = -0.76 \text{ V} \qquad \text{negative terminal}$$
$$Cu^{2+}(aq) + 2e^- \rightleftharpoons Cu(s) \qquad E^\ominus = +0.34 \text{ V} \qquad \text{positive terminal}$$

The half-equations below show the reactions that take place at each electrode in the cell.

- The more negative zinc system provides the electrons, so the movement of the zinc equilibrium is from right to left.
- The half-equations are added together and electrons cancelled to give the equation for the overall cell reaction:

$$Zn(s) \longrightarrow Zn^{2+}(aq) + 2e^-$$
$$Cu^{2+}(aq) + 2e^- \longrightarrow Cu(s)$$

Overall: $Zn(s) + Cu^{2+}(aq) \longrightarrow Cu(s) + Zn^{2+}(aq)$

The standard potentials can be used to work out the e.m.f. of the cell:
$$E^\ominus_{cell} = E^\ominus \text{ (positive terminal)} - E^\ominus \text{ (negative terminal)}$$
$$= 0.34 - (-0.76)$$
$$= 1.10 \text{ V}$$

Modern cells and batteries

Electrochemical cells are used as our modern-day cells and batteries. Cells can be divided into three main types:

1 *Non-rechargeable cells* – provide electrical energy until the chemicals have reacted to such an extent that the voltage falls. The cell is then 'flat' and is discarded.
2 *Rechargeable cells* – the chemicals in the cell react, providing electrical energy. The cell reaction can be reversed during recharging – the chemicals in the cell are regenerated and the cell can be used again. Common examples include:
   - nickel and cadmium (Ni-Cad) batteries, used in rechargeable batteries
   - lithium-ion and lithium polymer batteries, used in laptops.
3 *Fuel cells* – the cell reaction uses external supplies of a fuel and an oxidant, which are consumed and needs to be provided continuously. The cell will continue to provide electrical energy so long as there is a supply of fuel and oxidant.

Figure 1 Selection of cells and batteries

Examiner tip

In exams, you will not be expected to recall a specific storage cell. However, you might be expected to make predictions about a given cell. All relevant electrode potentials and other data will be supplied.

Fuel cells

Fuel cells have been around for over 150 years! In 1842 Sir William Grove, a Welsh physicist, invented the first fuel cell. He mixed hydrogen and oxygen in the presence of an electrolyte and produced electricity and water. The invention, which later became known as a fuel cell, didn't produce enough electricity to be useful at the time. Modern fuel cells are based on hydrogen, or hydrogen-rich fuels such as methanol, CH_3OH.

The hydrogen–oxygen fuel cell

A fuel cell uses energy from the reaction of a fuel with oxygen to create a voltage.
- The reactants flow in and products flow out while the electrolyte remains in the cell.
- Fuel cells can operate virtually continuously so long as the fuel and oxygen continue to flow into the cell. Fuel cells do not have to be recharged.

Figure 2 shows a simple hydrogen–oxygen fuel cell with an alkaline electrolyte.

The redox equilibria are shown below. The more negative of the two systems is the negative terminal of the cell.

$$2H_2O(l) + 2e^- \rightleftharpoons H_2(g) + 2OH^-(aq) \quad E^\ominus = -0.83 \text{ V} \quad \text{negative terminal}$$
$$\tfrac{1}{2}O_2(g) + H_2O(l) + 2e^- \rightleftharpoons 2OH^-(aq) \quad\quad E^\ominus = +0.40 \text{ V} \quad \text{positive terminal}$$

The more negative hydrogen system provides the electrons. This equilibrium is reversed when writing the half-equations at each electrode. The half equations are added together and electrons cancelled to give the equation for the overall cell reaction:

$$H_2(g) + 2OH^-(aq) \longrightarrow 2H_2O(l) + 2e^-$$
$$\tfrac{1}{2}O_2(g) + 2H_2O(l) + 2e^- \longrightarrow 2OH^-(aq)$$

Overall: $H_2(g) + \tfrac{1}{2}O_2(g) \longrightarrow H_2O(l)$

$$E^\ominus_{cell} = E^\ominus \text{ (positive terminal)} - E^\ominus \text{ (negative terminal)}$$
$$= 0.40 - (-0.83)$$
$$= 1.23 \text{ V}$$

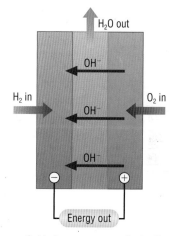

Figure 2 Hydrogen–oxygen fuel cell

Other fuel cells have been developed – these are based on hydrogen-rich fuels such as methanol, natural gas and petrol.

Figure 3 Space shuttle launch. The space shuttle's electricity is provided by hydrogen fuel cells – the water produced is used by the crew for drinking

Questions

1 What are the three main types of electrochemical cell?
2 What are the electrode reactions and the overall reaction in an alkaline hydrogen fuel cell?

191

By the end of this spread, you should be able to . . .

* Outline the development of fuel cell vehicles (FCVs) that use hydrogen gas and hydrogen-rich fuels.
* State the advantages of FCVs over conventional petrol- or diesel-powered vehicles.
* Understand how hydrogen might be stored in FCVs.
* Consider limitations of hydrogen fuel cells.
* Comment about the contribution of the 'hydrogen economy' to future energy and discuss its limitations.

Development of fuel cell vehicles

In the previous spread, we discussed the hydrogen fuel cell. In the future, a 'hydrogen economy' based on fuels cells may well contribute greatly to our ever-increasing energy needs. A current focus is the development of fuel cells as an alternative to the use of finite oil-based fuels in cars.

Scientists in the car industry are developing fuel cell vehicles (FCVs) – these use hydrogen gas or hydrogen-rich fuels.

Hydrogen-rich fuels include methanol, natural gas and petrol. These are mixed with water and converted into hydrogen gas by an onboard 'reformer'. The reformer operates at 250–300 °C to generate the hydrogen gas – for example:

$$CH_3OH + H_2O \longrightarrow 3H_2 + CO_2$$

The hydrogen can then be fed into fuel cells.

Alternatively, fuel cells have been developed in which methanol is the fuel rather than hydrogen. There are obvious advantages of a methanol fuel cell compared with use of hydrogen gas:

* a liquid fuel is easier to store than hydrogen gas
* methanol can be generated from biomass.

Methanol-based fuel cells are in the early days of development. They currently generate only a small amount of power. Note also that carbon dioxide is produced as a waste emission.

Advantages of fuel cell vehicles

Fuel cell vehicles have advantages over conventional petrol and diesel vehicles.

Less pollution and less CO₂

* Combustion of hydrocarbon fuels produces CO_2, which contributes to the greenhouse effect. Incomplete combustion produces toxic carbon monoxide, which must be removed by catalytic converters (see *AS Chemistry* spread 2.4.6).
* Hydrogen-rich fuels produce only small amounts of CO_2 and air pollutants.

Greater efficiency

* A petrol engine is less than 20% efficient in converting chemical energy by the combustion of petrol. Much of the chemical energy is wasted as heat.
* Hydrogen fuel cell vehicles are 40–60% efficient in converting the fuel's energy. This means that fuel consumption drops by more than half compared with a petrol- or diesel-powered car.

Figure 1 Liquid-hydrogen-powered car. Test car (BMW 735i) being filled with liquid hydrogen fuel – the fuel tank is in the car boot. A converted car can run for 300 kilometres on a full tank of liquid hydrogen

Figure 2 Methanol-powered bus in New York city. Methanol is a liquid used as a hydrogen carrier for fuel cells. Methanol fuel cells will reduce the dependence on petrol and improve urban air quality

Storage of hydrogen

The most obvious problem in storing hydrogen is that it is a gas with a very low boiling point. Gases are far more difficult to store than liquids such as petrol and diesel. Car manufacturers are developing several strategies for the storage of hydrogen.

- *Hydrogen can be stored as a liquid under pressure.* This in itself creates logistical problems. Even under pressure, a very low temperature is required and liquid hydrogen will need to be stored in a giant 'thermos flask' to prevent it from boiling.
- *Hydrogen can be adsorbed* onto the surface of a solid material in a similar way that a catalyst is able to hold gases to its surface.
- *Hydrogen can be absorbed* within some solid materials.

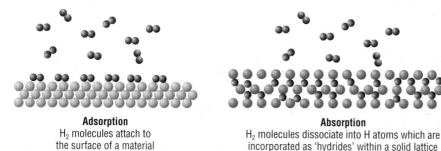

Adsorption
H_2 molecules attach to the surface of a material

Absorption
H_2 molecules dissociate into H atoms which are incorporated as 'hydrides' within a solid lattice

Figure 3 Storage of hydrogen by adsorption and absorption – large amounts of hydrogen can be stored in a small volume at low pressure and close to room temperature

Limitations of hydrogen fuel cells

There are logistical problems in the development and use of hydrogen fuel cells.

- The large-scale storage and transportation of hydrogen poses problems. A cost-effective and energy-efficient infrastructure needs to be in place to deliver large quantities of hydrogen fuel over long distances.
- The feasibility of storing a pressurised liquid.
- Current 'adsorbers' and 'absorbers' of hydrogen have a limited lifetime.
- Current fuel cells have a limited lifetime, requiring regular replacement and disposal following high production costs.
- Fuel cells use toxic chemicals in their production.

The hydrogen economy

The 'hydrogen economy' may contribute largely to future energy needs, but many limitations need to be resolved.

- The use of hydrogen as a fuel has to be accepted politically and by the general public.
- There are logistical problems in the handling and maintenance of hydrogen systems.
- Hydrogen is an 'energy carrier' and not an 'energy source'. Hydrogen must first be manufactured, either by electrolysis of water or by reacting methane (a finite fuel) with steam. The danger is that more energy may be used in making the hydrogen than is saved by its use. One strategy is to use renewable forms of energy, such as the wind or solar power, to generate the hydrogen.

Figure 4 Liquid-hydrogen filling station

Questions

1 How does a methanol fuel cell work?
2 Outline the main methods available for storing hydrogen in fuel cell vehicles.

Born–Haber cycles

Formation of an ionic lattice

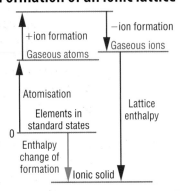

+ion formation
Gaseous atoms
−ion formation
Gaseous ions
Atomisation
Elements in standard states
Lattice enthalpy
0
Enthalpy change of formation
Ionic solid

Formation of a solution

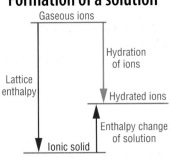

Gaseous ions
Hydration of ions
Lattice enthalpy
Hydrated ions
Enthalpy change of solution
Ionic solid

Entropy and free energy

$$\Delta S = \Sigma S(\text{products}) - \Sigma S(\text{reactants})$$

System more random: ΔS is positive
System more ordered: ΔS is negative

$$\Delta G = \Delta H - T\Delta S$$

For a reaction to be feasible $\Delta G < 0$

Electrode potentials

$$E^{\ominus}_{\text{cell}} = E^{\ominus}(\text{positive terminal}) - E^{\ominus}(\text{negative terminal})$$

The half cell with the more negative E value supplies electrons

The half cell with the less negative E value receives electrons

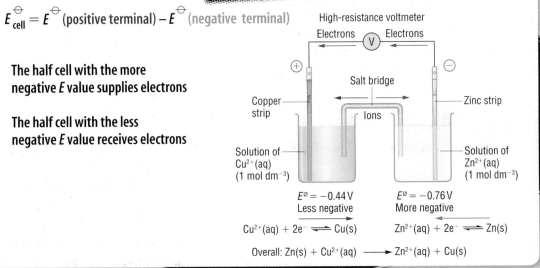

High-resistance voltmeter
Electrons — V — Electrons
Copper strip
Salt bridge
Zinc strip
Ions
Solution of $Cu^{2+}(aq)$ (1 mol dm^{-3})
Solution of $Zn^{2+}(aq)$ (1 mol dm^{-3})
$E^{\ominus} = -0.44\,V$ Less negative
$E^{\ominus} = -0.76\,V$ More negative
$Cu^{2+}(aq) + 2e^- \rightleftharpoons Cu(s)$
$Zn^{2+}(aq) + 2e^- \rightleftharpoons Zn(s)$

Overall: $Zn(s) + Cu^{2+}(aq) \longrightarrow Zn^{2+}(aq) + Cu(s)$

Practice questions

1 (a) Define the term lattice enthalpy.

(b) Write an equation for the change that accompanies lattice enthalpy of:

(i) potassium chloride;

(ii) lithium oxide;

(iii) aluminium fluoride.

2 The table below shows the enthalpy changes required to construct a Born–Haber cycle for the formation of potassium iodide in order to find its lattice enthalpy.

Name of enthalpy change	Equation	$\Delta H / \text{kJ mol}^{-1}$
	$K(s) \longrightarrow K(g)$	+89
Lattice enthalpy of potassium iodide	$K^+(g) + I^-(g) \longrightarrow KI(s)$	?
	$K(s) + \frac{1}{2}I_2(s) \longrightarrow KI(s)$	−328
	$I(g) + e^- \longrightarrow I^-(g)$	−295
	$\frac{1}{2}I_2(s) \longrightarrow I(g)$	+107
	$K(s) \longrightarrow K^+(g) + e^-$	+419

(a) Copy and complete the table with the names of the enthalpy changes.

(b) Using the information in the table, construct the Born–Haber cycle for potassium iodide.

(c) Calculate the lattice enthalpy for potassium iodide.

(d) Explain whether the lattice enthalpies of the following compounds would be more or less exothermic than that of potassium iodide:

(i) sodium iodide; **(ii)** calcium iodide.

3 You are provided with the following enthalpy changes.
$Mg^{2+}(g) + 2Cl^-(g) \longrightarrow MgCl_2(s)$ $\Delta H = -2493 \text{ kJ mol}^{-1}$
$MgCl_2(s) + aq \longrightarrow MgCl_2(aq)$ $\Delta H = -153 \text{ kJ mol}^{-1}$
$Cl^-(g) + aq \longrightarrow Cl^-(aq)$ $\Delta H = -363 \text{ kJ mol}^{-1}$

(a) Name each of these three enthalpy changes.

(b) Calculate the enthalpy change of hydration of Mg^{2+}.

4 (a) What is the sign of ΔS for the following reactions?

(i) $Mg(s) + H_2SO_4(aq) \longrightarrow MgSO_4(aq) + H_2(g)$

(ii) $2CO(g) + 2NO(g) \longrightarrow 2CO_2(g) + N_2(g)$

(b) You are provided with some standard entropies in the table below.

	$N_2(g)$	$H_2(g)$	$NH_3(g)$	$C_2H_5OH(g)$	$C_2H_4(g)$	$H_2O(g)$
$S^{\ominus}/$ $\text{J K}^{-1} \text{mol}^{-1}$	+192	+130	+192	+161	+220	+189

For each of the two reactions in **(i)** and **(ii)** below,
- calculate ΔS at 25 °C;
- calculate ΔG at 25 °C;
- find the temperature in °C at which $\Delta G = 0$ and comment on the significance of this value for the each reaction.

(i) $N_2(g) + 3H_2(g) \longrightarrow 2NH_3(g)$ $\Delta H = -92 \text{ kJ mol}^{-1}$

(ii) $C_2H_5OH(g) \longrightarrow C_2H_4(g) + H_2O(g)$ $\Delta H = +87 \text{ kJ mol}^{-1}$

5 (a) Construct redox equations from the following half-equations:

(i) $Pb(s) \longrightarrow Pb^{2+}(aq) + 2e^-$
$Ag^+(aq) + e^- \longrightarrow Ag(s)$

(ii) $Fe^{2+}(aq) \longrightarrow Fe^{3+}(aq) + e^-$
$Cr_2O_7^{2-}(aq) + 14H^+(aq) + 6e^- \longrightarrow Cr^{3+}(aq) + 7H_2O(l)$

(b) Use oxidation states to construct redox equations for the following reactions.

(i) In acid conditions silver metal, Ag, is oxidised to silver(I) ions, Ag^+, by NO_3^- ions, which are reduced to nitrogen(II) oxide, NO.

(ii) Aqueous thiosulfate ions, $S_2O_3^{2-}$, are oxidised to sulfate(VI) ions, SO_4^{2-}, by chorine gas, which is reduced to chloride ions.

6 (a) Define *standard electrode potential*.

(b) You are provided with the following standard electrode potentials:

$MnO_4^-(aq) + 8H^+(aq) + 5e^- \rightleftharpoons Mn^{2+}(aq) + 4H_2O(l)$
$E^{\ominus} = +1.52 \text{ V}$
$Fe^{3+}(aq) + e^- \rightleftharpoons Fe^{2+}(aq)$ $E^{\ominus} = +0.77 \text{ V}$
$Zn^{2+}(aq) + 2e^- \rightleftharpoons Zn(s)$ $E^{\ominus} = -0.76 \text{ V}$

Calculate the standard cell potential and cell reaction for the following cells, set up under standard conditions:

(i) MnO_4^-/Mn^{2+} and Fe^{3+}/Fe^{2+}

(ii) MnO_4^-/Mn^{2+} and Zn^{2+}/Zn

(iii) Zn^{2+}/Zn and Fe^{3+}/Fe^{2+}

7 Use the standard electrode potentials below to answer the questions which follow:

$Cr^{3+}(aq) + e^- \rightleftharpoons Cr^{2+}(aq)$ $E^{\ominus} = -0.41 \text{ V}$
$2H^+(aq) + 2e^- \rightleftharpoons H_2(g)$ $E^{\ominus} = 0.00 \text{ V}$
$O_2(g) + 4H^+(aq) + 4e^- \rightleftharpoons 2H_2O(l)$ $E^{\ominus} = +1.23 \text{ V}$

(a) Predict the reactions that could take place.

(b) Write an overall equation for each reaction in part **(a)**.

1 The table below shows the enthalpy changes needed to calculate the enthalpy change of formation of calcium oxide.

Process	Enthalpy change/kJ mol^{-1}
Lattice enthalpy for calcium oxide	−3459
First ionisation energy for calcium	+590
Second ionisation energy for calcium	+1150
First electron affinity for oxygen	−141
Second electron affinity for oxygen	+798
Enthalpy change of atomisation for oxygen	+249
Enthalpy change of atomisation for calcium	+178

(a) (i) Explain why the first ionisation energy of calcium is endothermic. [1]

   (ii) Explain why the first electron affinity for oxygen is exothermic. [1]

(b) (i) Draw a Born–Haber cycle for calcium oxide. Include
   • correct formulae and state symbols
   • energy changes in kJ mol^{-1}. [3]

   (ii) Use your Born–Haber cycle in (i) to calculate the enthalpy change of formation for calcium oxide. [2]

   (iii) The lattice enthalpy for iron(II) oxide is −3920 kJ mol^{-1}.
   Suggest a reason for the difference in lattice enthalpy between calcium oxide and iron(II) oxide. [1]

   [Total: 8]
   (OCR 2815/1 Jun07)

2 In this question, one mark is available for the quality of spelling, punctuation and grammar.
The lattice enthalpy of magnesium chloride, MgCl$_2$, can be determined using a Born–Haber cycle and the following enthalpy changes.

Name of process	Enthalpy change/kJ mol^{-1}
Enthalpy change of formation of MgCl$_2$	−641
Enthalpy change of atomisation of magnesium	+148
First ionisation energy of magnesium	+738
Second ionisation energy of magnesium	+1451
Enthalpy change of atomisation of chlorine	+123
Electron affinity of chlorine	−349

• Define, using an equation with MgCl$_2$ as an example, what is meant by the term *lattice enthalpy*.
• Construct a Born–Haber cycle for MgCl$_2$, including state symbols, and calculate the lattice enthalpy of MgCl$_2$.
• Explain why the lattice enthalpy of NaBr is much less exothermic than that of MgCl$_2$.

   [Total: 12]
   (OCR 2815/1 Jan07)

3 The Born–Haber cycle below can be used to calculate the lattice enthalpy for magnesium oxide.

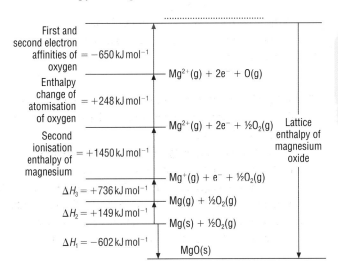

(a) (i) Write down the name for each of the following enthalpy changes.

ΔH_1

ΔH_2

ΔH_3 [3]

(ii) Write down the missing formulae at the **top** of the Born–Haber cycle. Include state symbols. [1]

(iii) The equations representing the first and second electron affinities for oxygen are shown below.

$O(g) + e^- \rightarrow O^-(g)$ $\Delta H = -141$ kJ mol^{-1}

$O^-(g) + e^- \rightarrow O^{2-}(g)$ $\Delta H = +791$ kJ mol^{-1}

Suggest why the enthalpy change for the second of these processes is positive. [1]

(b) (i) Use the Born–Haber cycle to calculate the lattice enthalpy of magnesium oxide. [2]

(ii) Describe how, and explain why, the lattice enthalpy of magnesium oxide differs from that of barium oxide. [3]

(c) Give **one** reason why magnesium oxide is a good material to make the lining of a furnace. [1]

[Total: 11]

(OCR 2815/1 Jan04)

4 When crystals of ammonium chloride, NH_4Cl, are dissolved in water, the temperature falls. The enthalpy change of solution of ammonium chloride, $NH_4^+Cl^-$, is +15 kJ mol^{-1}. The lattice enthalpy of $NH_4Cl(s) = -689$ kJ mol^{-1}. The enthalpy change of hydration of Cl^- is -364 kJ mol^{-1}.

(a) Draw a Born–Haber cycle to link together the lattice enthalpy, the enthalpy change of solution and the relevant enthalpy changes of hydration. [3]

(b) Calculate the enthalpy change of hydration of the ammonium ion, NH_4^+. [2]

(c) Explain why ammonium chloride dissolves spontaneously in water even though this process is endothermic. [2]

[Total: 7]

5 8.48 g of solid lithium chloride, LiCl, were dissolved in 100 cm^3 of water. The temperature of the solution increased from 22.0 °C to 34.5 °C. The specific heat capacity of the solution = 4.18 J g^{-1} K^{-1}.

(a) Assuming that there is no heat loss, calculate the enthalpy change of solution of lithium chloride. [3]

(b) The experiment was repeated using sodium chloride, NaCl. This time the temperature decreased and the enthalpy change of solution was calculated as +5.00 kJ mol^{-1}.

Explain why sodium chloride is soluble in water, despite its enthalpy change of solution being endothermic. [2]

(c) The diagram below shows an enthalpy cycle for the solution of lithium chloride.

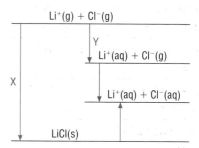

(i) Name the enthalpy changes **X** and **Y**. [2]

(ii) Suggest why the enthalpy change of solution of LiCl is exothermic while the enthalpy change of solution of NaCl is exothermic. [2]

[Total: 9]

6 The reaction between $NaHCO_3$ and HCl is endothermic. The equation for this reaction is shown below.

$NaHCO_3(aq) + HCl(aq) \rightarrow NaCl(aq) + H_2O(l) + CO_2(g)$

Equation 1

The reaction between sulfur dioxide and oxygen is exothermic. This is a reversible reaction. The equation for the forward reaction is shown below.

$2SO_2(g) + O_2(g) \rightarrow 2SO_3(g)$ Equation 2

(a) Predict, with reasons, the sign of ΔS in the two reactions above. [2]

(b) State the equation that links together ΔG, ΔH and ΔS. [1]

(c) Use the equation in (b) to help you explain the following.

(i) The reaction between $NaHCO_3$ and HCl, shown in Equation 1, takes place spontaneously at room temperature, despite the reaction being endothermic. [2]

(ii) The reaction between SO_2 and O_2, shown in Equation 2, does not take place at high temperatures, despite the forward reaction being exothermic. [3]

[Total: 8]

7 Hydrogen can be prepared by reacting together methane and steam. This is a reversible reaction.

$$CH_4(g) + H_2O(g) \rightleftharpoons CO(g) + 3H_2(g) \quad \Delta H = +210 \text{ kJ mol}^{-1}$$

(a) Predict the sign of ΔS for the forward reaction that produces hydrogen.
Explain your reasoning. [1]

(b) The standard entropies of the reactants and products in this reaction are shown below.

Species	$S^\ominus$/ J K^{-1} mol^{-1}
$CH_4(g)$	+186
$H_2O(g)$	+189
$CO(g)$	+198
$H_2(g)$	+131

Calculate ΔS for the forward reaction that produces hydrogen. [3]

(c) Show that the forward reaction does not take place at room temperature (25 °C). [2]

(d) Calculate the minimum temperature required for the forward reaction to take place. [2]

[Total: 8]

8 Some standard electrode potentials are shown below.

$$E^\ominus/V$$

$Ag^+ + e^- \rightleftharpoons Ag$	+0.80
$\frac{1}{2}Cl_2 + e^- \rightleftharpoons Cl^-$	+1.36
$Cu^{2+} + 2e^- \rightleftharpoons Cu$	+0.34
$Fe^{3+} + e^- \rightleftharpoons Fe^{2+}$	+0.77
$\frac{1}{2}I_2 + e^- \rightleftharpoons I^-$	+0.54

(a) Define the term *standard electrode potential*. [3]

(b) The diagram below shows an incomplete cell consisting of Cu/Cu^{2+} and Ag/Ag$^+$ half cells.

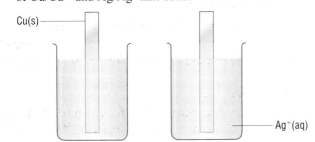

(i) Complete and label the diagram to show how the cell potential of this cell could be measured. [2]

(ii) On the diagram, show the direction of **electron** flow in the circuit if a current was allowed. [1]

(iii) Calculate the standard cell potential. [1]

(iv) Write the overall cell reaction. [1]

(c) Chlorine will oxidise Fe^{2+} to Fe^{3+} but iodine will not. Explain why, using the electrode potential data. [2]

[Total: 10]
(OCR 2815/6 Jun06)

9 Chlorine gas may be prepared in the laboratory by reacting hydrochloric gas with potassium manganate(VII). The following standard electrode potentials relate to this reaction.

$$\frac{1}{2}Cl_2 + e^- \rightleftharpoons Cl^- \qquad E^\ominus = +1.36 \text{ V}$$
$$MnO_4^- + 8H^+ + 5e^- \rightleftharpoons Mn^{2+} + 4H_2O \quad E^\ominus\ +1.52 \text{ V}$$

(a) Define the term *standard electrode potential*. [3]

(b) Determine the standard cell potential for a cell constructed from these two redox systems. [1]

(c) Use the half-equations above to:
(i) construct an ionic equation for the reaction between hydrochloric acid and potassium manganate(VII); [2]

(ii) determine the oxidation numbers of chlorine and manganese before and after the reaction has taken place; [2]

(iii) state what is oxidised and what is reduced in this reaction. [2]

(d) If potassium manganate(VII) and very dilute hydrochloric acid are mixed, there is no visible reaction. Suggest why there is no visible reaction in this case. [1]

[Total: 11]
(OCR 2815/6 Jun05)

10 An electrochemical cell was set up based on the following electrode reactions.

$$Cd^{2+}(aq) + 2e^- \rightleftharpoons Cd(s) \quad E^\ominus = -0.40 \text{ V}$$
$$Cr^{3+}(aq) + 3e^- \rightleftharpoons Cr(s) \quad E^\ominus = -0.74 \text{ V}$$

(a) (i) Draw a diagram of this cell working under standard conditions. [3]

(ii) Show on the diagram the direction of electron flow in the external circuit. [1]

(iii) Explain your answer to (ii). [2]

(b) Write a full ionic equation for the reaction taking place in this circuit. [1]

(c) (i) Calculate the standard cell potential of this cell. [1]

(ii) When water is added to the chromium half cell, the cell potential changes. Suggest **one** reason for this observation. [1]

[Total: 9]
(OCR 2815/6 Jun02)

11 The standard cell potential, $E^\ominus$, of $S_2O_8^{2-} + 2e^- \rightleftharpoons 2SO_4^{2-}$ is +2.01 V.
(a) Define the term *standard electrode potential*. [3]
(b) The standard cell potential, $E^\ominus$, of $S_2O_8^{2-} + 2e^- \rightleftharpoons 2SO_4^{2-}$ may be measured using the following apparatus.

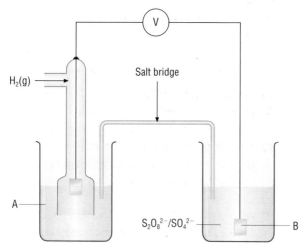

(i) What could be used for
   solution **A**
   solid **B**? [2]
(ii) The voltmeter is replaced by a lamp.
   On the diagram, show the direction of electron flow when the cell is used to light the lamp. [1]
(iii) Construct an equation to show the reaction taking place when the cell is used to light the lamp. [2]
(c) The solution of $S_2O_8^{2-}/SO_4^{2-}$ has a concentration of 1.00 mol dm^{-3} with respect to $S_2O_8^{2-}$ and SO_4^{2-}.
Calculate the masses needed to make up exactly 100 cm^3 of this solution, starting with solid $Na_2S_2O_8$ and Na_2SO_4. [3]
(d) A student made up the solution of $S_2O_8^{2-}/SO_4^{2-}$ at 1.00 mol dm^{-3} with respect to $S_2O_8^{2-}$ but 0.100 mol dm^{-3} with respect to SO_4^{2-}.
Suggest the effect this would have on the value of $E^\ominus$. Explain your answer. [2]
[Total: 13]
(OCR 2815/6 Jan08)

12 *In your answers to the questions that follow, show all of your working.*
At room temperature and pressure, RTP, 1 mol of gas molecules has a volume of 24 dm^3.
While digging his garden, a chemistry student found what appeared to be an ancient relic, possibly from the Bronze Age. The student knew that bronze is an alloy of copper with other metals including tin. He carried out a series of experiments on samples of the ancient relic.

(a) **Experiment 1**
The student determined the percentage of copper in the bronze relic by the following method.
He dissolved a small piece of the bronze relic, weighing 0.28 g, in concentrated (16 mol dm^{-3}) nitric acid, HNO_3. A blue solution **C** containing Cu^{2+} ions was formed together with a brown gas with the molecular formula NO_2. Solution **C** had a volume of 5.0 cm^3.
Equation **1** represents the equation for the reaction between copper and concentrated nitric acid.
$$Cu + 4H^+ + 2NO_3^- \longrightarrow Cu^{2+} + 2NO_2 + 2H_2O$$
Equation 1
The student analysed the blue colour from the Cu^{2+} ions in solution **C** using a colorimeter. He found out that the concentration of Cu^{2+} ions in solution **C** was 0.68 mol dm^{-3}.
The student concentrated the solution and obtained some blue crystals of a compound **A** with a percentage composition by mass of Cu, 26.29%; N, 11.60%; O 59.63%; H 2.48%. This composition included 3 waters of crystallisation.
(i) Calculate the percentage of copper in the bronze relic. [3]
(ii) Calculate the empirical formula of **A**. [2]
(iii) How would the formula of **A** normally be shown on a bottle of the chemical? [2]

(b) **Experiment 2**
The student dissolved another small piece of the bronze relic in dilute (8 mol dm^{-3}) nitric acid. A blue solution containing Cu^{2+} ions was again formed but this time a colourless gas was produced with the molecular formula NO.
Equation **2** represents the unbalanced equation for this second reaction.
$$Cu + H^+ + NO_3^- \longrightarrow Cu^{2+} + NO + H_2O$$
Equation 2
By considering oxidation numbers, balance equation **2**. [3]

(c) **Experiment 3**
The student heated a third small piece of the bronze relic with concentrated sulfuric acid. The copper in the bronze relic reacted to produce a blue solution and 90 cm^3 of a gas **B**, measured at RTP. The mass of the gas **B** collected was 0.24 g.
(i) Suggest a possible identity of gas **B**. [2]
(ii) Suggest a likely balanced equation for this reaction. [2]
[Total: 14]
(OCR 2816/1 Jun2005)

Module 3
Transition elements

Introduction

The transition elements have great importance in both biological and chemical systems. Iron is the key element in haemoglobin, the red pigment in blood which carries oxygen round the body, and plays a role as an important catalyst in the Haber process. Platinum, in *cis*-platin, is used therapeutically in the treatment of cancer and is also used with rhodium and palladium in catalytic converters.

One striking feature of the transition metals is their brightly coloured compounds. Rubies are red due to the presence of chromium(III) ions, blue glass is made by the addition of cobalt(II), and turquoise gemstones owe their colour to traces of copper(II) ions.

In this module, you will look more closely at the transition elements and explore their properties and uses. You will discover that reactions of transition metal ions in solution often result in changes in colour. Redox reactions of transition elements in titrations can be used to find the iron content of iron tablets and the copper content of brass.

Test yourself

1 Write the electron configuration of an Fe^{3+} ion.
2 Define the term coordinate bond.
3 What is a titration used for?
4 Name the platinum-containing drug used in cancer therapy.

① Transition metals

By the end of this spread, you should be able to ...

* Deduce the electron configurations of atoms and ions of the d-block elements.
* Describe the elements Ti–Cu as transition elements.

What are transition metals?

The d-block

The d-block elements are found sandwiched between Group 2 and Group 3 in the Periodic Table. Across the d-block, electrons are filling d-orbitals and the highest energy sub-shell is a d sub-shell (see spread 1.2.4 in *AS Chemistry*). In this spread we will consider the chemistry of the elements titanium to copper in the first row of the d-block.

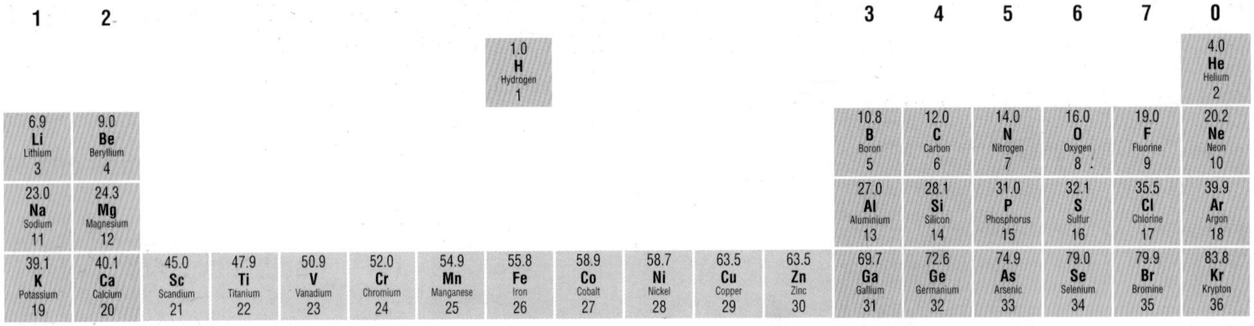

1	2											3	4	5	6	7	0
							1.0 H Hydrogen 1										4.0 He Helium 2
6.9 Li Lithium 3	9.0 Be Beryllium 4											10.8 B Boron 5	12.0 C Carbon 6	14.0 N Nitrogen 7	16.0 O Oxygen 8	19.0 F Fluorine 9	20.2 Ne Neon 10
23.0 Na Sodium 11	24.3 Mg Magnesium 12											27.0 Al Aluminium 13	28.1 Si Silicon 14	31.0 P Phosphorus 15	32.1 S Sulfur 16	35.5 Cl Chlorine 17	39.9 Ar Argon 18
39.1 K Potassium 19	40.1 Ca Calcium 20	45.0 Sc Scandium 21	47.9 Ti Titanium 22	50.9 V Vanadium 23	52.0 Cr Chromium 24	54.9 Mn Manganese 25	55.8 Fe Iron 26	58.9 Co Cobalt 27	58.7 Ni Nickel 28	63.5 Cu Copper 29	63.5 Zn Zinc 30	69.7 Ga Gallium 31	72.6 Ge Germanium 32	74.9 As Arsenic 33	79.0 Se Selenium 34	79.9 Br Bromine 35	83.8 Kr Krypton 36

Figure 1 First row of the d-block

Transition elements

A **transition element** is a d-block element that forms at least one ion with an incomplete d sub-shell. Scandium and zinc, the first and last members of this row, are not actually classed as transition metals because they do not have any ions with partially filled d-orbitals:

* scandium forms only the Sc^{3+} ion, in which d-orbitals are empty.
* zinc forms only the Zn^{2+} ion, in which the d-orbitals are completely full.

Writing electron configurations

The electron configuration of an element is the arrangement of the electrons in an atom of the element. Electrons occupy orbitals in order of energy level. Figure 2 shows the order of energy levels for the sub-shells making up the first four shells.

* The sub-shell energy levels in the third and fourth energy levels overlap.
* The 4s sub-shell fills before the 3d sub-shell.

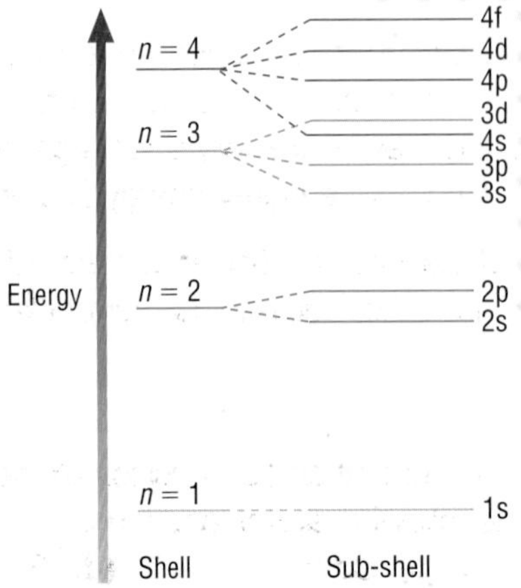

Figure 2 Energy-level diagram showing the overlap of the 3d and 4s sub-shells

The rules for filling orbitals are covered in detail in *AS Chemistry* spread 1.2.3.

From scandium to zinc, the 3d orbitals are being filled. The pattern is regular except for chromium and copper, which do not follow the Aufbau principle for placing electrons in orbitals:

- chromium – the 3d and 4s orbitals all contain one electron with no orbital being completely filled
- copper – the 3d orbitals are full, but there is only one electron in the 4s orbital.

In these two elements it is suggested that electron repulsions between the outer electrons are minimised, resulting in an increased stability of the chromium and copper atoms.

- In chromium atoms, the 4s and 3d orbitals are all half-filled.
- In copper atoms, the 3d orbitals are all filled.

The electron configurations of the elements from scandium to zinc are shown in Figure 3.

Element	Z	Electron configuration	Noble gas configuration	Electron in box diagram					
Scandium	21	$1s^2\,2s^2\,2p^6\,3s^2\,3p^6\,4s^2\,3d^1$	[Ar] $4s^2\,3d^1$	↑↓	↑				
Titanium	22	$1s^2\,2s^2\,2p^6\,3s^2\,3p^6\,4s^2\,3d^2$	[Ar] $4s^2\,3d^2$	↑↓	↑	↑			
Vanadium	23	$1s^2\,2s^2\,2p^6\,3s^2\,3p^6\,4s^2\,3d^3$	[Ar] $4s^2\,3d^3$	↑↓	↑	↑	↑		
Chromium	24	$1s^2\,2s^2\,2p^6\,3s^2\,3p^6\,4s^1\,3d^5$	[Ar] $4s^1\,3d^5$	↑	↑	↑	↑	↑	↑
Manganese	25	$1s^2\,2s^2\,2p^6\,3s^2\,3p^6\,4s^2\,3d^5$	[Ar] $4s^2\,3d^5$	↑↓	↑	↑	↑	↑	↑
Iron	26	$1s^2\,2s^2\,2p^6\,3s^2\,3p^6\,4s^2\,3d^6$	[Ar] $4s^2\,3d^6$	↑↓	↑↓	↑	↑	↑	↑
Cobalt	27	$1s^2\,2s^2\,2p^6\,3s^2\,3p^6\,4s^2\,3d^7$	[Ar] $4s^2\,3d^7$	↑↓	↑↓	↑↓	↑	↑	↑
Nickel	28	$1s^2\,2s^2\,2p^6\,3s^2\,3p^6\,4s^2\,3d^8$	[Ar] $4s^2\,3d^8$	↑↓	↑↓	↑↓	↑↓	↑	↑
Copper	29	$1s^2\,2s^2\,2p^6\,3s^2\,3p^6\,4s^1\,3d^{10}$	[Ar] $4s^1\,3d^{10}$	↑	↑↓	↑↓	↑↓	↑↓	↑↓
Zinc	30	$1s^2\,2s^2\,2p^6\,3s^2\,3p^6\,4s^2\,3d^{10}$	[Ar] $4s^2\,3d^{10}$	↑↓	↑↓	↑↓	↑↓	↑↓	↑↓

Figure 3 Electron configurations of the elements scandium to zinc

The electron configurations of d-block ions

In their reactions, transition element atoms lose electrons to form positive ions. Transition metals lose their 4s electrons before the 3d electrons. This seems surprising because the 4s orbitals are filled first. The 3d and 4s energy levels are very close together and, once electrons occupy the orbitals, the 4s electrons have a higher energy and are lost first.

When writing the electron configurations of the ions, it is easier to start with the electron configuration of the elements and then remove the required number of electrons.

Questions

1 What is the difference between a d-block element and a transition element?

2 Write the electronic configurations of the following atoms and ions:
(a) Cr; (b) Mn^{2+}; (c) Sc^{3+}; (d) Ti.

Examiner tip

When writing the electron configurations, the official way is to show all the $n = 3$ sub-shells before the $n = 4$ energy level. In this book, the energy levels have been shown in order of sub-shell filling with the 4s before the 3d. Either response is perfectly acceptable in exams, so cobalt can be shown as:

$1s^22s^22p^63s^23p^64s^23d^7$

or $1s^22s^22p^63s^23p^63d^74s^2$

Worked example

An iron atom has 26 electrons and has an electronic configuration of $1s^22s^22p^63s^23p^64s^23d^6$.

- To form the 3+ ion an iron atom has to lose three electrons.
- Fe^{3+} will have an electron configuration of $1s^22s^22p^63s^23p^63d^5$.
- Fe has lost two electrons from the 4s sub-shell and one electron from the d sub-shell.

A copper atom has 29 electrons and its electron configuration is $1s^22s^22p^63s^23p^64s^13d^{10}$.

- To form the 2+ ion a Cu atom has to lose two electrons – one from the 4s sub-shell and one from the 3d sub-shell.
- Cu^{2+} has an electron configuration of $1s^22s^22p^63s^23p^63d^9$.

Examiner tip

When forming ions in transition element chemistry, the 4s electrons are always removed before any electrons are taken from the 3d sub-shells.

By the end of this spread, you should be able to ...

* Illustrate the existence of more than one oxidation state for a transition element in its compounds.
* Illustrate the formation of coloured metal ions.

Physical properties

The transition elements are all metals and display all of the normal metallic properties. They are shiny in appearance and have high densities, high melting points and high boiling points. When solid, transition metals exist as giant metallic lattices containing delocalised electrons, which move to conduct electricity.

Metallic properties account for many of the uses of the transition elements. For example, nickel is alloyed with copper for extensive use throughout the world for making 'silver' coins; titanium is a component of joint replacement parts, including the ball-and-socket joints used in hip replacements; iron is cast to make telephone boxes and post boxes, or alloyed for use in the construction of buildings and bridges.

Figure 1 The Statue of Liberty is made of pure copper sheeting, hung on a framework of steel. The exception is the flame of the torch, which is coated with gold leaf. The green appearance of the copper comes from copper(II) carbonate, formed by reaction of copper with atmospheric gases

Figure 2 Transition metals are used in a wide variety of applications

In addition, transition elements form compounds in which the transition metal has different oxidation states. These compounds form coloured solutions when dissolved in water and frequently catalyse chemical reactions. These properties are a result of the electron configurations of the transition elements – in particular, partially filled d-orbitals.

Variable oxidation states

The transition elements from titanium to copper all form ions with two or more oxidation states. They all form compounds with ions in the +2 oxidation state. In most cases, this is the result of losing the two electrons from the 4s orbital. The 4s electrons are lost first because they are in the highest occupied energy level. However, because the 3d and 4s energy levels are so close in energy, the 3d electrons can also be lost when an atom forms a stable ion.

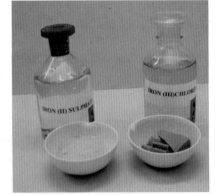

Figure 3 Coloured compounds of iron in different oxidation states

You need to be able to work out the oxidation numbers of the transition element ions found in compounds. To do this you need to use the rules for determining oxidation number introduced in *AS Chemistry* spread 1.1.14 – these rules are summarised in spread 2.2.9. The main oxidation numbers of the first row of the d-block elements are shown in Figure 4; the colours of the commonest ions are also shown.

Sc	Ti	V	Cr	Mn	Fe	Co	Ni	Cu	Zn
	+2	+2	+2	+2	+2	+2	+2	+2	+2
+3	+3	+3	+3	+3	+3	+3	+3	+3	
		+4	+4	+4	+4	+4	+4	+4	
		+5	+5	+5	+5	+5	+5		
			+6	+6	+6				
				+7					

Figure 4 Oxidation numbers and colours of the common d-block metal ions

The highest oxidation state of a transition element is often found in a strong oxidising agent. Manganese forms a compound called potassium permanganate, $KMnO_4$, a purple solid used as an oxidising agent in redox titrations. Chromium is found in potassium dichromate, $K_2Cr_2O_7$, an orange crystalline solid that acts as an oxidising agent in the preparation of aldehydes and ketones from alcohols. In these compounds, manganese and chromium have their maximum oxidation states of +7 and +6, respectively.

The systematic names of these compounds show the oxidation states:
* potassium manganate(VII), $KMnO_4$: oxidation number of Mn = +7
* potassium dichromate(VI), $K_2Cr_2O_7$: oxidation number of Cr = +6.

Coloured compounds

When white light passes through a solution containing transition metal ions, some of the wavelengths of visible light are absorbed. The colour that we observe is a mixture of the wavelengths of light that have *not* been absorbed. For example, a solution of copper(II) sulfate appears pale blue because the solution absorbs the red/orange region of the electromagnetic spectrum and reflects or transmits the blue.

Many coloured inorganic compounds contain transition metal ions. In fact, it is quite difficult to find coloured inorganic compounds that do not contain transition metal ions. Colour in inorganic chemistry is linked to the partially filled d-orbitals of transition metal ions.

Scandium(III) has the electron configuration $1s^2 2s^2 2p^6 3s^2 3p^6$ and is colourless in aqueous solution. Scandium(III) is formed from scandium ($1s^2 2s^2 2p^6 3s^2 3p^6 4s^2 3d^1$) by the loss of two 4s electrons and one 3d electron. There is no partially filled d-orbital and there is no colour.

Potassium dichromate   Potassium permanganate

Figure 5 Crystals of potassium manganate(VII) and potassium dichromate(VI)

Examiner tip

You are not required to *explain* the colour of transition metal ions in solutions. However, you should know some of the colours associated with the transition metal ions in solution.

Questions

1 State three properties of the transition elements.
2 Zinc is considered to be a d-block element and forms an ion Zn^{2+}. Suggest and explain the colour of solutions containing this ion.
3 What is the oxidation state of vanadium in V_2O_5 and in VO^{2+}?

By the end of this spread, you should be able to ...

* Illustrate the catalytic behaviour of the transition elements and/or their compounds.
* Describe the simple precipitation reactions of $Cu^{2+}(aq)$, $Co^{2+}(aq)$, $Fe^{2+}(aq)$ and $Fe^{3+}(aq)$ with aqueous sodium hydroxide.

Transition metals as catalysts

A catalyst is a substance that increases the rate of a chemical reaction by providing an alternative route for the reaction to follow – the alternative route has a lower activation energy. You will have already met some examples of catalysts during your AS chemistry course, and the basic principles of catalysis can be found in *AS Chemistry* spread 2.3.11.

Transition metals and their compounds are very effective catalysts. There are two main ways in which this catalysis takes place.

* Transition metals provide a surface on which a reaction can take place. Reactants are adsorbed onto the surface of the metal and held in place while a reaction occurs. After the reaction, the products are desorbed and the metal remains unchanged – see *AS Chemistry* spread 2.4.6.
* Transition metal ions have the ability to change oxidation states by gaining or losing electrons. They then bind to reactants forming intermediates as part of a chemical pathway with a lower activation energy.

Outlined below are some important chemical reactions that are catalysed by transition metals or their compounds.

Haber process

The Haber process is used to make ammonia, NH_3, from the reaction of nitrogen and hydrogen.

* $N_2(g) + 3H_2(g) \rightleftharpoons 2NH_3(g)$
* The catalyst is iron metal – it is used to increase the rate of reaction and to lower the temperature at which the reaction takes place.

Much of the ammonia produced in the Haber process is used in manufacturing fertilisers.

Contact process

The contact process is used to convert sulfur dioxide, SO_2, into sulfur trioxide, SO_3, in the manufacture of sulfuric acid, H_2SO_4.

* $2SO_2(g) + O_2(g) \rightleftharpoons 2SO_3(g)$
* The catalyst used is vanadium(V) oxide, V_2O_5, in which vanadium has the +5 oxidation state.

Sulfuric acid is an important inorganic chemical with many uses, including the production of fertilisers, detergents, adhesives and explosives, and as the electrolyte in car batteries.

Hydrogenation of alkenes

Hydrogen can be added across the C=C double bonds in unsaturated compounds to make saturated compounds – the process is called hydrogenation.

Figure 3 Hydrogenation of the C=C bond in alkenes

* The catalyst is nickel metal – it is used to lower the temperature and pressure needed to carry out the reaction.

Figure 1 View through the element of a catalytic converter from a car exhaust. The inner surface is coated with an alloy containing platinum, rhodium and palladium

Figure 2 Sample of vanadium(V) oxide – used as the catalyst in the contact process

This process is used in the hydrogenation of unsaturated vegetable oils to make spreadable margarines – see *AS Chemistry* spread 2.1.17.

Decomposition of hydrogen peroxide
Hydrogen peroxide decomposes slowly at room temperature and pressure into water and oxygen. A catalyst is added to increase the reaction rate.

* $2H_2O_2(aq) \longrightarrow 2H_2O(l) + O_2(g)$
* A suitable catalyst is manganese(IV) oxide, MnO_2, in which manganese has the +4 oxidation state. Manganese(IV) oxide is commonly called manganese dioxide.

The catalytic decomposition of hydrogen peroxide is often used in the laboratory as a simple and convenient preparation of oxygen gas.

Precipitation reactions
A precipitation reaction is one in which soluble ions, in separate solutions, are mixed together to produce an insoluble compound – this settles out of solution as a solid. The insoluble compound is called a precipitate.

Transition metal ions in aqueous solution react with aqueous sodium hydroxide, NaOH(aq), to form coloured precipitates. A blue solution of copper(II) ions reacts with aqueous sodium hydroxide to form a pale blue precipitate of copper(II) hydroxide, as shown in Figures 5 and 6. The equation for the reaction is:

$Cu^{2+}(aq) + 2OH^-(aq) \longrightarrow Cu(OH)_2(s)$
pale blue   pale blue
solution   precipitate

Cobalt(II), iron(II) and iron(III) ions also form precipitates with sodium hydroxide, and these reactions are detailed in the table below and in Figures 5 and 6.

Ion	Observation with NaOH	Equation
$Co^{2+}(aq)$	Pink solution containing $Co^{2+}(aq)$ reacts to form a blue precipitate – turning beige in the presence of air	$Co^{2+}(aq) + 2OH^-(aq) \longrightarrow Co(OH)_2(s)$ Precipitate is cobalt(II) hydroxide
$Fe^{2+}(aq)$	Pale green solution containing $Fe^{2+}(aq)$ forms a green precipitate – turning a rusty brown at its surface on standing in air*	$Fe^{2+}(aq) + 2OH^-(aq) \longrightarrow Fe(OH)_2(s)$ Precipitate is iron(II) hydroxide
$Fe^{3+}(aq)$	Pale yellow solution containing $Fe^{3+}(aq)$ forms a rusty-brown precipitate	$Fe^{3+}(aq) + 3OH^-(aq) \longrightarrow Fe(OH)_3(s)$ Precipitate is iron(III) hydroxide

* The precipitate changes colour because green Fe^{2+} ions are readily oxidised to rusty-brown Fe^{3+} ions.

Questions
1 What do you understand by the term *catalyst*?
2 Give an example of a chemical reaction that uses a transition metal compound as a catalyst.
3 A solution of an iron salt reacts with aqueous sodium hydroxide to form a red-brown precipitate. What is the iron ion present? What is its oxidation number?
4 Nickel is a transition element that forms an ion with an oxidation state of +2. Write an equation for the reaction of an aqueous solution containing this ion with aqueous sodium hydroxide.

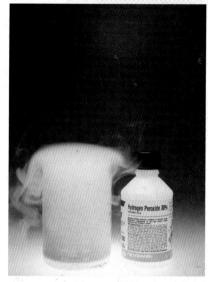

Figure 4 Hydrogen peroxide is rapidly decomposed into water and oxygen in the presence of manganese dioxide catalyst

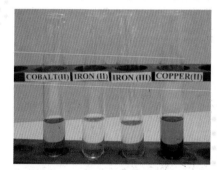

Figure 5 Solutions of $Co^{2+}(aq)$, $Fe^{2+}(aq)$, $Fe^{3+}(aq)$ and $Cu^{2+}(aq)$

Figure 6 Precipitates of $Co(OH)_2(s)$, $Fe(OH)_2(s)$, $Fe(OH)_3(s)$ and $Cu(OH)_2(s)$

By the end of this spread, you should be able to ...

* Explain the term *ligand* and its role in coordinate bonding.
* Describe and use the terms *complex ion* and *coordination number*.
* State and give examples of complexes with six-fold coordination with an octahedral shape.

Complex ions

In solution, the ions of transition metals form complex ions. A complex ion consists of a central metal ion surrounded by ligands.

* A **ligand** is a molecule or ion which donates a pair of electrons to the central metal ion to form a coordinate or dative covalent bond.
* A **coordinate bond** is one in which one of the bonded atoms provides both electrons for the shared pair. A coordinate bond is the name commonly used in transition metal chemistry for a dative covalent bond.
* The transition metal ion accepts the pair of electrons from the ligand in forming the coordinate bond.

An example of a complex ion is $[Cu(H_2O)_6]^{2+}$. The central metal ion is Cu^{2+} and the ligands are molecules of water, H_2O. Each water molecule donates a lone pair of electrons from its oxygen atom to the Cu^{2+} ion to form a coordinate bond.

In a complex ion, the **coordination number** indicates the number of coordinate bonds to the central metal ion. In the complex ion $[Cu(H_2O)_6]^{2+}$, the coordination number is 6 because there are six coordinate bonds in total from the ligands to the Cu^{2+} ion.

In the formula of the complex ion:
* square brackets group together the species making up the complex ion
* the overall charge is shown outside the brackets.

The overall charge is the sum of the individual charges of the transition metal ion and those of the ligands present in the complex. The six coordinate bonds in the complex ion $Fe(H_2O)_6]^{2+}$ are shown in Figure 1.

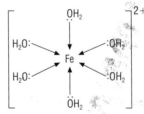

Figure 1 Complex ion $[Fe(H_2O)_6]^{2+}$ – each of the six H_2O ligands forms one coordinate (dative covalent) bond to the central metal ion. The shapes of six-coordinate complex ions are discussed on the next page

Common ligands

All ligands have one or more lone pairs of electrons in their outer energy level. Some ligands are neutral and carry no charge – other ligands are negatively charged. Each of the ligands in Figure 2 is a **monodentate ligand** – this means that the ligand donates just one pair of electrons to the central metal ion to form one coordinate bond.

> ### Key definitions
>
> A **complex ion** is a transition metal ion bonded to one or more ligands by coordinate bonds (dative covalent bonds).
>
> A **ligand** is a molecule or ion that can donate a pair of electrons with the transition metal ion to form a coordinate bond.
>
> The **coordination number** is the total number of coordinate bonds formed between a central metal ion and its ligands.

Name of ligand	Formula	Charge
Water	$:OH_2$	None – neutral ligand
Ammonia	$:NH_3$	None – neutral ligand
Thiocyanate	$:SCN^-$	−1
Cyanide	$:CN^-$	−1
Chloride	$:Cl^-$	−1
Hydroxide	$:OH^-$	−1

Worked example

State the formula and charge of the complex ion made from one titanium(III) ion and six water molecules.

$[Ti(H_2O)_6]^{3+}$

The formula shows that six H_2O ligands are bonded to one titanium(III) ion. As water is a neutral ligand, the overall charge on the ion is the same as the transition metal ion, which is +3.

Worked example

What is the oxidation number of the transition metal in the complex ion $[Co(H_2O)_5Cl]^+$?

This complex ion has five neutral water ligands with no charge, and one chloride ligand with a charge of -1.

Overall the complex ion has a 1+ positive charge.

Co	$(H_2O)_5$	Cl^-	Overall charge
?	5×0	-1	$+1$

In $[Co(H_2O)_5Cl]^+$, cobalt must have the oxidation state +2 to give the overall charge of +1.

Shapes of complex ions

Complex ions exist in a number of different shapes, but by far the most common is the octahedral shape. Octahedral complexes have six coordinate bonds attached to the central ion. The outer face of the shape is an eight-sided octahedron. In an octahedral complex, four of the ligands are in the same plane, one ligand is above the plane and the remaining ligand below the plane. All bond angles are 90°.

When cobalt(II) salts are dissolved in water, a pale pink solution is formed containing the octahedral complex ion $[Co(H_2O)_6]^{2+}$ – Figure 2 shows its shape.

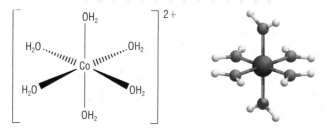

Figure 2 $[Co(H_2O)_6]^{2+}$ is an octahedral complex ion

In the diagram above:
- the solid wedge bonds come out of the plane of the paper, towards you
- the hatched wedge bonds go into the plane of the paper, away from you
- the ligands attached by the wedge bonds are at the corners of a square, with each bond separated by 90°
- the solid lines represent bonds in the plane of the paper
- the 'solid bonds' are separated from each wedge bond by 90°.

Questions

1   State the formula and charge of each of these complex ions:
    (a) one cobalt(II) ion and four chloride ions;
    (b) one iron(III) ion, five water molecules and a chloride ion.
2   In the complex ion $[Cu(NH_3)_4(H_2O)_2]^{2+}$, ammonia molecules and water molecules are both ligands. Why can water behave as a ligand?
3   Using any stated complex ion, show what is meant by the term *coordination number*.

By the end of this spread, you should be able to ...

* Describe stereoisomerism in transition metal complexes using examples of *cis–trans* and optical isomerism.
* Describe the use of *cis*-platin as an anti-cancer drug, and its action by binding to DNA.

What is a stereoisomer?

You met isomerism in *AS Chemistry* spread 2.1.7, and stereoisomerism and optical isomerism in spread 1.2.3 of this book. The focus there was on organic chemistry, but isomerism is also found among the complexes of the transition elements.

There are two types of stereoisomerism in transition element chemistry:
* *cis–trans* isomerism
* optical isomerism.

The definition for stereoisomerism must be broadened to accommodate both organic molecules and transition metal complexes:
* stereoisomers are molecules or complexes with the same structural formula but with a different spatial arrangement of these atoms.

In transition element chemistry, it is possible to have stereoisomers from complexes containing all types of ligands – monodentate, bidentate and multidentate. You will meet bidentate and multidentate ligands in spread 2.3.6.

Cis–trans isomerism

Some octahedral complex ions contain two different ligands, with four of one ligand and two of another. These complex ions can exist as *cis* and *trans* isomers.

The earliest examples of stereoisomerism involved Co^{3+} complex ions. In 1889, Sophus Mads Jorgensen, a Danish scientist, observed that some complex ions could form salts that had different colours.

However it was Alfred Werner, a Swiss chemist, who succeeded in explaining the existence of two isomers of $[Co(NH_3)_4Cl_2]^+$. Werner proposed that there were two isomers of this ion, one of which was purple and the other green. He suggested that the cobalt(III) ion is surrounded by four NH_3 ligands and two Cl^- ligands at the corners of an octahedron.

He called the purple isomer '*cis*':
* the two Cl^- ligands are at adjacent corners of the octahedron
* the two Cl^- ligands are at 90° to one another.

He called the green isomer '*trans*':
* the two Cl^- ligands are at opposite corners of the octahedron
* the two Cl^- ligands are at 180° to one another.

The *cis* and *trans* isomers of $[Co(NH_3)_4Cl_2]^+$ are shown in Figure 2.

Figure 1 Alfred Werner, one of the pioneers of transition element chemistry

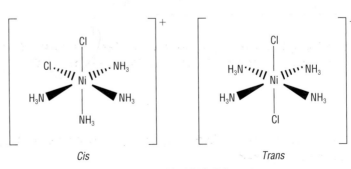

Figure 2 *Cis–trans* isomerism in $[Co(NH_3)_4Cl_2]^+$

Cis–trans isomerism in four-coordinate complexes

Cis–trans isomerism is also possible in some four-coordinate complexes with a square planar shape. In this shape, the ligands are arranged at the corners of a square. This structure is rather like an octahedral complex, but without the ligands above and below the plane. For *cis–trans* isomerism, the complex must contain two different ligands, with two of one ligand and two of another. Figure 3 shows the structures of the *cis* and *trans* isomers of $[NiCl_2(NH_3)_2]$.

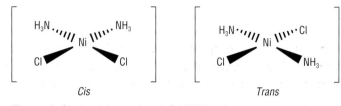

Figure 3 *Cis–trans* isomerism in $[NiCl_2(NH_3)_2]$

Transition metal complexes in medicine

Cis-platin is one of the most effective drugs against many forms of cancer. *Cis*-platin is the *cis*-isomer of a platinum complex, $[PtCl_2(NH_3)_2]$.

Cells divide to make more cells of the same type. Before cell division, a copy is made of the DNA present in the cell. After cell division, two new cells are formed, each containing the same DNA. Cancer is spread throughout an organism by cell division of a cancerous cell that can reproduce itself. It is believed that *cis*-platin acts by binding to the DNA of fast-growing cancer cells. This alters the DNA in the cancerous cells. It is generally believed that the cells are prevented from reproducing by these changes to the DNA structure. Activation of a cell's own repair mechanism eventually leads to the death of the cancer-containing cell. There are also several other theories for the action of *cis*-platin – showing that the science involved is currently uncertain.

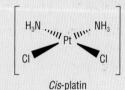

Cis-platin

Figure 4 Structure of *cis*-platin

The use of *cis*-platin is the basis for modern chemotherapy treatment of cancer. Unfortunately, treatment is associated with some unpleasant side effects.

Cis-platin is a very effective drug, but scientists have now developed a new generation of compounds with lower doses and fewer side effects. The most common of these is carboplatin, which was developed in the late 1980s to treat advanced ovarian cancer. It slows cancer growth by reacting with a cell's DNA.

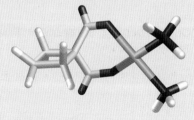

Figure 5 Structure of carboplatin. Can you spot where the platinum is? Do some research to find out what the other atoms are in the structure

Questions

1 What kind of stereoisomers can be formed by the complex ion $[Ru(H_2O)_4Cl_2]^+$? Draw diagrams to illustrate your answer.
2 Define the term *stereoisomers*.

By the end of this spread, you should be able to . . .

* Explain and use the term *bidentate ligand* (e.g. NH₂CH₂CH₂NH₂, 'en').
* Describe stereoisomerism in transition element multidentate complexes using examples of *cis–trans* and optical isomerism.

Bidentate ligands

Some ligands are **bidentate** – this means that the ligand can donate two lone pairs of electrons to the central metal ion to form two coordinate bonds. The most common bidentate ligand is ethane-1,2-diamine, $NH_2CH_2CH_2NH_2$, often shortened to 'en.' Each nitrogen atom in a molecule of $NH_2CH_2CH_2NH_2$ donates its lone pair to the metal ion.

Bidentate ligands can also form octahedral complexes. Figure 1 shows $[Ni(en)_3]^{2+}$, an example of an octahedral transition metal ion containing a bidentate ligand:

* each 'en' ligand forms two coordinate bonds to the central metal ion
* there are three 'en' ligands giving a total of $3 \times 2 = 6$ coordinate bonds
* the coordination number is 6.

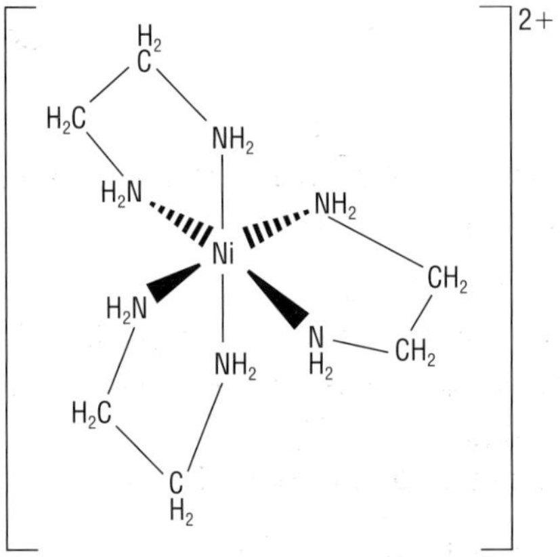

Figure 1 Complex ion $[Ni(en)_3]^{2+}$
(or $[Ni(NH_2CH_2CH_2NH_2)_3]^{2+}$)

Cis–trans isomerism from bidentate ligands

Octahedral complexes containing bidentate ligands can also show *cis–trans* isomerism. The complex ion $[Cr(C_2O_4)_2(H_2O)_2]^-$ has *cis* and *trans* isomers.

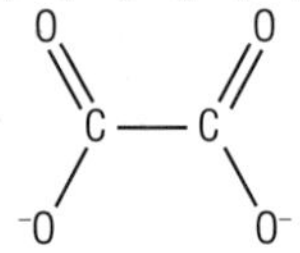

Figure 2 shows the ethanedioate ion, $C_2O_4^{2-}$, which is the bidentate ligand present in this complex ion.

The *cis* and *trans* isomers of $[Cr(C_2O_4)_2(H_2O)_2]^-$ are shown in Figure 3.

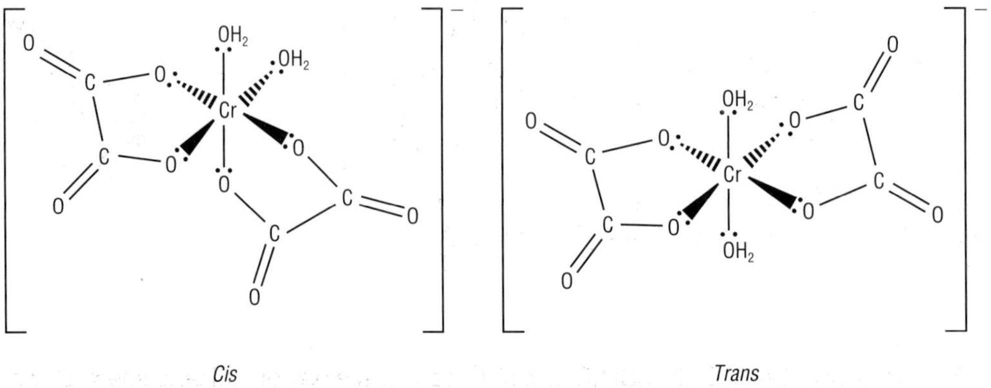

Figure 2 Structure of the ethanedioate ligand, $C_2O_4^{2-}$. Each O^- in the ion can form a coordinate bond with a transition metal ion

Cis   Trans

Figure 3 *Cis* and *trans* isomers of $[Cr(C_2O_4)_2(H_2O)_2]^-$

A hexadentate ligand

A hexadentate ligand has six lone pairs of electrons, each forming a coordinate bond to a metal ion in a complex ion. One example of a hexadentate ligand is ethylenediaminetetraacetic acid, shortened to EDTA. EDTA exists in complexes as the ion $EDTA^{4-}$.

EDTA is used to bind metal ions and is known as a chelating agent. This means that EDTA decreases the concentration of metal ions in solutions by binding them into a complex.

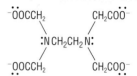

Figure 4 $EDTA^{4-}$ – note the six pairs of electrons, each of which can form a coordinate bond

Uses of EDTA

The ability of EDTA to bind to metals is exploited in industry, and EDTA is used in many different consumer products:
- in detergents – it binds to calcium and magnesium ions to reduce hardness in water
- in some foods – as a stabiliser to remove metal ions that might catalyse the oxidation of the product
- in medical applications – it is added to blood samples to prevent clotting and used to treat patients suffering from lead and mercury poisoning.

Optical isomers

Optical isomerism in transition element chemistry is associated with octahedral complexes that contain multidentate ligands.

The requirements for optical isomerism are:
- a complex with three molecules or ions of a bidentate ligand – e.g. $[Ni(en)_3]^{2+}$
- a complex with two molecules or ions of a bidentate ligand, and two molecules or ions of a monodentate ligand – e.g. $[Co(en)_2Cl_2]$
- a complex with one hexadentate ligand – e.g. $[Cu(EDTA)]^{2-}$

Optical isomers are non-superimposable mirror images of each other (see spread 1.2.3). Optical isomers rotate plane-polarised light differently – one of the isomers rotates the light clockwise and the other rotates the light anti-clockwise. A mixture containing equal amounts of the two isomers has no effect on plane-polarised light because the rotations cancel out.

Optical isomerism can be seen in the structures of the two isomers of $[Ni(NH_2CH_2CH_2NH_2)_3]^{2+}$, shown in Figure 5.

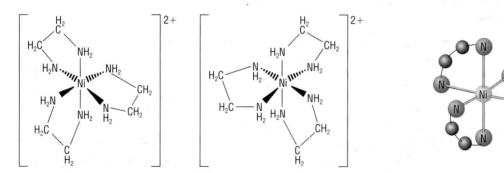

Figure 5 Optical isomers of $[Ni(NH_2CH_2CH_2NH_2)_3]^{2+}$

Question

1 The compound $[Co(H_2O)_2(NH_2CH_2CH_2NH_2)_2]^{2+}$ can form a pair of optical isomers. Draw diagrams to show the two isomers.

By the end of this spread, you should be able to . . .

* Describe the process of ligand substitution.
* Describe examples of ligand substitution using $[Cu(H_2O)_6]^{2+}$ and $[Co(H_2O)_6]^{2+}$ with ammonia and chloride ions.

Ligand substitution reactions

A **ligand substitution** reaction is one in which a ligand in a complex ion is replaced by another ligand. In many ligand substitution reactions, water molecules in an aqueous solution of a complex ion are replaced by another ligand.

The reaction of aqueous copper(II) ions and ammonia

An aqueous solution of copper(II) ions contains $[Cu(H_2O)_6]^{2+}$ complex ions which have a characteristic pale blue colour. When an excess of aqueous ammonia solution is added, the pale blue solution changes colour and a deep blue solution forms. This can be represented by the equilibrium equation:

$$[Cu(H_2O)_6]^{2+}(aq) + 4NH_3(aq) \rightleftharpoons [Cu(NH_3)_4(H_2O)_2]^{2+}(aq) + 4H_2O(l)$$
pale blue solution   deep blue solution

In the reaction, four of the water ligands are replaced by four ammonia ligands. Ligand substitution has taken place because one type of ligand has been replaced by another ligand.

The product, $[Cu(NH_3)_4(H_2O)_2]^{2+}$, has six ligands is an octahedral complex ion. The shape of the product is shown in Figure 1. The copper–oxygen bonds are longer than the copper–nitrogen bonds, so the shape is strictly described as a distorted octahedral shape.

If you carry out this experiment in the laboratory you will observe two changes. On addition of a small amount of ammonia, a pale blue precipitate of copper(II) hydroxide, $Cu(OH)_2$, forms. This is the precipitation reaction that you saw in spread 2.3.3. On addition of an excess of aqueous ammonia, the pale blue precipitate dissolves and a deep blue solution is formed containing $[Cu(NH_3)_4(H_2O)_2]^{2+}$ ions. These colour changes are shown in Figure 2.

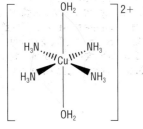

Figure 1 Octahedral shape of $[Cu(NH_3)_4(H_2O)_2]^{2+}$

$[Cu(H_2O)_6]^{2+}$ $Cu(OH)_2$ $[Cu(NH_3)_4(H_2O)_2]^{2+}$

Figure 2 Stepwise reaction of aqueous copper(II) ions with aqueous ammonia

The reaction of copper(II) ions and hydrochloric acid

When concentrated hydrochloric acid is added to an aqueous solution containing copper(II) ions, the pale blue solution initially forms a green solution before finally turning yellow. This reaction exists in equilibrium and can be reversed by adding water to the yellow solution to return it to the original blue colour.

This can be represented by the equilibrium equation:

$$[Cu(H_2O)_6]^{2+}(aq) + 4Cl^-(aq) \rightleftharpoons [CuCl_4]^{2-}(aq) + 6H_2O(l)$$
$$\text{pale blue solution} \qquad\qquad \text{yellow solution}$$

Note that the $[Cu(H_2O)_6]^{2+}$ complex ion has six ligands, but the $[CuCl_4]^{2-}$ complex ion has only four ligands. This occurs because the chloride ligands are larger than the water ligands and have stronger repulsions, so fewer chloride ligands can fit around the central metal ion. $[CuCl_4]^{2-}$ has a tetrahedral shape, as in a CH_4 molecule.

Because the reaction is an equilibrium system, you can apply le Chatelier's principle to predict how the colour of the solution changes when the reaction conditions are modified.

The colours in this equilibrium are shown in Figure 3.

$$[Cu(H_2O)_6]^{2+} + 4Cl^- \rightleftharpoons CuCl_4^{2-} + 6H_2O$$

Figure 3 Equilibrium reaction of aqueous copper(II) ions with chloride ions

The reaction of cobalt(II) ions and concentrated hydrochloric acid

An aqueous solution of cobalt(II) ions contains $[Co(H_2O)_6]^{2+}$ complex ions, which have a characteristic pale pink colour. If concentrated hydrochloric acid is added, the solution forms a dark blue solution. The concentrated hydrochloric acid providing a high concentration of chloride ions. The six water molecules in the complex ion are replaced by four chloride ions.

The reaction is reversible and can be represented by the equilibrium equation:

$$[Co(H_2O)_6]^{2+}(aq) + 4Cl^-(aq) \rightleftharpoons [CoCl_4]^{2-}(aq) + 6H_2O(l)$$
$$\text{pink solution} \qquad\qquad \text{blue solution}$$

The changes in this equilibrium are shown in Figure 4.

$$[Co(H_2O)_6]^{2+} + 4Cl^- \rightleftharpoons CoCl_4^{2-} + 6H_2O$$

Figure 4 Equilibrium reaction of aqueous cobalt(II) ions with chloride ions

> **Examiner tip**
>
> The green solution often observed at the end of the reaction is the mixing of the blue and yellow colours in the equilibrium system. Some students are impatient and fail to add the chloride to excess. If excess chloride ions are added, the solution will appear yellow – and that is the correct observation!

Questions

1 What do you understand by the term *ligand substitution*?

2 A solution of ammonia was slowly added to an aqueous solution containing copper(II) ions until the ammonia was in excess. Initially a pale blue precipitate formed, followed by the formation of a deep blue solution.
   (a) Identify the pale blue precipitate and write an equation for its formation.
   (b) Write the formula of the complex ion in the deep blue solution.

3 A test tube holding the equilibrium system shown below was placed in a beaker of ice and the colour of the solution changed from blue to pink.

$$[Co(H_2O)_6]^{2+}(aq) + 4Cl^-(aq) \rightleftharpoons [CoCl_4]^{2-}(aq) + 6H_2O(l)$$

On heating, the solution turned blue. Using these observations, state whether the forward reaction is exothermic or endothermic. Explain your answer.

2.3 (8) Ligand substitution and stability constants

By the end of this spread, you should be able to ...

* Explain the biochemical importance of iron in haemoglobin, including ligand substitution.
* Use and understand the term *stability constant*, K_{stab}.
* Deduce expressions for K_{stab} of ligand substitutions and understand that a large K_{stab} results in formation of a stable complex ion.

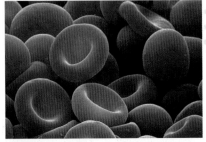

Figure 1 Scanning electron micrograph of red blood cells – these biconcave, disc-shaped cells transport oxygen from the lungs to all the cells in the body

Examiner tip

For information only: a prosthetic group is a non-protein component of a complete protein that is important for the protein's biological activity.

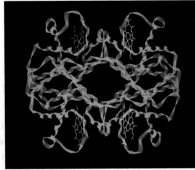

Figure 2 Haemoglobin is an oxygen-carrying molecule found in red blood cells. It consists of four polypeptide chains (blue); each chain carries a haem complex (purple) capable of reversibly binding oxygen. At its centre, the haem group contains an iron(II) ion, which binds one oxygen molecule

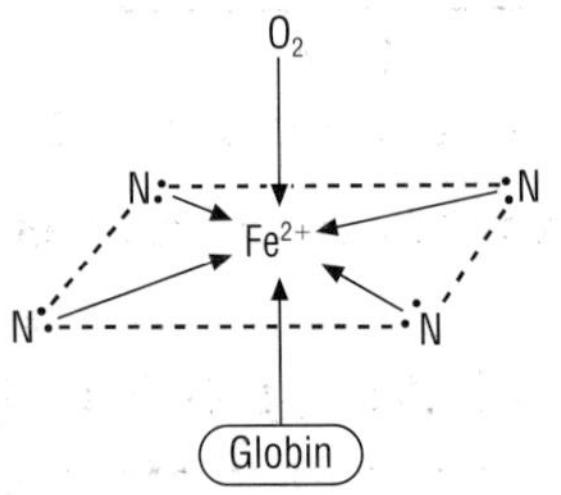

Figure 3 Part of the structure of haemoglobin showing the coordinate bonds formed to the central metal ion

Haemoglobin and ligand substitution

A tiny drop of blood contains millions of red blood cells, which are responsible for carrying oxygen around the body. Red blood cells carry oxygen efficiently because they contain haemoglobin, a complex protein composed of four polypeptide chains.

Each protein chain contains four non-protein components called haem groups. Each haem group has an Fe^{2+} ion at its centre – oxygen can bind to the Fe^{2+} ion. This allows haemoglobin to carry out its function of transporting oxygen around the body.

When the Fe^{2+} ion in each haem group is bound to oxygen, the haem group is red in colour. As blood passes through the lungs, the haemoglobin picks up oxygen and carries it to the cells, where it can be released. Red blood cells can also pick up carbon dioxide as a waste product to be transported back to the lungs, where it is released.

Figure 3 shows the coordinate bonds formed around the Fe^{2+} ions in a haem group.

- There are four coordinate bonds between the Fe^{2+} ion and the nitrogen atoms in the haem structure.
- A further coordinate bond is formed to the protein globin.
- A final coordinate bond can form to an oxygen molecule, which is then transported.

Carbon monoxide – the silent killer?

Carbon monoxide and oxygen can both bind to haemoglobin at the same binding site. Unfortunately, carbon monoxide binds more strongly to the haemoglobin than oxygen. If both carbon monoxide and oxygen are present in the lungs, carbon monoxide will bind to the haemoglobin – leaving fewer haemoglobin molecules to bind to oxygen molecules. Tissues can then be starved of oxygen because carbon monoxide rather than oxygen is being carried round the body. The reaction is an example of ligand substitution because carbon monoxide molecules can replace oxygen molecules in haemoglobin.

To make matters worse, this ligand substitution is irreversible – once on a haem group, a carbon monoxide molecule cannot be removed and the haemoglobin is useless.

Carbon monoxide is formed during the incomplete combustion of carbon-containing fuels. It is a colourless and odourless gas, so we have no warning if it is present in the air around us. Sensors, rather like smoke detectors, are available to give some warning of this silent killer. Low levels of carbon monoxide in the blood can cause headaches, nausea and potential suffocation. Burning tobacco releases the gas and this is one of the reasons why long-term smokers become short of breath.

Stability constants

In spread 2.1.8, you learned how to write expressions for the equilibrium constant, K_c, in terms of concentration. When chloride ions are added to an aqueous solution of some transition metal ions, an equilibrium is set up. This example involves $[Co(H_2O)_6]^{2+}$ ions:

$$[Co(H_2O)_6]^{2+}(aq) + 4Cl^-(aq) \rightleftharpoons [CoCl_4]^{2-}(aq) + 6H_2O(l)$$

An expression for K_c can be written for this reaction. However, we use the term **stability constant, K_{stab}**, rather than K_c.

$$K_{stab} = \frac{[CoCl_4^{2-}]}{[[Co(H_2O)_6]^{2+}][Cl^-]^4}$$

Note that water has been left out of the expression for K_{stab}. This is common practice because all the species are dissolved in water, which is in large excess and its concentration is virtually constant.

Worked example 1

Write an expression for the stability constant, K_{stab}, for the following equilibrium and deduce the units for K_{stab}.

$$[Cu(H_2O)_6]^{2+}(aq) + 4NH_3(aq) \rightleftharpoons [Cu(NH_3)_4(H_2O)_2]^{2+}(aq) + 4H_2O(l)$$

$$K_{stab} = \frac{[[Cu(NH_3)_4(H_2O)_2]^{2+}]}{[[Cu(H_2O)_6]^{2+}][NH_3]^4} \qquad \frac{mol\ dm^{-3}}{(mol\ dm^{-3})(mol\ dm^{-3})^4} = dm^{12}\ mol^{-4}$$

Key definition

The **stability constant**, K_{stab}, is the equilibrium constant for an equilibrium existing between a transition metal ion surrounded by water ligands and the complex formed when the same ion has undergone a ligand substitution.

The value of K_{stab}

A large value of K_{stab} indicates that the position of an equilibrium lies to the right. Complex ions with high stability constants are more stable than those with lower values. A large value of a stability constant shows that the ion is easily formed.

Tables of K_{stab} values are available. These usually compare the stability of complexes with those containing H_2O ligands.

It is possible to calculate values for K_{stab} and for the concentrations of the ions in a solution given appropriate data.

Worked example 2

Two complexes $[Co(NH_3)_6]^{3+}$ and $[Ni(NH_3)_6]^{2+}$ have stability constants of 1.34×10^{23} and 8.96×10^8 dm^{18} mol^{-6}, respectively. Explain which of the complexes is the more stable.

$[Co(NH_3)_6]^{3+}$ is the more stable complex because it has the larger stability constant. This means that $[Co(NH_3)_6]^{3+}$ is more easily formed than the $[Ni(NH_3)_6]^{2+}$ complex.

Worked example 3

The following system is set up and allowed to reach equilibrium:

$$[Cu(H_2O)_6]^{2+}(aq) + 4Cl^-(aq) \rightleftharpoons [CuCl_4]^{2-}(aq) + 6H_2O(l)$$

At 25 °C, the equilibrium mixture contained 1.17×10^{-5} mol dm^{-3} $[Cu(H_2O)_6]^{2+}$ and 0.800 mol dm^{-3} Cl^-. $K_{stab} = 4.17 \times 10^5$ dm^{12} mol^{-4}.

Write an expression for K_{stab} and calculate the concentration of the $[CuCl_4]^{2-}$ at equilibrium.

$$K_{stab} = \frac{[[CuCl_4]^{2-}]}{[[Cu(H_2O)_6]^{2+}][Cl^-]^4}$$

$$4.17 \times 10^5\ dm^{12}\ mol^{-4} = \frac{[[CuCl_4]^{2-}]}{1.17 \times 10^{-5}\ mol\ dm^{-3} \times (0.800\ mol\ dm^{-3})^4}$$

$\therefore$ concentration of $[CuCl_4]^{2-} = 2.00$ mol dm^{-3}

STRETCH and CHALLENGE

Why are multidentate complexes so stable?
Bidentate and multidentate ligands, such as EDTA, form particularly stable complex ions. The additional stability of these complex ions results from an increase in entropy.

Ligand substitution by EDTA is accompanied by a large increase in the total number of moles, which increases the entropy:

$$[Cu(H_2O)_6]^{2+} + edta^{4-} \rightleftharpoons [Cu(edta)]^{2-} + 6H_2O$$

2 mol	7 mol

$K_{stab} = 6.3 \times 10^{18}$ dm^3 mol^{-1} or log K = 18.8 at 25 °C

The K_{stab} value is much greater than ligand substitution with NH_3.

$$[Cu(H_2O)_6]^{2+} + 4NH_3(aq) \rightleftharpoons [Cu(NH_3)_4(H_2O)_2]^{2+} + 4H_2O(l)$$

5 mol	5 mol

$K_{stab} = 1.2 \times 10^{13}$ dm^{12} mol^{-4} or log K = 13.1 at 25 °C

Questions

1 How many molecules of oxygen can one molecule of haemoglobin carry?

2 Write an expression for K_{stab} for the following equilibrium system:

$$[Ni(H_2O)_6]^{2+}(aq) + 6NH_3(aq) \rightleftharpoons [Ni(NH_3)_6]^{2+}(aq) + 6H_2O(l)$$

3 Which of the following complexes is the more stable?

$[Co(NH_3)_6]^{2+}$, $K_{stab} = 1.34 \times 10^5$ dm^{18} mol^{-6},
or $[Ni(NH_3)_6]^{2+}$, $K_{stab} = 8.96 \times 10^8$ dm^{18} mol^{-6}

⑨ Redox titrations

By the end of this spread, you should be able to ...

✳ Describe redox behaviour in transition elements using suitable examples.
✳ Carry out redox titrations and carry out structured calculations involving MnO_4^-.

Oxidation and reduction in transition element chemistry

In spread 2.3.2 you saw that transition elements form compounds with variable oxidation states. This means that transition metal ions can take part in oxidation and reduction reactions as their ions change from one oxidation state to another.

MnO_4^- and Fe^{2+}

Iron has two common oxidation states in its compounds, Fe^{2+} and Fe^{3+}. The iron(II) oxidation state is less stable than the iron(III) oxidation state. In the presence of air, or when in contact with another oxidising agent, iron(II) is readily oxidised to iron(III):

$$Fe^{2+}(aq) \longrightarrow Fe^{3+}(aq) + e^-$$

Manganese exists in a number of oxidation states, but the most important are the +2 and +7 oxidation states. MnO_4^- is a strong oxidising agent. In acidic solution, MnO_4^- can be reduced to form Mn^{2+}:

$$MnO_4^-(aq) + 8H^+(aq) + 5e^- \longrightarrow Mn^{2+}(aq) + 4H_2O(l)$$

MnO_4^- is commonly used to oxidise solutions containing iron(II), but is also used as an oxidising agent in many other chemical reactions. The reaction between iron(II) and acidified manganate(VII) is shown below:

$$MnO_4^-(aq) + 8H^+(aq) + 5Fe^{2+}(aq) \longrightarrow Mn^{2+}(aq) + 5Fe^{3+}(aq) + 4H_2O(l)$$

We can use these and other redox reactions to estimate the concentration of a transition metal ion in an unknown solution, or to calculate the percentage composition of a metal in a solid sample of a compound or alloy.

Carrying out redox titrations

Redox titrations are carried out in a similar way to acid–base titrations.

- An acid–base reaction involves the transfer of protons from the acid to the base and the formation of water and a salt.
- A redox titration involves the transfer of electrons from one species to another.

Just as an acid can be titrated against a base, an oxidising agent can be titrated against a reducing agent. You can find out more about titration techniques and the calculations that follow in *AS Chemistry* spread 1.1.13.

In an acid–base reaction, the end point is identified using an indicator, which appears as one colour in the acid solution and as a different colour in the alkaline solution. Indicators can also be used in some redox titrations – but in some reactions there is a colour change that provides the end point without the need for an indicator.

Aqueous potassium manganate(VII) contains $MnO_4^-(aq)$ and is a common oxidising agent that is self-indicating:

- $MnO_4^-(aq) \longrightarrow Mn^{2+}(aq)$
  purple   almost colourless

$Mn^{2+}(aq)$ ions are very pale pink in colour. In manganate(VII) titrations, the solutions used are very dilute and the $Mn^{2+}(aq)$ ions are present in low concentrations. This means that the very pale pink colour from $Mn^{2+}(aq)$ cannot be seen and the solution appears colourless. The $MnO_4^-(aq)$ ions have such a deep purple colour that this masks any other colour present.

Figure 1 Colours of solutions containing $MnO_4^-(aq)$ (left) and $Mn^{2+}(aq)$ ions

Redox titrations involving MnO_4^- ions

MnO_4^- is typically used to oxidise solutions containing iron(II) ions. The solution is acidified with sulfuric acid. Hydrochloric acid cannot be used because it reacts with MnO_4^-. This would not only invalidate the titration results but would also cause you some distress – chlorine gas is given off!

Typically these reactions are carried out with potassium manganate(VII) solution in the burette and the iron(II) solution in the conical flask. As the manganate(VII) solution is added to the iron(II) solution, it is decolourised. The end point of the titration is when excess MnO_4^- ions are present – this shows up as the first hint of a permanent pink colour in the solution in the conical flask.

Figure 2 Measuring the concentration of iron(II) ions in a solution of iron(II) sulfate. You can see the purple colour forming as the MnO_4^- runs into the acidified $Fe^{2+}(aq)$ solution in the conical flask

Worked example

$25.0~cm^3$ of a solution of an iron(II) salt required $23.00~cm^3$ of $0.0200~mol~dm^{-3}$ potassium manganate(VII) for complete oxidation in acid solution.
(a) Calculate the amount, in moles, of manganate(VII), MnO_4^-, used in the titration.
(b) Deduce the amount, in moles, of iron(II), Fe^{2+}, in the solution that was titrated.
(c) Calculate the concentration, in mol dm^{-3}, of Fe^{2+} in the solution that was titrated.

(a) Calculate the amount, in mol, of MnO_4^- that reacted.

$$n(MnO_4^-) = \frac{c \times V}{1000} = 0.0200 \times \frac{23.00}{1000} = 4.60 \times 10^{-4}~mol$$

(b) Calculate the amount, in mol, of Fe^{2+}, that was used in the titration.

equation	$MnO_4^-(aq)$ reacts with 5 $Fe^{2+}(aq)$	
moles	1 mol	5 mol
actual moles	4.60×10^{-4} mol	2.30×10^{-3} mol

(c) Calculate the concentration, in mol dm^{-3}, of Fe^{2+} in the solution.

$$c(Fe^{2+}) = \frac{n \times 1000}{V} = \frac{2.30 \times 10^{-3} \times 1000}{25.0} = 0.0920~mol~dm^{-3}$$

Examiner tip

For **(a)** we use amount, $n = c \times \dfrac{V}{1000}$

For **(b)** we use the balanced equation (see previous page) to work out the reacting quantities of MnO_4^- and Fe^{2+}:

$$MnO_4^-(aq) \equiv 5Fe^{2+}(aq)$$

For **(c)**, we rearrange: $n = c \times \dfrac{V}{1000}$

to get $c = \dfrac{n \times 1000}{V}$

Questions

1 MnO_4^- is used as an oxidising agent in redox titrations.
   (a) What do you understand by the term *redox reaction*?
   (b) What is the oxidation state of Mn in MnO_4^-?
2 $25.0~cm^3$ of $0.150~mol~dm^{-3}~Fe^{2+}$ required $32.50~cm^3$ of potassium manganate(VII) for complete oxidation in acid solution.
   Calculate the concentration of the potassium manganate(VII) solution used in the reaction.

By the end of this spread, you should be able to ...

✳ **Understand how to carry out redox titrations and carry out structured calculations involving MnO_4^- and $I_2/S_2O_3^{2-}$.**

✳ **Perform non-structured titration calculations, based on experimental results.**

Common examination questions based on MnO_4^- and Fe^{2+} titrations include:

- calculating the molar mass and formula of an iron(II) salt
- finding the percentage of iron in iron tablets
- calculating the percentage purity of iron samples.

You may need to use the relevant half-equations and the overall equation introduced in the previous spread:

$Fe^{2+}(aq) \longrightarrow Fe^{3+}(aq) + e^-$

$MnO_4^-(aq) + 8H^+(aq) + 5e^- \longrightarrow Mn^{2+}(aq) + 4H_2O(l)$

$MnO_4^-(aq) + 8H^+(aq) + 5Fe^{2+}(aq) \longrightarrow Mn^{2+}(aq) + 5Fe^{3+}(aq) + 4H_2O(l)$

In this section we will look at some of these types of questions.

Worked example: calculating the molar mass and formula of an iron(II) salt

A student weighed out 2.950 g of hydrated iron(II) sulfate, $FeSO_4 \cdot xH_2O$, and dissolved it in 50 cm³ of sulfuric acid. This solution was poured into a 250 cm³ volumetric flask and made up to the mark with distilled water. 2.50 cm³ of this solution was titrated with 0.0100 mol dm⁻³ potassium manganate(VII), $KMnO_4$, and 21.20 cm³ of the MnO_4^- solution was used.

(a) Calculate the amount, in moles, of manganate(VII), MnO_4^-, used in the titration.

(b) Deduce the amount, in moles, of iron(II), Fe^{2+}, in the 25.0 cm³ solution used.

(c) Deduce the amount, in moles, of iron(II), Fe^{2+}, in 250 cm³ of the solution.

(d) Determine the molar mass of the iron(II) salt.

(e) Determine the value of x and the formula of the hydrated iron(II) salt.

(a) Calculate the amount, in mol, of MnO_4^- that reacted:

$$n(MnO_4^-) = c \times \frac{V}{1000} = 0.0100 \times \frac{21.20}{1000} = 2.12 \times 10^{-4}\ mol$$

(b) Calculate the amount, in mol, of Fe^{2+}, that was used in the titration:

equation	$MnO_4^-(aq)$ reacts with $5Fe^{2+}(aq)$	
moles from equation	1 mol	5 mol
actual moles	2.12×10^{-4} mol	1.06×10^{-3} mol

(c) Calculate the amount, in mol, Fe^{2+}, that was used to make up the 250 cm³ solution:

25.0 cm³ $Fe^{2+}(aq)$ contains 1.06×10^{-3} mol

So the 250 cm³ solution contains $10 \times 1.06 \times 10^{-3} = 0.0106$ mol

(d) Calculate the molar mass of the iron(II) salt:

$$M(FeSO_4 \cdot xH_2O) = \frac{mass}{n} = \frac{2.950}{0.0106} = 278.3\ g\ mol^{-1}$$

(e) Determine the value of x in the iron(II) salt:

Mass of $FeSO_4$ in 1 mole of $FeSO_4 \cdot xH_2O = 55.8 + 32.1 + (16 \times 4) = 151.9$ g

Mass of H_2O in 1 mole of $FeSO_4 \cdot xH_2O = 278.3 - 151.9 = 126.4$ g

∴ value of $x = \dfrac{126.4}{18.0} = 7$. So the formula of the hydrated salt is $FeSO_4 \cdot 7H_2O$

Worked example: calculate the percentage mass of iron in some tablets

A multi-vitamin tablet contains vitamins A, B, C and D along with some iron. 0.325 g of a powdered tablet was dissolved in water and a little dilute sulfuric acid added. 12.10 cm^3 of 0.002 00 mol dm^{-3} potassium manganate(VII) solution was titrated until a permanent pale pink colour was observed. Calculate the percentage mass of iron in the tablets.

$$\text{Amount of MnO}_4^- = c \times \frac{V}{1000} = 0.002\,00 \times \frac{12.10}{1000}$$
$$= 2.42 \times 10^{-5} \text{ mol}$$

Amount of Fe^{2+} = 5 × 2.42 × 10^{-5} = 1.21 × 10^{-4} mol

$$n = \frac{\text{mass}}{M}$$

∴ mass = $n \times M$ = 1.21 × 10^{-4} × 55.8 = 0.006 75 g

∴ % iron = $\dfrac{\text{mass of iron}}{\text{mass of tablet}} \times 100 = \dfrac{0.006\,75}{0.325} = 2.08\%$

Figure 1 Some multi-vitamins have traces of iron, which can be analysed by titration

Examiner tip

This is an unstructured titration calculation so you will need to think about the steps required. To find the percentage of iron you will need to find the mass of iron in the tablets. The titration result will enable you to calculate the number of moles, and the mass should then be calculated easily.

Examiner tip

Remember that
$$\text{amount, } n = \frac{\text{mass, } m}{\text{molar mass, } M}$$

The molar mass of Fe^{2+} has the same numerical value as the relative atomic mass of Fe. (The mass of the two missing electrons is negligible.)

Sometimes you will be expected to apply your knowledge of titration calculations to unfamiliar situations. You will normally be given some additional information to help you.

Worked example: oxidising hydrogen peroxide

A 25.0 cm^3 portion of hydrogen peroxide was poured into a 250 cm^3 volumetric flask and made up to the mark with distilled water. A 25.0 cm^3 portion of this solution was acidified and titrated against 0.0200 mol dm^{-3} potassium manganate(VII) solution. 38.00 cm^3 of the MnO$_4^-$ solution was required for complete oxidation. Calculate the concentration of the original hydrogen peroxide solution.

The following equation represents the oxidation of hydrogen peroxide:

$$H_2O_2(aq) \longrightarrow O_2(g) + 2H^+(aq) + 2e^-$$

Step 1: You will need to write the fully balanced equation for the reaction to find the molar relationship between the two reactants (see spread 2.2.9 on combining half-equations).

The equation for MnO$_4^-$ ions is:

$$MnO_4^-(aq) + 8H^+(aq) + 5e^- \longrightarrow Mn^{2+}(aq) + 4H_2O(l)$$

The fully balanced equation is:

$$2MnO_4^-(aq) + 6H^+(aq) + 5H_2O_2(aq) \longrightarrow 2Mn^{2+}(aq) + 8H_2O(l) + 5O_2(g)$$

Step 2: Calculate the amount of MnO$_4^-$ that reacted.

$$n(MnO_4^-) = c \times \frac{V}{1000} = 0.0200 \times \frac{38.00}{1000} = 7.60 \times 10^{-4} \text{ mol}$$

Step 3: Deduce the amount of H$_2$O$_2$ that was used in the titration.

equation	2MnO$_4^-$ (aq) reacts with	5 H$_2$O$_2$(aq)
moles from equation	2 mol	5 mol
actual moles	7.60 × 10^{-4} mol	1.90 × 10^{-3} mol

Step 4: Calculate the concentration, in mol dm^{-3}, of the hydrogen peroxide solution.

$$c = \frac{n \times 1000}{V} = \frac{1.90 \times 10^{-3} \times 1000}{25.0} = 0.0760 \text{ mol dm}^{-3}$$

Step 5: Remember that the solution was diluted by a factor of 10 at the start of the experiment, so the concentration of the original hydrogen peroxide solution is 0.760 mol dm^{-3}.

Questions

1 A sample of metal of mass 2.50 g was thought to contain iron. The metal was reacted with sulfuric acid and the resulting solution made up to 250 cm^3. 25.0 cm^3 of this solution was titrated against 0.0180 mol dm^{-3} potassium manganate(VII) solution and 24.80 cm^3 of the solution was required. Calculate the percentage by mass of iron in the metal sample.

2 The equation below shows the reduction of potassium dichromate(VI):

$$Cr_2O_7^{2-}(aq) + 6e^- + 14H^+(aq) \longrightarrow Cr^{3+}(aq) + 7H_2O(l)$$

What volume of 0.002 00 mol dm^{-3} iron(II) sulfate will be oxidised by 20.00 cm^3 of 0.0150 mol dm^{-3} potassium dichromate(VI) in acid solution?

By the end of this spread, you should be able to ...

✳ Carry out structured calculations involving $I_2/S_2O_3^{2-}$.
✳ Perform unstructured titration calculations based on experimental results.

The reaction between iodine, I_2, and thiosulfate ions, $S_2O_3^{2-}$, is a redox reaction that is useful in chemical analysis:

$$2S_2O_3^{2-}(aq) + I_2(aq) \longrightarrow 2I^-(aq) + S_4O_6^{2-}(aq)$$

In this reaction, aqueous iodine is reduced to iodide ions, I^-, by aqueous sodium thiosulfate, which forms the tetrathionate ion, $S_4O_6^{2-}$.

The concentration of iodine in a solution can be determined by titration with a solution of sodium thiosulfate of known concentration. This titration can be used to determine the concentration of a solution of an oxidising agent that reacts with iodide ions, I^-, to produce iodine, I_2. Examples of oxidising agents that can be analysed using this method include copper(II), Cu^{2+}, dichromate(VI), $Cr_2O_7^{2-}$, and chlorate(I), ClO^-.

The method for analysing an oxidising agent is outlined below.

- Iodide ions are added to the oxidising agent under investigation – a redox reaction takes place, producing iodine.
- The iodine is titrated against a sodium thiosulfate solution of known concentration.
- From the results, you can calculate the amount of iodine, and hence the concentration of the oxidising agent.

Figure 1 Household bleach contains 'sodium hypochlorite', which is the common name for sodium chlorate(I), NaClO. A redox titration can be used to estimate the concentration of chlorate(I) ions, ClO^-, in bleach

Worked example

30.0 cm³ of bleach was added to an excess of iodide ions, I^-, in dilute sulfuric acid. Bleach contains chlorate(I) ions, $ClO^-(aq)$. In the presence of acid, chlorate(I) ions oxidise iodide ions, I^-, to iodine, I_2:

$$ClO^-(aq) + 2I^-(aq) + 2H^+(aq) \longrightarrow Cl^-(aq) + I_2(aq) + H_2O(l) \qquad \text{Equation 1}$$

The iodine formed was titrated against 0.200 mol dm⁻³ sodium thiosulfate and 29.45 cm³ of the thiosulfate was needed:

$$2S_2O_3^{2-}(aq) + I_2(aq) \longrightarrow 2I^-(aq) + S_4O_6^{2-}(aq) \qquad \text{Equation 2}$$

Calculate the concentration of chlorate(I) ions in the bleach.

(a) Calculate the amount, in mol, of $S_2O_3^{2-}$ that reacted.

$$n(S_2O_3^{2-}) = c \times \frac{V}{1000} = 0.200 \times \frac{29.45}{1000} = 5.89 \times 10^{-3} \text{ mol}$$

(b) Calculate the amount, in mol, of I_2, that was used in the titration.

Equation 2	$2S_2O_3^{2-}(aq)$	$+ I_2(aq)$
moles from equation	2 mol	1 mol
actual moles	5.890×10^{-3} mol	2.945×10^{-3} mol

(c) Deduce the amount, in mol, of chlorate(I) ions, ClO^-, needed to produce this amount of I_2.

Equation 1	$ClO^-(aq)$	$\longrightarrow$	$I_2(aq)$
moles from equation	1 mol		1 mol
actual moles	2.945×10^{-3} mol		2.945×10^{-3} mol

(d) Work out the concentration, in mol dm⁻³, of ClO^-, in the bleach.

$$c = \frac{2.945 \times 10^{-3} \times 1000}{30.0} = 0.0982 \text{ mol dm}^{-3}$$

Estimating the copper content of solutions and alloys

If a solution containing $Cu^{2+}(aq)$ ions is mixed with aqueous iodide ions, $I^-(aq)$:
- iodide ions are oxidised to iodine, $2I^- \longrightarrow I_2 + 2e^-$
- copper(II) ions are reduced to copper(I) ions, $Cu^{2+} + e^- \longrightarrow Cu^+$

$$2Cu^{2+}(aq) + 4I^-(aq) \longrightarrow 2CuI(s) + I_2(aq)$$

- This produces a light brown/yellow solution and a white precipitate of copper(I) iodide – the precipate appears to be light brown due to the iodine in the solution.
- This mixture is then titrated against sodium thiosulfate of known concentration. As the iodine reacts, the iodine colour gets paler during the titration. When the colour is a pale straw, a small amount of starch is added to help with the identification of the end point. A blue/black colour forms.
- The blue/black colour disappears sharply at the end point because all the iodine has reacted.

You can use the results to determine the concentration of iodine in the titration mixture, and then the concentration of copper(II) ions in the original solution can be determined. You can use this method to estimate the concentration of copper(II) ions in an unknown solution or the percentage of copper in alloys such as brass and bronze. The alloy is first reacted with nitric acid. Brass and bronze can both be reacted to produce a solution containing copper(II) ions. This can then be reacted with potassium iodide solution. The iodine that forms is then titrated against sodium thiosulfate.

The worked example below is an unstructured titration calculation – you are provided with the titration results and relevant equations, but you need to work out how to tackle the problem yourself.

Figure 2 The famous bronze Champions Statue commemorating four of the players who won the 1966 World Cup

---

Worked example

A sample of bronze was analysed to find the proportion of copper it contained. 0.500 g of the bronze was reacted with nitric acid to give a solution containing Cu^{2+} ions. The solution was neutralised and reacted with iodide ions to produce iodine. The iodine was titrated with 0.200 mol dm^{-3} sodium thiosulfate, and 22.40 cm^3 were required.
Calculate the percentage of copper in the bronze.

$$Cu(s) \longrightarrow Cu^{2+}(aq) + 2e^-$$
$$2Cu^{2+}(aq) + 4I^-(aq) \longrightarrow 2CuI(s) + I_2(aq)$$
$$2S_2O_3^{2-}(aq) + I_2(aq) \longrightarrow 2I^-(aq) + S_4O_6^{2-}(aq)$$

Step 1: Calculate the amount of thiosulfate used.
$$n(S_2O_3^{2-}) = c \times \frac{V}{1000} = 0.200 \times \frac{22.40}{1000} = 4.48 \times 10^{-3} \text{ mol}$$

Step 2: Calculate the reacting amounts.
2 mol Cu^{2+} produces 1 mol I_2 which reacts with 2 mol $S_2O_3^{2-}$.
1 mol Cu reacted with nitric acid to produce 1 mol Cu^{2+}
So 1 mol Cu is equivalent to 1 mol $S_2O_3^{2-}$

Step 3: Calculate the % copper.
$$n(Cu) = n(S_2O_3^{2-}) = 4.48 \times 10^{-3} \text{ mol}$$
Mass of copper $= n \times M = 4.48 \times 10^{-3} \times 63.5 = 0.28448$ g
$$\% \text{ copper in bronze} = \frac{0.28448}{0.500} \times 100 = 56.9\%$$

Examiner tip

Note that we have not rounded anything until the final answer, which is given to three significant figures to match the smallest number of significant figures given in the question. If we had rounded the mass of copper to 0.284 g, we would have found the percentage of copper to be 56.8%.

Questions

1 A piece of impure copper of mass 0.900 g was reacted with nitric acid. The copper(II) solution formed was neutralised and reacted with iodide ions to produce iodine. The iodine was titrated with 0.500 mol dm^{-3} sodium thiosulfate, and 23.70 cm^3 were required. What is the percentage of copper in the sample?

2 A student weighed 5.95 g of an unknown copper(II) salt and dissolved it in distilled water. This solution was added to a 250 cm^3 volumetric flask and was made up to the mark with distilled water. 25.0 cm^3 of the solution was transferred into a conical flask and 10 cm^3 of potassium iodide (an excess) was added. The copper(II) ions reacts with iodide ions, I^-.
The resulting solution was titrated against 0.100 mol dm^{-3} sodium thiosulfate, and 24.05 cm^3 of sodium thiosulfate was used.
Calculate the molar mass of the copper(II) salt.

Definition

- Ions with partially filled d-orbitals

Complex ions

- Coordination numbers
- Central metal ions
- Ligands

Transition metals ions

NaOH(aq) →

Precipitation

- Fe^{2+} Green
- Fe^{3+} Brown
- Cu^{2+} Blue
- Co^{2+} Pink

Ligand substitution

- A ligand is an electron pair donor to a metal ion
- Substitution: replacement of one ligand by another

Shape: octahedral

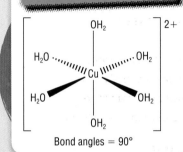

Bond angles = 90°

Properties

- Coloured compounds
- Catalytic properties
- Variable oxidation states

Isomerism

Cis–trans ←————————————→ Optical

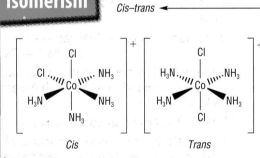

Cis   *Trans*

Stability constants

$$[Cu(H_2O)_6]^{2+}(aq) + 4Cl^-(aq) \rightleftharpoons [CuCl_4]^{2-}(aq) + 6H_2O(l)$$

$$K_{stab} \frac{[[Cu(Cl_4)]^{2-}]}{[[Cu(H_2O)_6]^{2+}][Cl^-]^4}$$

Redox titrations

Reduction is gain of electrons or a decrease in oxidation number

$$MnO_4^-(aq) + 8H^+(aq) + 5e^- \longrightarrow Mn^{2+}(aq) + 4H_2O(l)$$

Oxidation is loss of electrons or an increase in oxidation number

$$Fe^{2+}(aq) \longrightarrow Fe^{3+}(aq) + e^-$$

Practice questions

(1) Write the electron configurations of the following atoms and ions:

(a) Fe; **(b)** Fe^{2+};
(c) Cu^+; **(d)** Co.

(2) What is the oxidation state of chromium in the following species?

(a) $K_2Cr_2O_7$; **(b)** $Cr_2O_2Cl_2$; **(c)** $[Cr(NH_3)_6]^{3+}$.

(3) Iron is a typical transition element because it shows more than one oxidation state in its compounds and has catalytic properties.

(a) Write the electron configuration for iron(III) and use it to explain why iron is a transition metal.

(b) State one use of iron or one of its compounds in catalysis. State the name of the catalyst and write an equation for the reaction that is catalysed.

(4) **(a)** **(i)** What is meant by the term *d-block element*?

(ii) What is meant by the term *transition element*?

(iii) Write the electron configuration of a vanadium atom.

(b) Aqueous transition metal ions can react with aqueous hydroxide ions, OH^-.
Copy and complete the table below.

Metal ion	Formula and state symbol of the product formed with OH^-(aq)	Colour of product
Fe^{2+}		
Fe^{3+}		

(5) Copper ions, Cu^{2+}, take part in a number of different ligand substitution reactions.

(a) What is meant by the terms:

(i) *ligand*;

(ii) *complex ion*;

(iii) *ligand substitution*?

(b) Copper(II) salts form aqueous solutions containing $[Cu(H_2O)_6]^{2+}$ ions. A ligand substitution reaction involving $[Cu(H_2O)_6]^{2+}$ is shown below:

$$[Cu(H_2O)_6]^{2+}(aq) \xrightarrow{\text{reagent Y}} [Cu(H_2O)_2(NH_3)_4]^{2+}(aq)$$

(i) Identify reagent Y.

(ii) Write a balanced equation for this ligand substitution.

(iii) State the colours of the two complex ions involved in this ligand substitution.

(iv) Draw the shape of the $[Cu(H_2O)_6]^{2+}$ complex ion, showing clearly any bond angles.

(6) **(a)** Explain what is meant by the term *coordination number* in a complex ion.

(b) State the formula and shape of the complex containing:

(i) Co(II) and six H_2O ligands;

(ii) Ni^{2+} and three $(NH_2CH_2CH_2NH_2)$ ligands.

(c) Sketch and label the two isomers of:

(i) $[Cr(H_2O)_4Cl_2]^+$; **(ii)** $[NiCl_2(NH_3)_2]$.

(d) Sketch and label all the isomers of the complex $[Ni(NH_2CH_2CH_2NH_2)_2Cl_2]$.

(7) The concentration of ethanedioic acid, $(COOH)_2$, can be determined by its reaction with acidified manganate(VII) ions, MnO_4^-. Ethanedioic acid is oxidised to carbon dioxide as shown in the equation:

$$(COOH)_2\,(aq) \longrightarrow 2CO_2(g) + 2H^+(aq) + 2e^-$$

(a) Write the half-equation for the reduction of acidified manganate(VII) ions.

(b) Construct a balanced equation for the reaction between ethanedioic acid and acidified manganate(VII) ions.

(c) $25.0\ cm^3$ of an aqueous solution of ethanedioic acid reacted exactly with $15.0\ cm^3$ of $0.0200\ mol\ dm^{-3}$ manganate (VII) ions.
Calculate the concentration, in $mol\ dm^{-3}$, of the solution of ethanedioic acid.

(8) A solution containing aqueous MnO_4^- ions was titrated against $25.0\ cm^3$ of $0.0500\ mol\ dm^{-3}$ Fe^{2+}(aq) ions in acid solution:

$$MnO_4^-(aq) + 8H^+(aq) + 5Fe^{2+}\,(aq) \longrightarrow$$
$$Mn^{2+}(aq) + 5Fe^{3+}(aq) + 4H_2O(l)$$

The volume of the solution containing aqueous manganate(VII) ions required to reach the end point was $12.30\ cm^3$.

(a) State the colour change at the end point.

(b) Calculate the concentration, in $mol\ dm^{-3}$, of the aqueous manganate (VII) ion solution used in the titration.

(9) A sample of copper(II) sulfate was dissolved in water to produce $250\ cm^3$ of solution. An excess of aqueous potassium iodide, KI(aq), was added to $25.0\ cm^3$ of this solution and iodine was formed in the reaction:

$$2Cu^{2+}(aq) + 4I^-(aq) \longrightarrow 2CuI(s) + I_2(aq)$$

The iodine produced was titrated with $0.200\ mol\ dm^{-3}$ sodium thiosulfate, $Na_2S_2O_3$(aq):

$$I_2(aq) + 2S_2O_3^{2-}(aq) \longrightarrow 2I^-(aq) + S_4O_6^{2-}(aq)$$

The average titre obtained was $20.15\ cm^3$.

Calculate the concentration of the copper(II) sulfate solution.

1 The compound $FeSO_4 \cdot 7H_2O$ can be used to kill moss in grass. Iron(II) ions in a solution of $FeSO_4 \cdot 7H_2O$ are slowly oxidised to form iron(III) ions.

(a) Describe a test to show the presence of iron(III) ions in a solution of $FeSO_4 \cdot 7H_2O$. [1]

(b) The percentage purity of an impure sample of $FeSO_4 \cdot 7H_2O$ can be determined by titration against potassium dichromate(VI), $K_2Cr_2O_7$, under acid conditions, using a suitable indicator.

During the titration, $Fe^{2+}(aq)$ ions are oxidised to $Fe^{3+}(aq)$ ions.

Stage 1 – A sample of known mass of the impure $FeSO_4 \cdot 7H_2O$ is added to a conical flask.

Stage 2 – The sample is dissolved in an excess of dilute sulfuric acid.

Stage 3 – The contents of the flask are titrated against $K_2Cr_2O_7(aq)$.

(i) The reduction half-equation for acidified dichromate(VI) ions, $Cr_2O_7^{2-}$, is as follows.

$$Cr_2O_7^{2-}(aq) + 14H^+(aq) + 6e^- \rightarrow 2Cr^{3+}(aq) + 7H_2O(l)$$

Construct the balanced equation for the redox reaction between $Fe^{2+}(aq)$, $Cr_2O_7^{2-}$ and $H^+(aq)$. [2]

(ii) In stage 1, a student uses a 0.655 g sample of impure $FeSO_4 \cdot 7H_2O$.

In stage 3, the student uses 19.6 cm³ of 0.0180 mol dm⁻³ $Cr_2O_7^{2-}$ to reach the end point.

One mol of $Cr_2O_7^{2-}$ reacts with 6 moles of Fe^{2+}.

Calculate the percentage purity of the impure sample of $FeSO_4 \cdot 7H_2O$. [4]

[Total: 7]

(OCR 2815/1 Jan06)

2 Dilute aqueous copper(II) sulfate contains $[Cu(H_2O)_6]^{2+}$ ions.

(a) Concentrated hydrochloric acid is added drop by drop to a small volume of dilute aqueous copper(II) sulfate. The equation for the reaction taking place is as follows.

$$Cu(H_2O)_6]^{2+}(aq) + 4Cl^-(aq) \rightleftharpoons [CuCl_4]^{2-}(aq) + 6H_2O(l)$$

(i) Describe the observations that would be made during the addition of the concentrated hydrochloric acid. [1]

(ii) Describe the bonding within the complex ion, $[CuCl_4]^{2-}$. [2]

(b) Concentrated aqueous ammonia is added drop by drop to aqueous copper(II) sulfate until present in excess. Two reactions take place, one after the other, to produce the complex ion $[Cu(NH_3)_4(H_2O)_2]^{2+}(aq)$.

Describe the observations that would be made during the addition of concentrated aqueous ammonia. [2]

(c) Ammonia is a simple molecule. The H–N–H bond angle in an isolated ammonia molecule is 107°.

The diagram shows part of the $[Cu(NH_3)_4(H_2O)_2]^{2+}$ ion and the H–N–H bond angle in the ammonia ligand.

Explain why the H–N–H bond angle in the ammonia ligand is 109.5° rather than 107°. [3]

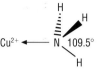

[Total: 8]

(OCR 2815/1 Jan06)

3 In this question, one mark is available for the quality of use and organisation of scientific terms.

Copper and iron are typical transition elements. One of the characteristic properties of a transition element is that it can form complex ions.

Explain in terms of electronic configuration why copper is a transition element.

Give an example of a complex ion that contains copper. Draw the three-dimensional shape of the ion and describe the bonding within this complex ion.

Transition elements show typical metallic properties. Describe three other typical properties of transition elements. Illustrate each property using copper or iron or their compounds. [12]

[Total: 12]

(OCR 2815/1 Jun05)

4 A moss killer contains iron(II) sulfate.

Some of the iron(II) sulfate gets oxidised to form iron(III) sulfate. During the oxidation iron(II) ions, Fe^{2+}, react with oxygen, O_2, and hydrogen ions to make water and iron(III) ions, Fe^{3+}.

(a) Complete the electronic configuration for Fe^{3+} and use it to explain why iron is a transition element.
Fe^{3+}: $1s^2 2s^2 2p^6$ [2]

(b) State **two** typical properties of compounds of a transition element. [2]

(c) Describe how aqueous sodium hydroxide can be used to distinguish between aqueous iron(II) sulfate and aqueous iron(III) sulfate. [2]

(d) Construct the equation for the oxidation of acidified iron(II) ions by oxygen. [2]

(e) The percentage by mass of iron in a sample of moss killer can be determined by titration against acidified potassium manganate(VII).

Stage 1 – A sample of moss killer is dissolved in excess sulfuric acid.

Stage 2 – Copper turnings are added to the acidified sample of moss killer and the mixture is boiled carefully for five minutes. Copper reduces any iron(III) ions in the sample to give iron(II) ions.

Stage 3 – The reaction mixture is filtered into a conical flask to remove excess copper.

Stage 4 – The contents of the flask are titrated against aqueous potassium manganate(VII).

(i) Suggest why it is important to remove all the copper in stage 3 before titrating in stage 4. [1]

(ii) The ionic equation for the redox reaction between acidified MnO_4^- and Fe^{2+} is given below.

$$MnO_4^-(aq) + 8H^+(aq) + 5Fe^{2+}(aq) \rightarrow$$
$$Mn^{2+}(aq) + 4H_2O(l) + 5Fe^{3+}(aq)$$

Explain, in terms of electron transfer, why this reaction involves both oxidation and reduction. [2]

(iii) A student analyses a 0.675 g sample of moss killer using the method described. In stage 4, the student uses 22.5 cm^3 of 0.0200 mol dm^{-3} MnO_4^- to reach the end point.

Calculate the percentage by mass of iron in the moss killer. [4]

[Total: 15]
(OCR 2815/1 Jan05)

5 Aqueous copper(II) sulfate contains $[Cu(H_2O)_6]^{2+}$ ions. Aqueous ammonia is added drop by drop to a small volume of aqueous copper(II) sulfate. Two reactions take place, one after the other, as shown in the equations.

$$[Cu(H_2O)_6]^{2+}(aq) + 2OH^-(aq) \rightarrow Cu(OH)_2(s) + 6H_2O(l)$$
$$Cu(OH)_2(s) + 2H_2O(l) + 4NH_3(aq) \rightarrow$$
$$[Cu(NH_3)_4(H_2O)_2]^{2+}(aq) + 2OH^-(aq)$$

(a) Describe the observations that would be made as ammonia is added drop by drop until it is in an excess. [2]

(b) Draw the shape for the $[Cu(H_2O)_6]^{2+}$ ion. Include the bond angles in your diagram. [2]

(c) Water is a simple molecule. The H–O–H bond angle in an isolated water molecule is 104.5°.

The diagram shows part of the $[Cu(H_2O)_6]^{2+}$ ion and the H–O–H bond angle in the water ligand.

Explain why the H–O–H bond angle in the water ligand is 107° rather than 104.5°. [3]

[Total: 7]
(OCR 2815/1 Jan05)

6 Copper is a typical transition element.
It forms coloured compounds.
It forms complex ions.
It has more than one oxidation state in its compounds.

(a) State **one** other typical property of a transition element. [1]

(b) Dilute aqueous copper(II) sulfate is a blue solution containing $[Cu(H_2O)_6]^{2+}$ ions. A ligand substitution involving $[Cu(H_2O)_6]^{2+}$ is shown below.

$$[Cu(H_2O)_6]^{2+} \xrightarrow{\text{reagent X}} [CuCl_4]^{2-}$$
blue solution \qquad yellow

(i) Suggest a shape for the $[CuCl_4]^{2-}$ ion. Include the bond angles in your diagram. [2]

(ii) State the **formula** of the ligand in $[CuCl_4]^{2-}$. [1]

(iii) State the name or formula of reagent X. [1]

(iv) Explain, with the aid of a balanced equation, what is meant by the term *ligand substitution*. [2]

[Total: 7]
(OCR 2815/1 Jan04)

7 Aqueous hydrogen peroxide, H_2O_2, is used to sterilise contact lenses. H_2O_2 decomposes to make oxygen and water as shown in the equation.

$$2H_2O_2(aq) \rightarrow 2H_2O(l) + O_2(g)$$

(a) Decomposition of hydrogen peroxide is a redox reaction. Use oxidation numbers to show that oxidation and reduction take place. [2]

(b) The concentration of an aqueous solution of hydrogen peroxide can be determined by titration. Aqueous potassium manganate(VII), $KMnO_4$, is titrated against a solution of hydrogen peroxide in the presence of acid. The half-equation for the oxidation of H_2O_2 is as follows:

$$H_2O_2(aq) \rightarrow O_2(g) + 2H^+(aq) + 2e^-$$

The half-equation for the reduction of acidified MnO_4^- is as follows:

$$MnO_4^-(aq) + 8H^+(aq) + 5e^- \rightarrow Mn^{2+}(aq) + 4H_2O(l)$$

(i) Construct the equation for the reaction between H_2O_2, MnO_4^- ions and H^+ ions. [2]

(ii) A student takes a 25.0 cm^3 sample of aqueous hydrogen peroxide and places this into a conical flask and then adds sulfuric acid to acidify the hydrogen peroxide.

The student titrates this sample of acidified hydrogen peroxide against a solution containing 0.0200 mol dm^{-3} $MnO_4^-(aq)$ ions. For complete reaction with the acidified hydrogen peroxide, the student uses 17.50 cm^3 of this solution containing $MnO_4^-(aq)$ ions.

Calculate the concentration, in mol dm^{-3}, of the aqueous hydrogen peroxide.

2 mol MnO_4^- reacts with 5 mol H_2O_2. [3]

(c) Acidified hydrogen peroxide oxidises Fe^{2+}(aq) to Fe^{3+}(aq). Describe a simple chemical test to show the presence of Fe^{3+}(aq).

Name of reagent used ……

Observation …… [2]

[Total: 9]

(OCR 2815/1 Jan04)

8 Ruthenium (Ru) is a metal in the second transition series. It forms complex ions with the following formulae.

A = $[Ru(H_2O)_6]^{3+}$

B = $[Ru(H_2O)_5Cl]^{2+}$

C = $[Ru(H_2O)_4Cl_2]^+$

(a) (i) What is the oxidation number of ruthenium in **B**? [1]

(ii) One of the complex ions, **A**, **B** or **C**, shows stereoisomerism.

Draw diagrams to show the structures of the two isomers. [2]

(iii) Name this type of stereoisomerism. [1]

(b) The complex ion $[Ru(H_2O)_6]^{3+}$ can be converted into $[Ru(H_2O)_5Cl]^{2+}$.

(i) Suggest a suitable reagent for this conversion. [1]

(ii) What type of reaction is this? [1]

[Total: 6]

(OCR 2815/6 Jan07)

9 Platinum forms complexes with a co-ordination number of 4.

(a) (i) Explain the term *co-ordination number*. [1]

(ii) State the shape of these platinum complexes. [1]

(b) The tetrachloroplatinate(II) ion readily undergoes the following reaction.

$[PtCl_4]^x + 2NH_3 \rightleftharpoons [Pt(NH_3)_2Cl_2]^y + 2Cl^-$

(i) What type of reaction is this? [1]

(ii) Suggest values for x and y in the equation. [2]

(c) The complex $[Pt(NH_3)_2Cl_2]^y$ exists in two isomeric forms.

(i) Draw diagrams to show the structures of these isomers. [2]

(ii) What type of isomerism is this? [1]

(iii) One of the isomers of $[Pt(NH_3)_2Cl_2]^y$ is an important drug used in the treatment of cancer.

How does this drug help in the treatment of cancer? [2]

[Total: 10]

(OCR 2815/6 Jun06)

10 (a) A complex ion contains one Fe^{3+} ion, four molecules of ammonia and two chloride ions.

(i) What is the formula of this complex ion? [1]

(ii) This complex shows *cis–trans* isomerism. Draw diagrams to show the structures of the *cis* and *trans* isomers. [3]

(iii) What is the co-ordination number of this complex ion? [1]

(b) Describe the role of *cis*-platin as an important therapeutic drug. [2]

[Total: 7]

(OCR 2815/6 Jan03)

11 The following is an account of a laboratory experiment.

• A solution was prepared by dissolving some copper(II) sulfate to give 250 cm^3 of aqueous solution.

• 25.0 cm^3 of this solution was treated with an excess of aqueous potassium iodide, KI.

$2Cu^{2+}(aq) + 4I^-(aq) \rightarrow 2CuI(s) + I_2(aq)$

• The iodine produced was titrated with 0.100 mol dm^{-3} sodium thiosulfate.

$I_2(aq) + 2S_2O_3^{2-}(aq) \rightarrow 2I^-(aq) + S_4O_6^{2-}(aq)$

The average titre obtained was 22.00 cm^3 of the thiosulfate solution.

(a) State the oxidation number of S in $S_2O_3^{2-}$. [1]

(b) Calculate the amount of $S_2O_3^{2-}$ ions in the titre. [1]

(c) Calculate the amount of I_2 produced. [1]

(d) Calculate the amount of Cu^{2+} ions in 25.0 cm^3 of solution. [1]

(e) Calculate the concentration of the aqueous copper(II) sulfate in mol dm^{-3}. [1]

[Total: 5]

(OCR 2815/6 Jun02)

12 This question is about complex ions of cobalt.

(a) The equilibrium below exists between two complex ions of cobalt(II).

$$[Co(H_2O)_6]^{2+} + 4Cl^- \rightleftharpoons [CoCl_4]^{2-} + 6H_2O$$

   (i) What colour change occurs from left to right? [1]

   (ii) Suggest the shape of the $[CoCl_4]^{2-}$ ion? [1]

   (iii) What type of reaction is this? [1]

(b) Cobalt(III) forms a complex ion of formula $[Co(en)_3]^{3+}$, where 'en' is ethane-1,2-diamine, $H_2NCH_2CH_2NH_2$. Draw a displayed formula for ethane-1,2-diamine, showing clearly any lone pairs of electrons. [1]

(c) The $[Co(en)_3]^{3+}$ ion shows a type of stereoisomerism.

   (i) What type of stereoisomerism does it show? [1]

   (ii) Draw 3D diagrams to show the two stereoisomers of $[Co(en)_3]^{3+}$.

   You may use 'en' to represent ethane-1,2-diamine. [2]

[Total: 7]
(OCR 2815/6 Jun08)

13 Iron is a transition element. It forms compounds in which the oxidation number of iron is +2, +3 or +6.

(a) Give the electronic configuration of Fe^{2+} and use it to explain why iron is a transition element. [1]

(b) Transition elements form complex ions.

Choose an example of a complex ion in which iron has the +2 oxidation state.

   (i) Write the formula for your chosen complex ion formed by iron. [1]

   (ii) Draw the shape of your chosen complex ion. Indicate clearly the three-dimensional shape and the bond angles. [2]

   (iii) Describe the bonding within your chosen complex ion. [2]

(c) Complex ions containing transition elements often undergo ligand substitution reactions.

Describe one ligand substitution reaction of a complex ion containing iron in the +3 oxidation state. [2]

(d) Describe how aqueous sodium hydroxide can be used to distinguish between aqueous solutions of iron(II) chloride and iron(III) chloride.

Write an ionic equation, with state symbols, for one of the reactions you describe. [2]

(e) FeO_4^{2-} reacts with Cr^{3+} ions as shown in the following equation.

$$FeO_4^{2-} + Cr^{3+} \longrightarrow CrO_4^{2-} + Fe^{3+}$$

Use oxidation numbers to explain why this reaction involves both oxidation and reduction. [2]

(f) FeO_4^{2-} decomposes in the presence of hydrogen ions, forming iron(III) ions, oxygen and water. Construct the ionic equation for this reaction. [2]

[Total: 14]
(OCR 2815/1 Jan08)

14 This question is about molybdenum and iron. Molybdenum steel is extremely hard.

Molybdenum is made by heating molybdenum(VI) oxide and reducing to molybdenum metal by aluminium.

(a) Construct an equation to show the reduction of molybdenum(VI) oxide to molybdenum metal by aluminium. [1]

(b) Molybdenum has the electronic configuration $[Kr]4d^55s^1$, where $[Kr]$ represents the electronic configuration for krypton.

Complete the electronic configuration for Mo^{3+} and use it to explain why molybdenum is a transition element. [1]

(c) Molybdenum(IV) oxide, MoO_2, can be oxidised by dichromate(VI) ions, $Cr_2O_7^{2-}$, under acidic conditions. The relevant half-equations are as follows.

$$MoO_2(s) + 2H_2O(l) \longrightarrow MoO_4^{2-}(aq) + 4H^+(aq) + 2e^-$$
$$Cr_2O_7^{2-}(aq) + 14H^+(aq) + 6e^- \longrightarrow 2Cr^{3+}(aq) + 7H_2O(l)$$

Construct an equation for the oxidation of MoO_2 by $Cr_2O_7^{2-}$ ions under acidic conditions. [2]

(d) Iron can form the ferrate(VI) ion, FeO_4^{2-}.

   (i) What is the formula of potassium ferrate(VI)? [1]

   (ii) Aqueous ferrate(VI) ions can be made by the oxidation of iron(III) oxide by chlorine in alkaline conditions.

   $$Fe_2O_3 + 3Cl_2 + 10OH^- \longrightarrow 2FeO_4^{2-} + 6Cl^- + 5H_2O$$

   A 1.00 g sample of Fe_2O_3 is added to 10.0 cm^3 of 4.00 mol dm^{-3} KOH.

   Which reagent, Fe_2O_3 or KOH, is in excess? Explain your answer. [3]

[Total: 8]
(OCR 2815/1 Jun08)

Answers

Spread answers

1.1.1 Introduction to aromatic chemistry

1 Any compound with the benzene ring is classified as an aromatic compound.
2 Crude oil and coal.
3 Any chemical known or believed to cause cancer in humans.
4 Ethylbenzene, phenol, cyclohexane, styrene.
5 The empirical formula is the simplest whole-number ratio of atoms of each element present in a compound; the molecular formula shows the number of atoms of each element present in a molecule.

1.1.2 The structure of benzene

1 The bromine would be decolourised.
$C_6H_{10} + Br_2 \longrightarrow C_6H_{10}Br_2$
2 Electrophilic addition.
3 X-ray crystallography.

1.1.3 The delocalised model of benzene

1 Trigonal planar and 120°
2 The p-orbitals on the carbon atoms overlap sideways to produce a delocalised π-cloud of electron density containing six electrons above and below the plane of the carbon atoms.
3 These are electrons which do not belong to one specific atom, but are shared between more than two atoms.
4 Substitution maintains the stability of the benzene ring, whereas the product from an addition reaction would be less stable because the delocalisation in the ring structure would be lost.

1.1.4 Benzene and its reactions

1 Concentrated nitric acid and concentrated sulfuric acid.
2 $C_6H_6 + HNO_3 \longrightarrow C_6H_5NO_2 + H_2O$
3 Aluminium chloride, iron(III) chloride or iron.
4 An atom or group of atoms that is attracted to an electron-rich centre of another molecule or atom, where it accepts a pair of electrons to form a new covalent bond.

1.1.5 Substitution reactions of benzene

1 $C_6H_6 + Cl_2 \longrightarrow C_6H_5Cl + HCl$
Catalyst: $AlCl_3$, $FeCl_3$ or Fe.
2 Electrophilic substitution.

3 $Cl_2 + FeCl_3 \longrightarrow Cl^+ + FeCl_4^-$

$FeCl_4^- + H^+ \longrightarrow FeCl_3 + HCl$

1.1.6 The reactivity of alkenes and benzene

1 This is an atom or group of atoms that is attracted to an electron-rich centre or atom, where it accepts a pair of electrons to form a new covalent bond.
2 (a) The bromine will be decolourised with cyclohexene but not with benzene.
(b)

1.1.7 Phenols

1 $C_6H_5OH + KOH \longrightarrow C_6H_5O^-K^+ + H_2O$; organic product is potassium phenoxide.
2 There would be effervescence and a gas would be evolved; the potassium would disappear.
3 (a) is a phenol – it is the only structure with an –OH group attached directly to the ring. (b) and (c) are alcohols as the –OH group in connected to a side chain.

1.1.8 Brominaton and uses of phenols

1 The bromine is decolourised and a white precipitate is formed.
2 Bromine reacts more rapidly with phenol than with benzene because a pair of non-bonding electrons (occupying a p-orbital) on the oxygen atom in the phenol group is drawn into the benzene ring; this creates a higher electron density in the ring structure and activates the ring. The increased electron density polarises bromine molecules, which are then attracted more strongly towards the ring structure than in benzene.
3 Surfactants, detergents, antiseptics, epoxy resins for paints.

1.1.9 An introduction to carbonyl compounds

1 (a)

(b)

(c)

2 In an aldehyde the functional group (C=O) is on the end of the carbon chain, whereas in a ketone it is at a point somewhere along the carbon chain with a C atom on each side.

3 This is the part of the organic molecule responsible for its chemical reactions.

4 (a) hexanal;　　**(b)** butanal;　　**(c)** octan-2-one.

1.1.10　Oxidation of alcohols and aldehydes

1 (a) For example,

Primary

Secondary

Tertiary

(b) The answers given here are subject to the original isomers drawn. The primary alcohol above would be oxidised to form 2-methylbutanal; or if in excess oxidising agent to 2-methylbutanoic acid. The secondary alcohol above would form pentan-3-one. The tertiary alcohol would not be oxidised.

(c) Oxidising agent: sulfuric acid and potassium dichromate(VI)
Conditions: heat and distil product for an aldehyde; heat under reflux for a ketone or carboxylic acid.

2 You would need to distil the product as it is made in order to separate the propanal from the rest of the reaction mixture and to prevent further oxidation to propanoic acid.

3 [O] represents an oxidising agent.

1.1.11　Reactions of aldehydes and ketones

1 (a) $CH_3COCH_3 + 2[H] \longrightarrow CH_3CH(OH)CH_3$
(b) $CH_3CH_2CH_2CH_2CHO + 2[H] \longrightarrow$
$$CH_3CH_2CH_2CH_2CH_2OH$$
(c) $(CH_3)_2CHCOCH_2CH_2CH_3 + 2[H] \longrightarrow$
$$(CH_3)_2CHCH(OH)CH_2CH_2CH_3$$

2 A nucleophile is an atom or group of atoms that is attracted to an electron-deficient centre, where it donates a pair of electrons to form a new covalent bond.

3

4

1.1.12　Chemical tests on carbonyl compounds

1 (a) Propanone and propanal.
(b) 2,4-dinitrophenylhydrazine (2,4-DNP) – when 2,4-DNP is added to an aldehyde or a ketone, a deep yellow or orange precipitate is formed.
(c) Tollens' reagent – gives a silver mirror with an aldehyde but not with a ketone.

2 The solid would be purified by recrystallisation and filtered to produce a fine, dry crystalline solid. The melting point of this solid would be taken and compared to a data table for identification.

3 Condensation.

1.1.13　Carboxylic acids

1 (a) butanoic acid;
(b) 2-methylpropanoic acid;
(c) 2,3-dimethyloctanoic acid.

2 (a) $2C_3H_7COOH + 2Na \longrightarrow 2C_3H_7COO^-Na^+ + H_2$
(b) $C_3H_7COOH + NaOH \longrightarrow C_3H_7COO^-Na^+ + H_2O$
(c) $2C_3H_7COOH + MgCO_3 \longrightarrow$
$$(C_3H_7COO^-)_2Mg^{2+} + CO_2 + H_2O$$

1.1.14　Esters

1 $CH_3CH_2COOH + CH_3OH \longrightarrow CH_3CH_2COOCH_3 + H_2O$
$(CH_3CH_2CO)_2O + CH_3OH \longrightarrow$
$$CH_3CH_2COOCH_3 + CH_3CH_2COOH$$

2 (a)

CH₃(CH₂)₄—C(=O)—O—(CH₂)₃CH₃

(b) $CH_3(CH_2)_4COOCH_2CH_2CH_2CH_3 + H_2O \longrightarrow$
$CH_3(CH_2)_4COOH + CH_3CH_2CH_2CH_2OH$

$CH_3(CH_2)_4COOCH_2CH_2CH_2CH_3 + OH^- \longrightarrow$
$CH_3(CH_2)_4COO^- + CH_3CH_2CH_2CH_2OH$

3 Flavourings, solvents, perfumes.

1.1.15 Fats and oils – building triglycerides

1 (a)

H—C(H)(H)—C(H)(H)—C(H)(H)—C(H)(H)—C(H)(H)—C(H)(H)—C(H)(H)—C(H)(H)—C(H)(H)—C(=O)—O—H

(b) Carboxylic acid or carboxyl group.

(c)

H—C(H)—O—C(=O)—C₉H₁₉
|
H—C(H)—O—C(=O)—C₉H₁₉
|
H—C(H)—O—C(=O)—C₉H₁₉

2 Fats and oils are very similar substances. They differ in their physical states because of their melting points. When the melting point of the ester is above room temperature the substance exists as a solid and is called a fat; when the melting point is below room temperature, the substance exists as an oil.

3 Simple triglycerides contain three molecules of the same fatty acid; mixed triglycerides have two or three different fatty acids in their structures.

4 18:3 (9,12,15)

1.1.16 Triglycerides, diet and health

1 A saturated fat contains no double bonds; an unsaturated fat contains at least one double C=C bond.

2

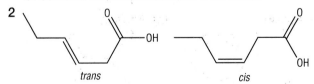

trans   *cis*

3 Meat products, meat pies, sausages, butter, lard, pastry, cakes, biscuits.

4 A crop that is used to produce a carbon-neutral fuel absorbs the same amount of CO_2 when it grows as is released when the fuel is combusted.

1.1.17 Amines

1 (a) C_2H_7N
(b) C_2H_7N
(c) $(CH_3)_2NH$

2 (a) A (structure) $NH_3^+NO_3^-$ **B** (structure) $NH_3^+Cl^-$

(b) (i) propylammonium chloride
(ii) $CH_3CH_2CH_2NH_2 + HCl \longrightarrow CH_3CH_2CH_2NH_3^+Cl^-$

1.1.18 Amines and their reactions

1 $CH_3CH_2CH_2CH_2Br + 2NH_3 \longrightarrow$
$CH_3CH_2CH_2CH_2NH_2 + NH_4Br$
The reaction is carried out using ethanol as the solvent.

2 $C_6H_6 + HNO_3 \longrightarrow C_6H_5NO_2 + H_2O$
using concentrated HNO_3 and concentrated H_2SO_4 at 50 °C;
$C_6H_5NO_2 + 6[H] \longrightarrow C_6H_5NH_2 + 2H_2O$
using concentrated HCl and Sn under reflux.

3 (a) $C_6H_5NH_2 + HNO_2 + HCl \longrightarrow C_6H_5N_2^+Cl^- + 2H_2O$
(b) The reaction is carried out below 10 °C.
(c)

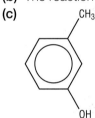

1.2.1 Amino acids

1 (a) The pH at which a particular molecule carries no net electrical charge.
(b) A dipolar ionic form of an amino acid that is formed by the donation of a hydrogen ion from the carboxyl group to the amino group. As both + and – charges are present there is no overall charge.

2 (a)

H—N⁺(H)(H)—C(H)(CH₃)—C(=O)—O⁻

(b)

⁺H₃N—C(H)(CH₃)—C(=O)—O⁻ + NaOH ⟶ H₂N—C(H)(CH₃)—C(=O)—O⁻Na⁺ + H₂O

3 (a)

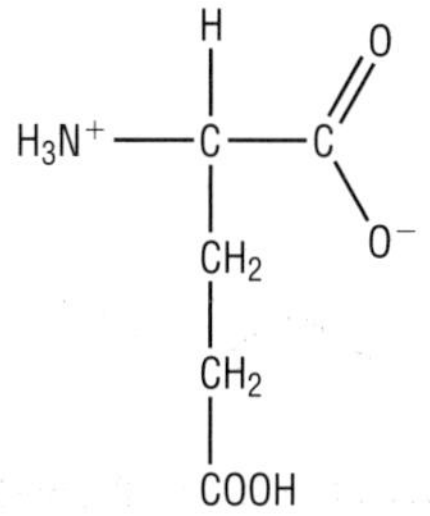

(b)

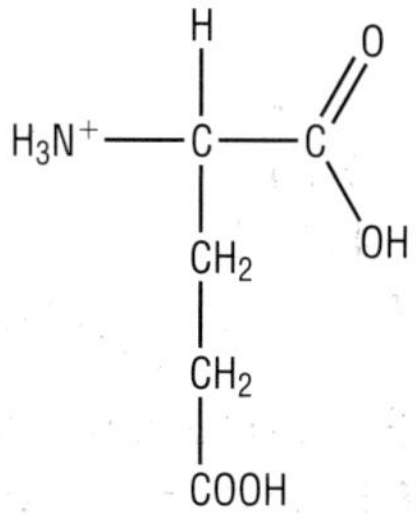

(c)

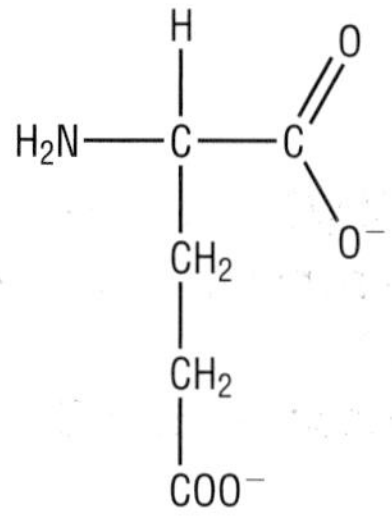

1.2.2 Polypeptides and proteins

1 (a)

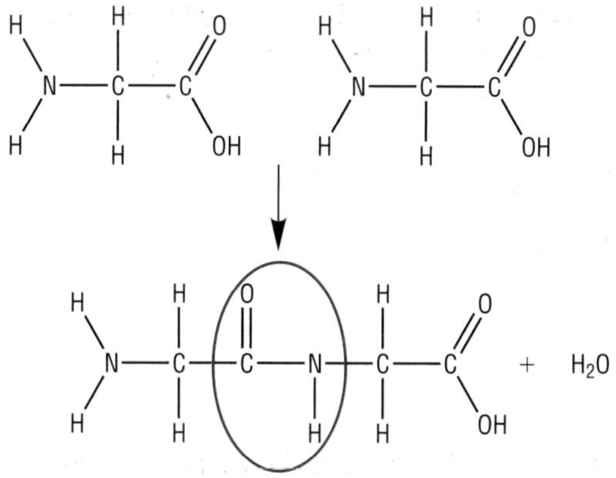

(b) Amine and carboxylic acid (or carboxyl group).

2

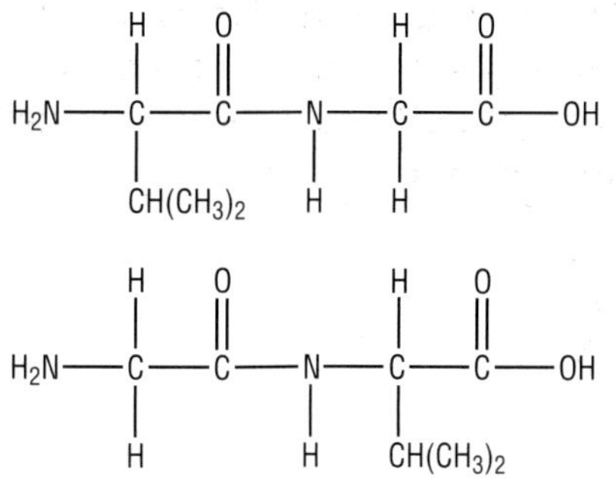

3 (a) 5

(b) HCl(aq), reflux.

(c)

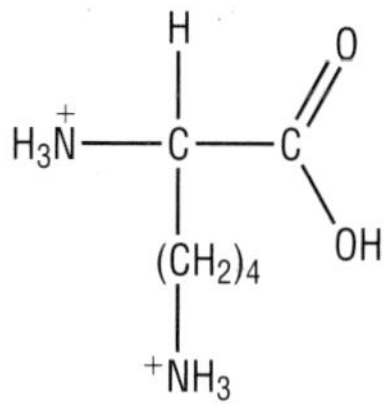

1.2.3 Optical isomerism

1 (a) A carbon atom attached to four different atoms or groups of atoms.

(b)

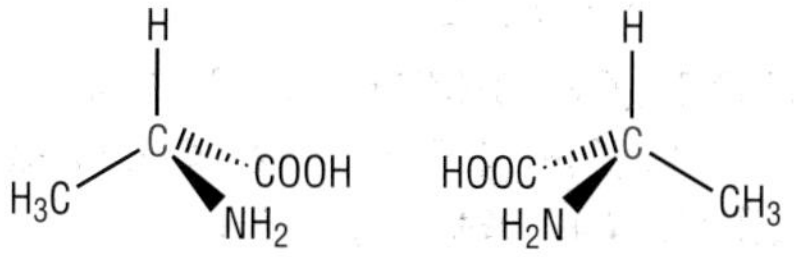

(c) 109.5°

2 Only **A** – it is the only one with a chiral carbon.

3 Two pairs because it has two chiral carbons.

1.2.4 Condensation polymerisation – polyesters

1 Carboxylic acid (or carboxyl group) and an alcohol.

2

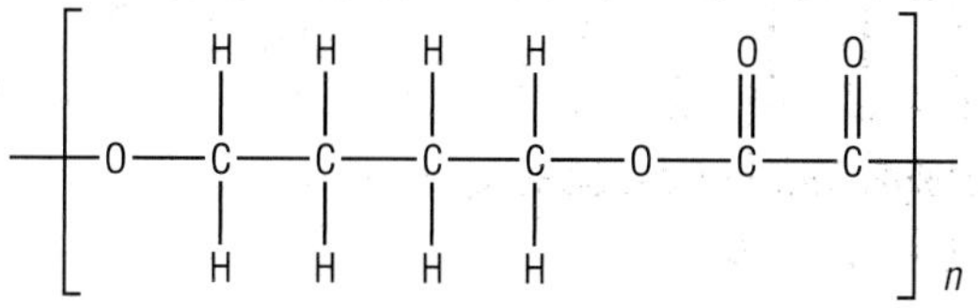

3 Terylene and poly(lactic acid).

1.2.5 Condensaton polymerisation – polyamides

1 Carboxylic acid (or carboxyl group) and an amine.

2

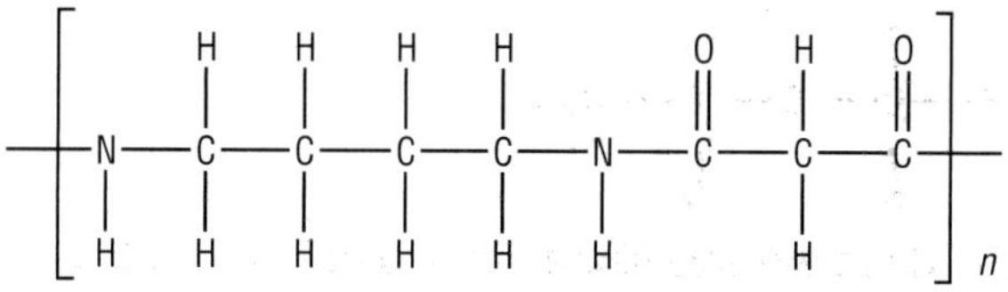

3 Kevlar and nylon-6,6.

1.2.6 Addition and condensation polymerisation

1 For example:

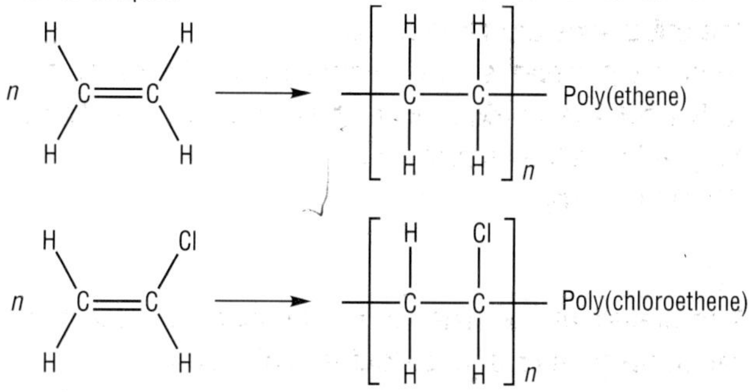
Poly(ethene)
Poly(chloroethene)

2 (a) Addition

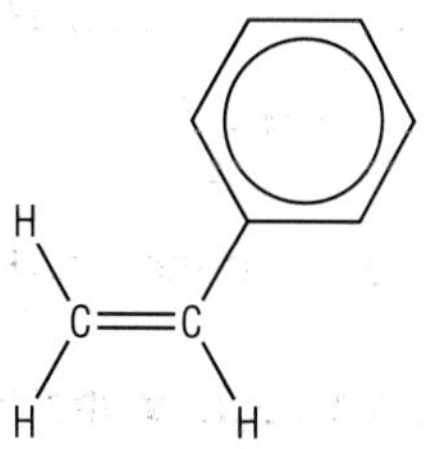

(b) Addition

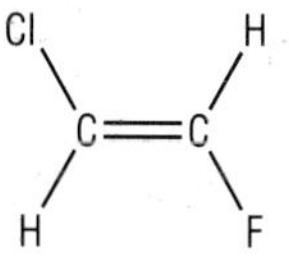

(c) Condensation

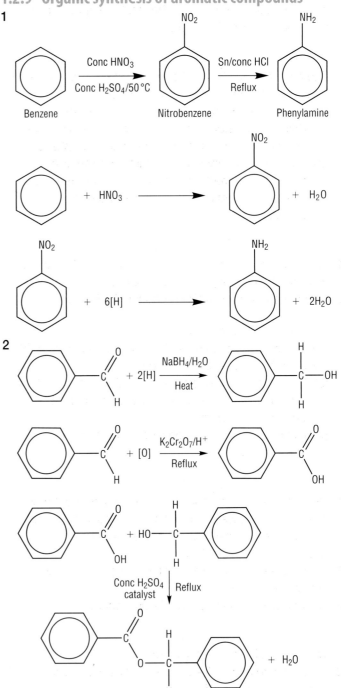

1.2.7 Breaking down condensation polymers

1

2 A polymer is considered to be biodegradable if, under carefully controlled conditions of temperature, humidity and the presence of microorganisms, it breaks down completely into carbon dioxide and water. A polymer is considered to be degradable if it breaks down into smaller fragments when exposed to light, heat or moisture.

3 Synthetic polymers that are designed to become weak and brittle when exposed to sunlight for prolonged periods.

1.2.8 Organic synthesis of aliphatic compounds

1 (a)

$$Cl-CH_2-CH_2-OH \xrightarrow[C_2H_5OH]{NH_3} H_2N-CH_2-CH_2-OH$$

$$\xrightarrow[]{K_2Cr_2O_7/H^+ \ \ Reflux}$$

$$H_2N-CH_2-COOH$$

(b)

1.2.9 Organic synthesis of aromatic compounds

1

Benzene → (Conc HNO₃, Conc H₂SO₄/50 °C) → Nitrobenzene → (Sn/conc HCl, Reflux) → Phenylamine

$$+ \ HNO_3 \longrightarrow \quad + \ H_2O$$

$$+ \ 6[H] \longrightarrow \quad + \ 2H_2O$$

2

$$+ \ 2[H] \xrightarrow[Heat]{NaBH_4/H_2O}$$

$$+ \ [O] \xrightarrow[Reflux]{K_2Cr_2O_7/H^+}$$

$$+ \ HO- $$ → (Conc H₂SO₄ catalyst, Reflux) → $$+ \ H_2O$$

1.2.10 Chirality in pharmaceutical synthesis

1 A carbon atom attached to four different atoms or groups of atoms.

2 Compounds with the same structural formula but with different arrangements of the atoms in space.

3 A larger dosage will be required as only one stereoisomer is pharmacologically active. The inactive stereoisomer may lead to harmful side effects.

1.3.1 Separation by chromatography

1 The mobile phase is the phase that moves; the stationary phase is the phase that does not move.
2 (a) A solid stationary phase separates by adsorption. Different components in a mixture bind to different degrees with the surface of the stationary phase.
(b) A liquid stationary phase separates by relative solubility. Different components in a mixture have different solubilities in the liquid of the stationary phase.

1.3.2 Thin-layer chromatography

1 The stationary phase is a solid material, usually silica or alumina; the mobile phase is a liquid solvent.
2 Component **A**: $R_f = 0.38$; component **B**: $R_f = 0.07$.

1.3.3 Gas chromatography

1 The stationary phase is a liquid or solid on an inert solid support; the mobile phase is a gas.
2 (a) Methanol: 0.8 min; ethanal: 0.9 min; ethanol: 1.1 min; 2-methylpropan-1-ol: 1.3 min; propanone: 1.6 min; propan-1-ol: 1.8 min.
(b) Evidence for blood alcohol content – likely to be linked to drink driving.

1.3.4 Gas chromatography–mass spectrometry

1 GC separates the components; each component then passes into a mass spectrometer; the resulting mass spectrum is matched against a spectral database to identify the component.
2 Forensic analysis, environmental analysis, airport security and space probes.
3 The comet 67P Churyumov–Gerasimenko.

1.3.5 What is nuclear magnetic resonance?

1 Tetramethylsilane (TMS), $(CH_3)_4Si$.
2 $CDCl_3$; deuterated solvents are used because they do not produce a proton NMR signal.

1.3.6 Carbon-13 NMR spectroscopy

1 (a) Two peaks; CH_3CH_2OH, $\delta = 5$–55 ppm; CH_3CH_2OH, $\delta = 50$–70 ppm.
(b) Two peaks; $CH_3CH_2NH_2$, $\delta = 5$–55 ppm; $CH_3CH_2NH_2$, $\delta = 35$–60 ppm.

(c) Four peaks; $CH_3COCH_2CH_3$, $\delta = 5$–55 ppm; $CH_3COCH_2CH_3$, $\delta = 160$–185 ppm $CH_3COCH_2CH_3$, $\delta = 5$–55 ppm; $CH_3COCH_2CH_3$, $\delta = 5$–55 ppm.
2 (a) A: **C**–Br; B: 2 × **CH$_3$**–C
(b) D: –**C**OO; E: O–**C**H$_3$; F: **C**H$_3$–CO

1.3.7 Analysis of carbon-13 NMR spectra

1 (a) $CH_3CH_2CH_2CHO$; $CH_3CH_2COCH_3$; $(CH_3)_2CHCHO$
(b) $CH_3CH_2CH_2CHO$ has four carbon environments; $CH_3CH_2COCH_3$ has four carbon environments; $(CH_3)_2CHCHO$ has three carbon environments.
(c) In the carbon-13 NMR spectrum, there are three peaks. Therefore there are three carbon environments. Therefore $(CH_3)_2CHCHO$ produced the spectrum.
(d) The other two isomers could not be distinguished as they both have four carbon environments and similar chemical shift values.

1.3.8 Proton NMR spectroscopy

1 (a) Three peaks; 3:2:1
CH_3CH_2CHO, $\delta = 0.5$–2.0 ppm;
CH_3CH_2CHO, $\delta = 2.0$–2.9 ppm;
CH_3CH_2CHO, $\delta = 9.0$–10.0 ppm.
(b) One peak; CH_3COCH_3, $\delta = 2.0$–2.9 ppm.
(c) Three peaks; 1:2:3;
$HCOOCH_2CH_3$, $\delta = 9.0$–10.0 ppm;
$HCOOCH_2CH_3$, $\delta = 3.0$–4.3 ppm;
$HCOOCH_2CH_3$, $\delta = 0.5$–2.0 ppm.
(d) Three peaks; 6:1:1;
$(CH_3)_2CHCHO$, $\delta = 0.5$–2.0 ppm;
$(CH_3)_2CHCHO$, $\delta = 2.0$–2.9 ppm;
$(CH_3)_2CHCHO$, $\delta = 9.0$–10.0 ppm.
(e) Three peaks; 5:2:3;
$C_6H_5CH_2CH_3$, $\delta = 6.5$–8.0 ppm;
$C_6H_5CH_2CH_3$, $\delta = 2.0$–2.9 ppm;
$C_6H_5CH_2CH_3$, $\delta = 0.5$–2.0 ppm.
2 (a) A: **C**H–Br; B: $(CH_3)_2$–C
(b) C: –O**C**H$_2$; D: **C**H$_3$CO; E: **C**H$_3$–C

1.3.9 Spin–spin coupling in proton NMR spectra

1 (a) CH_3–CH_2; **(b)** CH_3–CH; **(c)** CH_2–CH; **(d)** CH_2–CH_2
2 (a) triplet–quartet; **(b)** doublet–quartet;
(c) singlet, triplet–triplet.

1.3.10 NMR spectra of –OH and –NH protons

1 (a) Without D_2O: three peaks in the ratio 1(HO):2(CH_2):1(COOH);
With D_2O: one peak only (CH_2).

(b) Without D_2O: three peaks in the ratio
2(H_2N) : 2(CH_2) : 1(COOH);
With D_2O: one peak only (CH_2).

(c) Without D_2O: four peaks in the ratio
3(CH_3) : 1(CH) : 1(HO) : 1(COOH);
With D_2O: two peaks in the ratio 3(CH_3) : 1(CH).
CH_3 peak split as a doublet; CH peak split as a quartet.

2 COOH at δ = 11.0 ppm; NH_2 at δ = 5.2 ppm;
CH at δ = 3.8 ppm and CH_3 at δ = 1.2 ppm.
Structure = CH_3CHNH_2COOH

1.3.11 Spin–spin coupling examples

1

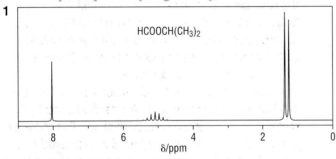

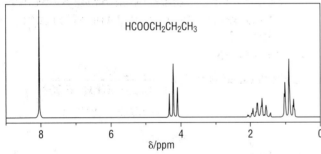

1.3.12 NMR in medicine

1 Patients might associate 'nuclear' with the use of radioactive elements.

2 Patients with ferromagnetic metal implants; patients with heart pacemakers.

3 A computer processes the information from the scan by taking 'slices' over time.

4 Bones produce a dark image, whereas fatty tissue produces a much brighter image. Brain disorders show up from differences in the water content of tissues.

1.3.13 Combined techniques

1 From left to right: 2(OCH_2) : 1(OH) : 2(CH_2) : 3(CH_3).

1.3.14 Combined techniques problems

1 IR spectrum
IR absorption at 1730 cm^{-1} for C=O.
Mass spectrum
Molecular ion peak at m/z = 102
Empirical formula mass = 102,
so molecular formula = $C_5H_{10}O_2$

Fragment ion at m/z = 71 for loss of 31. OCH_3?
Fragment ion at m/z = 43 for CH_3CO^+ or $C_3H_7^+$?
^{13}C NMR
4 types of C
C at δ = 173 ppm for **C**OO.
^{1}H NMR
3 types of H in relative amounts 3:1:6
From formula, 10H in total
δ = 3.7 ppm for 3Hs of type: OC**H**
δ = 2.6 ppm for 1H of type C**H**C=O
δ = 1.2 ppm for 6H of type CC**H**
δ = 3.7 ppm is a singlet which supports OC**H**$_3$
δ = 2.6 ppm is a heptet so there must be 6 Hs on adjacent carbons, as (CH_3)$_2$
δ = 1.2 ppm is a doublet so there must be one H on an adjacent carbon
From splitting, there must be (CH_3)$_2$CH.
Structure
Combining all the information,
compound = (CH_3)$_2$CHCOOCH_3.

2 IR spectrum
IR absorption at 1700 cm^{-1} for C=O.
Mass spectrum
Molecular ion peak at m/z = 134
Empirical formula mass = 134,
so molecular formula = $C_9H_{10}O_2$.
^{13}C NMR
7 types of C
4 C at δ = 150–127 ppm are aromatic. Together these account for 6 C atoms.
This pattern matches a 1,4 disubstituted benzene ring with different groups attached to 1 and 4 positions.
C at δ = 190 ppm for carbonyl **C**O (in aldehyde or ketone)
2 remaining C atoms as **C**–C, C–**C**=O or **C**–O.
^{1}H NMR
4 types of H, in relative amounts 1:4:2:3
δ = 7–7.9 ppm for 4 aromatic protons
δ = 9.8 ppm for 1H of type **C**HO
δ = 1.2 ppm for 3H of type **C**H**C**H
δ = 2.7 ppm for 2H of type **C**HC=O of **C**HAr
δ = 1.2 ppm is a triplet adjacent to a carbon with 2H;
δ = 2.7 ppm is a quartet adjacent to a carbon with 3H.
From triplet/quartet splitting there must be CH_3CH_2.
Structure
Combining all the information,
compound = $CH_3CH_2C_6H_4CHO$:

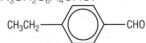

The aromatic protons δ = 7–7.9 ppm are split into two doublets.

2.1.1 Rate of reaction

1. $mol\ dm^{-3}\ s^{-1}$
2. A tangent is drawn to the curve at the required time, and the gradient of the tangent is measured.
3. The change in concentration of a reactant or product at the start of the reaction, when $t = 0$.

2.1.2 Measuring reaction rates

1. (a), (b)

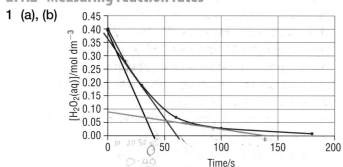

(c) $t = 0$ s,   rate = approximately $0.01\ mol\ dm^{-3}\ s^{-1}$
   $t = 20$ s, rate = approximately $0.006\ mol\ dm^{-3}\ s^{-1}$
   $t = 90$ s, rate = approximately $0.00065\ mol\ dm^{-3}\ s^{-1}$

2.1.3 Orders and the rate equation

1. (a) Rate quadruples;
   (b) rate triples;
   (c) rate increases by 27 times.
2. (a) R: zero order; S: 2nd order; T: 1st order
   (b) $rate = k[S]^2[T]$
3. $k = 8.3 \times 10^4\ dm^6\ mol^{-2}\ s^{-1}$

2.1.4 Half-lives

1. $0.20\ mol\ dm^{-3}$
2. 1520 s

2.1.5 Orders from rate–concentration graphs

1. (a) P: zero order; Q: 2nd order.
   (b) $rate = k[Q]^2$
2. (a) R: 2nd order; S: 1st order.
   (b) Reactant R

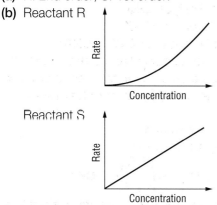

   Reactant S

2.1.6 Initial rates and the rate constant

1. From experiment 1, $k = \dfrac{(1.82 \times 10^{-6})}{(0.001\,00)^2(0.001\,00)}$
   $= 1820\ dm^6\ mol^{-2}\ s^{-1}$

   From experiment 3, $k = \dfrac{(7.28 \times 10^{-6})}{(0.002\,00)^2(0.001\,00)}$
   $= 1820\ dm^6\ mol^{-2}\ s^{-1}$

2. (a) Comparing experiment 1 with experiment 2, [A] and [C] have been kept constant and [B] has increased by 2 times; the rate has increased by $4 = 2^2$ times; so the reaction is 2nd order for B.
   Comparing experiment 2 with experiment 3, [B] and [C] have been kept constant and [A] has decreased by 0.5 times; the rate has decreased by 0.5 times; so the reaction is 1st order for A.
   Comparing experiment 1 with experiment 4, only [B] has been kept constant and [A] has increased by 3 times; A is 1st order, so this change will also increase the rate by 3 times; the rate has also increased by 3 times; [C] has decreased by 0.5 times, but there is no further change in the rate, so the reaction is zero order for C.
   (b) $rate = k[A][B]^2$
   (c) From experiment 1, $k = \dfrac{2.16 \times 10^{-7}}{(0.001\,00)(0.003\,00)^2}$
   $= 24.0\ dm^6\ mol^{-2}\ s^{-1}$

2.1.7 Rate-determining step

1. $rate = k[X]^2$
2. (a) The slowest step in the reaction mechanism of a multi-step reaction.
   (b) The rate-determining step involves two molecules of NO.
   (c) $2NO(g) \longrightarrow N_2O_2(g)$
   $N_2O_2(g) + O_2(g) \longrightarrow NO_2(g)$

2.1.8 The equilibrium constant, K_c

1. (a) $K_c = \dfrac{[NO_2(g)]^2}{[NO(g)]^2\ [O_2(g)]}$; units are $dm^3\ mol^{-1}$
   (b) $K_c = \dfrac{[HBr(g)]^2}{[H_2(g)]\ [Br_2(g)]}$; no units
   (c) $K_c = \dfrac{[NH_3(g)]^2}{[N_2(g)]\ [H_2(g)]^3}$; units are $dm^6\ mol^{-2}$

2.1.9 Calculations using K_c

1. $K_c = \dfrac{[NH_3(g)]^2}{[N_2(g)]\ [H_2(g)]^3} = 7.99 \times 10^{-2}\ dm^6\ mol^{-2}$

2 (a) $2.00 - 0.43 = 1.57$ mol

(b) $[CH_3COOH]$: $0.43/V$ mol dm^{-3}

$[C_2H_5OH]$: $1.43/V$ mol dm^{-3}

$[CH_3COOC_2H_5]$: $1.57/V$ mol dm^{-3}

and $[H_2O]$: $1.57/V$ mol dm^{-3}

(c) $K_c = 4.01$

2.1.10 The equilibrium position and K_c

1 Reaction A is exothermic; K_c decreases with increasing temperature.

Reaction B is endothermic; K_c increases with increasing temperature.

2.1.11 The equilibrium constant, K_c, and the rate constant, k

1 (a) $K_c = \dfrac{[NO_2(g)]^2}{[NO(g)]^2 \, [O_2(g)]}$

(b) An increase in concentration of NO(g) increases the term on the bottom of the expression. The equilibrium moves to restore K_c by increasing the top and decreasing the bottom – so the equilibrium moves from left to right.

(c) There are more concentration terms on the bottom of the expression. A decrease in pressure will decrease the term on the bottom of the expression by more than the top. The equilibrium moves to restore K_c by increasing the bottom and decreasing the top – so the equilibrium moves from right to left.

2 A change in pressure changes the top of the K_c expression by the same as the bottom of the expression. The relative concentrations still give the correct K_c value, so the system is still at equilibrium and does not move.

2.1.12 The road to acids

1 Identified: H_2SO_4, HNO_3; missed: HCl, HBr.

2 Arrhenius acid: any substance that dissociates in water to form protons, H$^+$.

Arrhenius base: any substance that dissociates in water to form hydroxide ions, OH$^-$.

Brønsted–Lowry acid: any substance that is a proton donor, irrespective of the solvent.

Brønsted–Lowry base: any substance that is a proton acceptor, irrespective of the solvent.

Lewis acid: any substance that is an electron-pair acceptor.

Lewis base: any substance that is an electron-pair donor.

2.1.13 The role of H$^+$ in reactions of acids

1 (a) $HNO_3(aq) \longrightarrow H^+(aq) + NO_3^-(aq)$

(b) $H_2CrO_4(aq) \longrightarrow H^+(aq) + HCrO_4^-(aq)$

$HCrO_4^-(aq) \rightleftharpoons H^+(aq) + CrO_4^{2-}(aq)$

2 (a) full: $H_2SO_4(aq) + MgCO_3(s) \longrightarrow$
$MgSO_4(aq) + CO_2(g) + H_2O(l)$

ionic: $2H^+(aq) + MgCO_3(s) \longrightarrow$
$Mg^{2+}(aq) + CO_2(g) + H_2O(l)$

(b) full: $H_2SO_4(aq) + K_2CO_3(aq) \longrightarrow$
$K_2SO_4(aq) + CO_2(g) + H_2O(l)$

ionic: $2H^+(aq) + CO_3^{2-}(aq) \longrightarrow CO_2(g) + H_2O(l)$

(c) full: $2HCl(aq) + CaO(s) \longrightarrow CaCl_2(aq) + H_2O(l)$

ionic: $2H^+(aq) + CaO(s) \longrightarrow Ca^{2+}(aq) + H_2O(l)$

(d) full: $HNO_3(aq) + NaOH(aq) \longrightarrow NaNO_3(aq) + H_2O(l)$

ionic: $H^+(aq) + OH^-(aq) \longrightarrow H_2O(l)$

(e) full: $6HCl(aq) + 2Al(s) \longrightarrow 2AlCl_3(aq) + 3H_2(g)$

ionic: $6H^+(aq) + 2Al(s) \longrightarrow 2Al^{3+}(aq) + 3H_2(g)$

2.1.14 Conjugate acid–base pairs

1 (a) acid 1: HIO_3, base 1: IO_3^-; acid 2: H_3O^+, base 2: H_2O

(b) acid 1: CH_3COOH, base 1: CH_3COO^-;
acid 2: H_3O^+, base 2: H_2O

(c) acid 1: H_2O, base 1: OH^-; acid 2: NH_4^+, base 2: NH_3

2 $HF(aq) + H_2O(l) \rightleftharpoons H_3O^+(aq) + F^-(aq)$

2.1.15 What is pH?

1 pH 2 has 100 times more H$^+$ ions than pH 4.

2 (a) 2.48; **(b)** 3.33; **(c)** 11.62; **(d)** –0.36.

3 (a) 2.95×10^{-7} mol dm^{-3}; **(b)** 1.35×10^{-3} mol dm^{-3}

(c) 2.63×10^{-10} mol dm^{-3}; **(d)** 1.32 mol dm^{-3}.

2.1.16 Strong and weak acids

1 (a) $K_a = \dfrac{[H^+(aq)] \, [HCOO^-(aq)]}{[HCOOH(aq)]}$;

(b) $K_a = \dfrac{[H^+(aq)] \, [CH_3CH_2COO^-(aq)]}{[CH_3CH_2COOH(aq)]}$

2 (a) $pK_a = 0.64$; **(b)** $pK_a = 2.60$; **(c)** $pK_a = 10.32$

3 (a) $K_a = 1.26 \times 10^{-3}$ mol dm^{-3};

(b) $K_a = 6.31 \times 10^{-8}$ mol dm^{-3};

(c) $K_a = 2.51 \times 10^{-11}$ mol dm^{-3}.

2.1.17 Calculating pH for strong and weak acids

1 (a) 2.48

(b) 4.39

2 (a) $K_a = 7.02 \times 10^{-7}$ mol dm^{-3}; $pK_a = 6.15$

(b) 5.34×10^{-10} mol dm^{-3}; $pK_a = 9.27$

2.1.18 The ionisation of water

1 $K_w = [H^+(aq)][OH^-(aq)]$

2 (a) $[OH^-(aq)] = 10^{-8}$ mol dm^{-3};

(b) $[OH^-(aq)] = 10^{-12}$ mol dm^{-3};

(c) $[OH^-(aq)] = 10^{-3}$ mol dm^{-3}.

3 (a) $[H^+(aq)] = 10^{-1}$ mol dm^{-3};
(b) $[H^+(aq)] = 10^{-10}$ mol dm^{-3};
(c) $[H^+(aq)] = 10^{-4}$ mol dm^{-3}.

2.1.19 pH values of bases

1 (a) $[H^+(aq)] = 5.0 \times 10^{-12}$ mol dm^{-3}; pH = 11.30
(b) $[H^+(aq)] = 1.75 \times 10^{-14}$ mol dm^{-3}; pH = 13.76.
2 (a) pH = 11.70;   **(b)** pH = 12.55.
3 (a) 2.69×10^{-2} mol dm^{-3};   **(b)** 6.61×10^{-1} mol dm^{-3}.

2.1.20 Buffer solutions

1 A buffer solution is a system that minimises pH changes on addition of small amounts of an acid or a base.
2 The buffer equilibrium system is
$CH_3CH_2COOH(aq) \rightleftharpoons H^+(aq) + CH_3CH_2COO^-(aq)$
Added acid, $H^+(aq)$, reacts with the conjugate base, $CH_3CH_2COO^-(aq)$:
$H^+(aq) + CH_3CH_2COO^-(aq) \longrightarrow CH_3CH_2COOH(aq)$
Added alkali, $OH^-(aq)$, reacts with the $H^+(aq)$:
$H^+(aq) + OH^-(aq) \longrightarrow H_2O(l)$
$H^+(aq)$ ions are replaced by the equilibrium moving to the right.

2.1.21 pH values of buffer solutions

1 $[H^+(aq)] = K_a \times \dfrac{[HCOOH(aq)]}{[HCOO^-(aq)]} = 1.6 \times 10^{-4} \times \dfrac{0.15}{0.065}$
$= 3.7 \times 10^{-4}$ mol dm^{-3}
pH = $-\log [H^+(aq)] = -\log (3.7 \times 10^{-4}) = 3.43$.
2 An increase in $H^+(aq)$ ions in the blood is removed by $HCO_3^-(aq)$, forming carbonic acid:
$H^+(aq) + HCO_3^-(aq) \longrightarrow H_2CO_3(aq)$
The carbonic acid is converted into aqueous carbon dioxide by an enzyme. In the lungs the dissolved carbon dioxide is converted into carbon dioxide gas, which is then exhaled.

2.1.22 Neutralisation – titration curves

1 (a) The equivalence point is the point in a titration at which the volume of one solution has reacted exactly with the volume of the second solution.
(b) The end point of an indicator is the point at which there are equal concentrations of the weak acid and conjugate base forms of the indicator.
2 Bromocresol green: strong acid–strong base and strong acid–weak base.
Thymolphthalein: strong acid–strong base and weak acid–strong base.

2.1.23 Neutralisation – enthalpy changes

1 $\Delta H = -57.5$ kJ mol^{-1}

2 (a) Strong acids are completely dissociated. The actual reaction taking place has the same ionic equation: $H^+(aq) + OH^-(aq) \longrightarrow H_2O(l)$. The other ions are *spectator ions* that do not take part in the reaction.
(b) Weak acids are only partially dissociated. Although the same neutralisation reaction takes place, most of the weak acid molecules must first dissociate to release $H^+(aq)$ and this process requires energy. Consequently the enthalpy change recorded is slightly less exothermic.

2.2.1 Lattice enthalpy

1 Lattice enthalpy is defined as the enthalpy change that accompanies the formation of one mole of an ionic compound from its gaseous ions under standard conditions.
2 Lattice enthalpies cannot be measured directly because it is impossible to use gaseous ions to form one mole of an ionic lattice experimentally.
3 (a) $Li(g) \longrightarrow Li^+(g) + e^-$
(b) $Ca(s) + Cl_2(g) \longrightarrow CaCl_2(s)$
(c) $Ca^{2+}(g) + 2Cl^-(g) \longrightarrow CaCl_2(s)$

2.2.2 Constructing Born–Haber cycles

1 (a) $Mg(s) + Br_2(l) \longrightarrow MgBr_2(s)$; enthalpy change of formation
(b) $Mg^{2+}(g) + 2Br^-(g) \longrightarrow MgBr_2(s)$; lattice enthalpy
2 Energy is required to break the bond between the two chlorine atoms in a chlorine molecule.
3 Lattice enthalpies cannot be measured directly because it is impossible to use gaseous ions to form one mole of an ionic lattice from gaseous ions experimentally.

2.2.3 Born–Haber cycle calculations

1 $\Delta H_{LE} = -435 - 81 - 121 - 403 + 346 = -694$ kJ mol^{-1}

2.2.4 Further examples of Born–Haber cycles

1 (a)

(b) (i) The enthalpy change involved in adding one electron to each atom in one mole of gaseous atoms to form one mole of gaseous 1– ions.

(ii) $-642 = +150 + 242 + 736 + 1450 + (-2 \times \Delta H_{EA})$ $+ (-2493)$

$\Delta H_{EA} = -363.5$ kJ mol^{-1}

2.2.5 Enthalpy change of solution

1 (a) Lattice enthalpy; enthalpy change of hydration of sodium ions; enthalpy change of solution of sodium fluoride.

(b) $F^-(g) + aq \longrightarrow F^-(aq)$

(c) $\Delta H = -457$ kJ mol^{-1}

2.2.6 Understanding hydration and lattice enthalpies

1 Cs^+ ions are larger than Na^+ ions so the electrostatic forces of attraction between Na^+ and Cl^- ions are stronger than between Cs^+ and Cl^- ions.

2 (a) Lattice enthalpy is defined as the enthalpy change that accompanies the formation of one mole of an ionic compound from its gaseous ions under standard conditions.

(b) As the ionic radius increases, the attraction between the ions decreases. This leads to the lattice enthalpy becoming less negative and the ionic bonds weaker. As ionic charge increases, attraction between ions increases. Therefore the lattice enthalpy becomes more negative and the ionic bonds stronger.

3 Small ions have a greater attraction for water molecules than do large ions – so when they become hydrated more energy is released. As Na^+ is smaller than Rb^+, the hydration enthalpy of Na^+ ions will be more exothermic than that of Rb^+ ions.

2.2.7 Entropy

1 (a) +ve;   **(b)** –ve;   **(c)** –ve.

2 (a) –323 J K^{-1} mol^{-1};   **(b)** –216.5 J K^{-1} mol^{-1}

2.2.8 Free energy

1 –806 kJ mol^{-1}

2 347 °C

2.2.9 Redox

1 (a) $2Al + 3Cu^{2+} \longrightarrow 2Al^{3+} + 3Cu$

(b) $10Br^- + 2MnO_4^- + 16H^+ \longrightarrow 5Br_2 + 2Mn^{2+} + 8H_2O$

2 (a) $2HBr + H_2SO_4 \longrightarrow Br_2 + SO_2 + 4H_2O$

(b) $3H_2S + 2HNO_3 \longrightarrow 3S + 2NO + 4H_2O$

2.2.10 Cells and half cells

1 (a) It consists of 1 mol dm^{-3} hydrochloric acid as the source of $H^+(aq)$, hydrogen gas, $H_2(g)$, at 100 kPa (1 atmosphere) pressure, and an inert platinum electrode.

(b) (i) along a wire;   **(ii)** through a salt bridge.

2.2.11 Cell potentials

1 (a) (i) 1.02 V;   **(ii)** 0.30 V;   **(iii)** 2.10 V

(b) $Cu(s) + Cl_2(g) \longrightarrow Cu^{2+}(aq) + 2Cl^-(aq)$
$3Fe^{2+}(aq) + 2Cr(s) \longrightarrow 3Fe(s) + 2Cr^{3+}(aq)$
$Cr(s) + 1\frac{1}{2}Cl_2(g) \longrightarrow Cr^{3+}(aq) + 3Cl^-(aq)$

2.2.12 The feasibility of reactions

1 (a) Ni(s) with $Fe^{3+}(aq)$; Ni(s) with $Br_2(aq)$; Ni(s) with $O_2(g)$ and $H^+(aq)$
$Fe^{2+}(aq)$ with $Br_2(aq)$; $Fe^{2+}(aq)$ with $O_2(g)$ and $H^+(aq)$
$Br^-(aq)$ with $O_2(g)$ and $H^+(aq)$

(b) $Ni(s) + 2Fe^{3+}(aq) \longrightarrow Ni^{2+}(aq) + 2Fe^{2+}(aq)$
$Ni(s) + Br_2(aq) \longrightarrow Ni^{2+}(aq) + 2Br^-(aq)$
$2Ni(s) + O_2(g) + 4H^+(aq) \longrightarrow 2Ni^{2+}(aq) + 2H_2O(l)$
$2Fe^{2+}(aq) + Br_2(aq) \longrightarrow 2Fe^{3+}(aq) + 2Br^-(aq)$
$4Fe^{2+}(aq) + O_2(g) + 4H^+(aq) \longrightarrow 4Fe^{3+}(aq) + 2H_2O(l)$
$4Br^-(aq) + O_2(g) + 4H^+(aq) \longrightarrow 2Br_2(aq) + 2H_2O(l)$

2.2.13 Storage and fuel cells

1 Non-rechargeable cells; rechargeable cells; fuel cells.

2 Negative electrode: $H_2(g) + 2OH^-(aq) \longrightarrow 2H_2O(l) + 2e^-$
Positive electrode: $\frac{1}{2}O_2(g) + 2H_2O(l) + 2e^- \longrightarrow 2OH^-(aq)$
Overall cell reaction: $H_2(g) + \frac{1}{2}O_2(g) \longrightarrow H_2O(l)$

2.2.14 Hydrogen for the future

1 It is reacted with water in an onboard reformer to produce hydrogen:
$CH_3OH(l) + H_2O(l) \longrightarrow 3H_2(g) + CO_2(g)$
The hydrogen is then fed into conventional fuel cells.

2 As a liquid under pressure; adsorbed onto the surface of a solid material; absorbed within a solid material.

2.3.1 Transition metals

1 (a) A d-block element is one in which electrons are filling d-orbitals and the highest energy sub-shell is a d sub-shell. A transition element is a d-block element which forms at least one ion with an incomplete d sub-shell.

2 (a) $1s^22s^22p^63s^23p^64s^13d^5$;

(b) $1s^22s^22p^63s^23p^63d^5$;

(c) $1s^22s^22p^63s^23p^6$;

(d) $1s^22s^22p^63s^23p^64s^23d^2$.

2.3.2 Properties of transition metal compounds

1 They form coloured compounds; have variable oxidation states; can be used as catalysts.
2 Colourless because the Zn^{2+} ion does not have a partially filled d-orbital.
3 +5 and +4.

2.3.3 Catalysis and precipitation

1 A catalyst increases the rate of a chemical reaction without undergoing any permanent change itself. It does this by providing an alternative route of lower activation energy.
2 e.g. The Haber process uses iron as a catalyst;
$N_2(g) + 3H_2(g) \rightleftharpoons 2NH_3(g)$
3 Fe^{3+}; +3
4 $Ni^{2+}(aq) + 2OH^-(aq) \longrightarrow Ni(OH)_2(s)$

2.3.4 Transition metals and complex ions

1 (a) $[CoCl_4]^{2-}$;   (b) $[Fe(H_2O)_5Cl]^{2+}$
2 A water molecule has a lone pair of electrons which can be donated to the central metal ion. The copper accepts these electrons and a coordinate bond is formed. A ligand is an electron-pair donor to a metal ion.
3 The coordination number is the total number of coordinate bonds formed between the central metal ion and any ligands; in the complex ion $[Co(H_2O)_6]^{2+}$ there are six coordinate bonds, one from each of the water ligands surrounding Co^{2+} – the coordination number is 6.

2.3.5 Stereoisomerism in complex ions

1 *cis–trans*

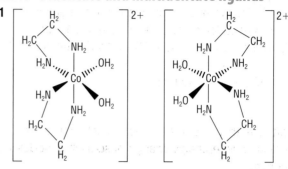

| Cis | Trans |

2 Stereoisomers are molecules or complexes with the same structural formula but a different spatial arrangement of these atoms.

2.3.6 Bidentate and multidentate ligands

1

2.3.7 Ligand substitution in complexes

1 This is a reaction in which a ligand in a complex ion is replaced by another ligand.
2 (a) $Cu(OH)_2$ precipitates
$Cu^{2+}(aq) + 2OH^-(aq) \longrightarrow Cu(OH)_2(s)$
(b) $[Cu(NH_3)_4(H_2O)_2]^{2+}$
3 Increasing the temperature of the system causes the equilibrium to move to the right; the forward reaction is endothermic.

2.3.8 Ligand substitution and stability constants

1 Four (one for each of the 4 haem groups).
2 $K_{stab} = \dfrac{[[Ni(NH_3)_6]^{2+}]}{[[Ni(H_2O)_6]^{2+}][NH_3]^6}$
3 $[Ni(NH_3)_6]^{2+}$ is more stable because its K_{stab} is larger than that for $[Co(NH_3)_6]^{2+}$.

2.3.9 Redox titrations

1 (a) A reaction in which one species is oxidised at the same time as another species is reduced.
(b) +7
2 Amount of Fe^{2+} = 0.003 75 mol;
amount of MnO_4^- = 0.000 750 mol.
Concentration of MnO_4^- = 0.0231 mol dm^{-3}.

2.3.10 Examples of redox titrations

1 Amount of MnO_4^- = 0.000 446 mol;
amount of Fe^{2+} = 0.002 23 mol.
Amount of Fe^{2+} in 250 cm^3 = 0.0223 mol;
mass of Fe = 1.245 g; %Fe = 49.8%.
2 900 cm^3

2.3.11 Redox titrations – iodine and thiosulfate

1 Amount of $S_2O_3^{2-}$ = 0.011 85 mol;
amount of Cu^{2+} = 0.011 85 mol;
% copper = 83.6%.
2 Amount of $S_2O_3^{2-}$ = 2.405 × 10^{-3} mol
Amount of $Cu^{2+}(aq)$ in 25.0 cm^3 = 2.405 × 10^{-3} mol
Amount of salt in 250 cm^3 solution = 2.405 × 10^{-2} mol
Molar mass of copper salt = 247 g mol^{-1}

Practice answers

1.1 Rings, polymers and analysis

1 Benzene is a cyclic hydrocarbon with six carbon atoms and six hydrogen atoms. The six carbon atoms are arranged in a planar hexagonal ring. Each of the carbon atoms is joined to two other carbon atoms and a hydrogen atom. The shape around each carbon atom is trigonal planar with a bond angle of 120°.

Each carbon atom has a fourth outer-shell electron in a 2p orbital above and below the plane of the carbon atoms. These p electrons on each carbon atom overlap with the electrons in the p-orbitals on either side. This results in the formation of a ring of electron density above and below the plane of the carbon atoms. This extensive overlapping produces a delocalised system of π-bonds which spread over all six carbon atoms. Each carbon–carbon bond length and bond enthalpy is between the values for a C–C single bond and a C=C double bond.

2 (a) Dyes and pharmaceuticals.
(b) Concentrated nitric and concentrated sulfuric acid are needed, with the mixture being heated to 50 °C.
(c) $C_6H_6 + HNO_3 \longrightarrow C_6H_5NO_2 + H_2O$
(d) (i)

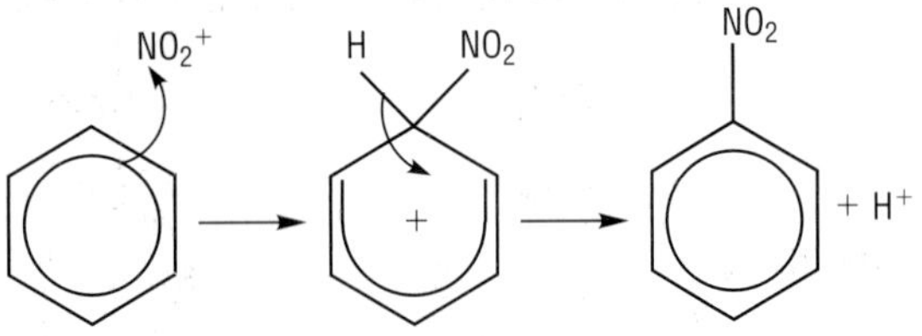

(ii) The movement of a pair of electrons to either make or break a bond.
(iii) Nitryl cation (or nitronium ion); reacts as an electrophile.

3 (a) (i)

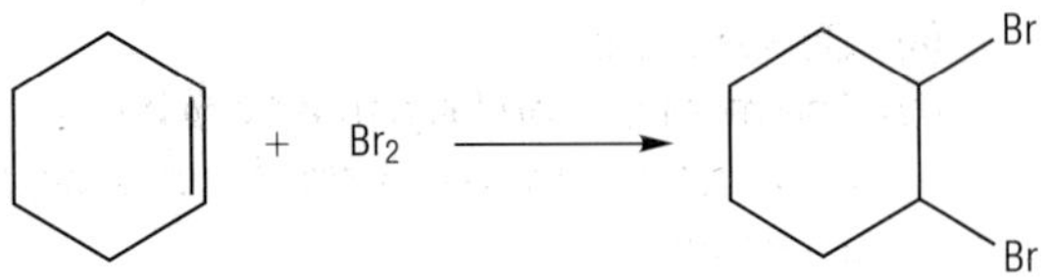

(ii) 1,2-dibromocyclohexane.
(iii) The bromine water would be decolourised.
(iv) Electrophilic addition.
(b) (i) $C_6H_6 + Br_2 \longrightarrow C_6H_5Br + HBr$
(ii) Fe, $AlBr_3$ or $FeBr_3$
(iii) Bromobenzene.
(iv) The bromine would be decolourised.
(v) Electrophilic substitution.

4 (a)

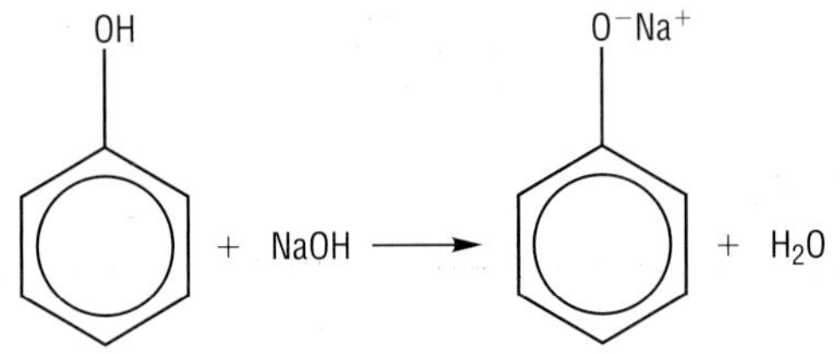

Sodium phenoxide

(b)

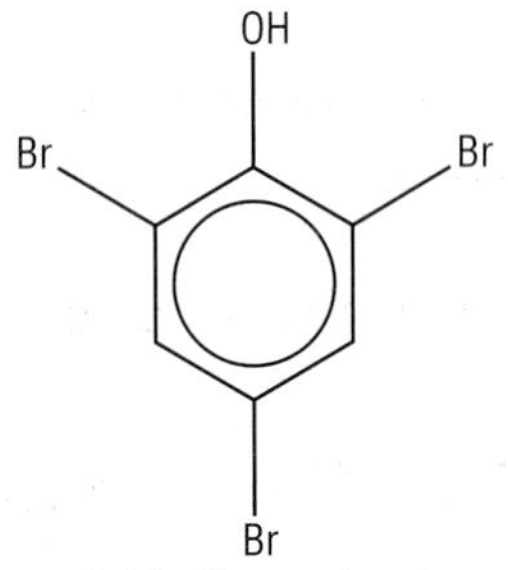

2,4,6-tribromophenol

(c) B is sodium phenoxide, $C_6H_5O^-Na^+$ (structure in **(a)**); D is hydrogen.
(d) Antiseptics, making dyes, making pharmaceuticals, making resins.

5 (a)

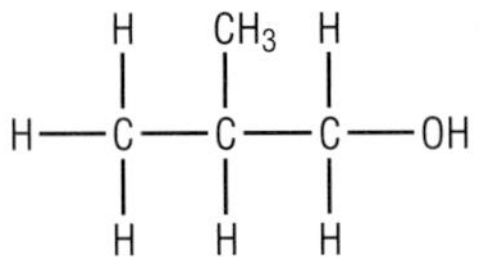

Primary
2-methylpropan-1-ol

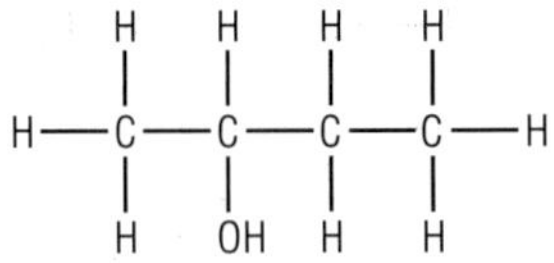

Secondary
butan-2-ol

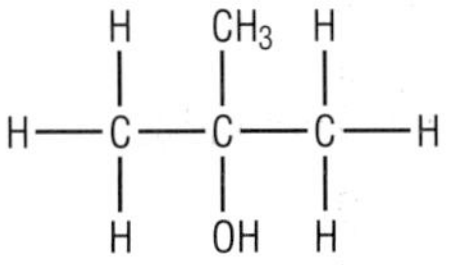

Tertiary
2-methylpropan-2-ol

(b) (i) 2-methylpropan-1-ol and butan-2-ol would react and the orange colour of dichromate(VI) would change to a green colour; 2-methylpropan-2-ol would not react.

(ii)

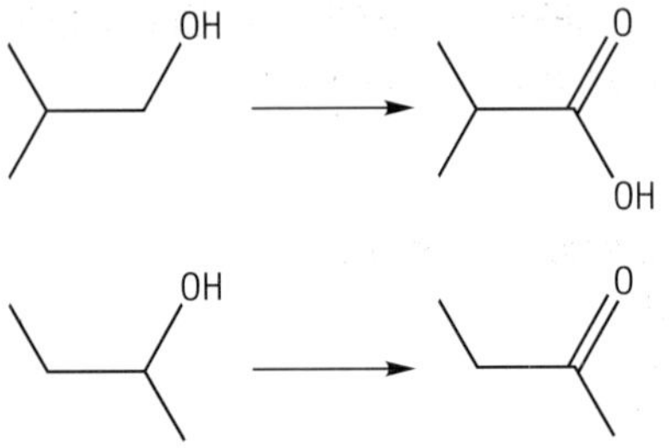

6 (a)

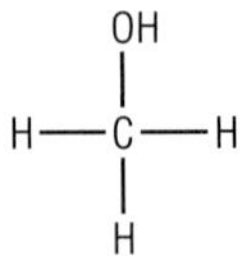

(b)

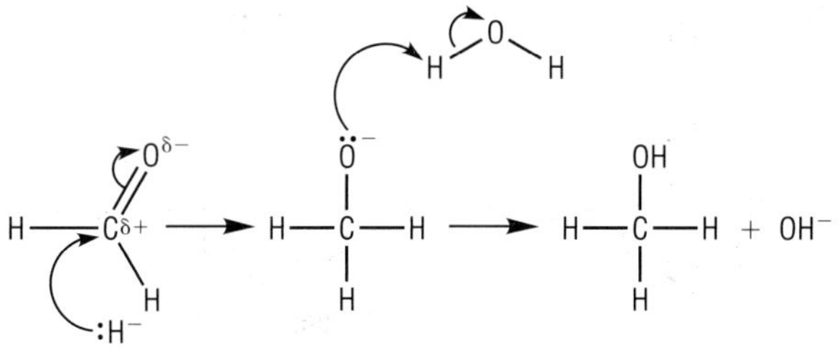

7 (a) Potassium propanoate

(b) Methanol and concentrated sulfuric acid, reflux;
$CH_3CH_2COOH + CH_3OH \longrightarrow$
$CH_3CH_2COOCH_3 + H_2O$

(c) Effervescence because carbon dioxide gas is evolved.

8 (a)

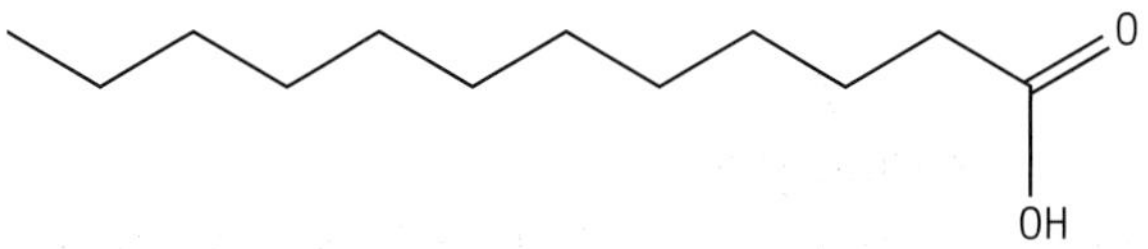

(b)

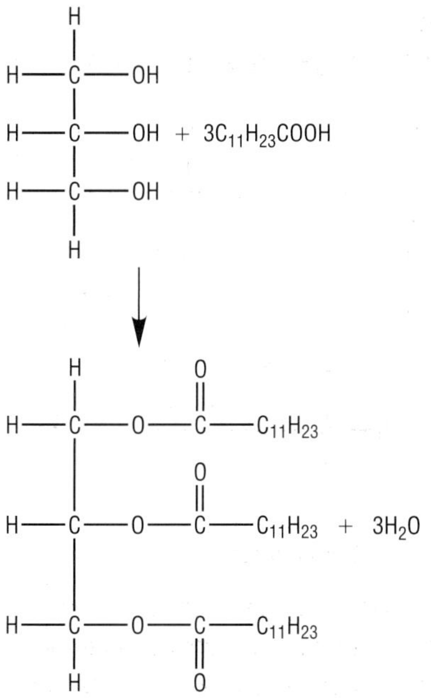

9 (a) A base is defined as proton acceptor.

(b) (i) $CH_3CH_2CH_2NH_2 + HCl \longrightarrow CH_3CH_2CH_2NH_3^+Cl^-$

(ii) Propylammonium chloride.

1.2 Polymers and synthesis

1 (a) Carboxylic acid and amine.

(b)

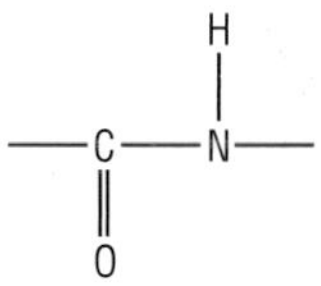

(c) (i) A zwitterion ion is a dipolar ionic form of an amino acid formed by donation of a hydrogen ion from the carboxyl group to the amino group.

(ii) Isoelectric point.

(iii)

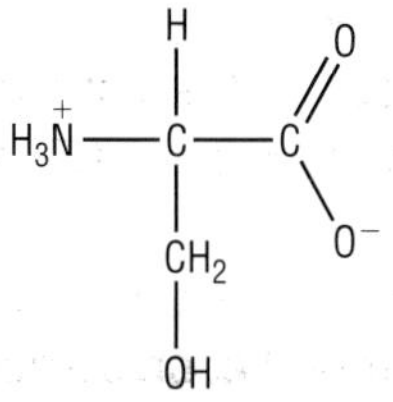

(d)

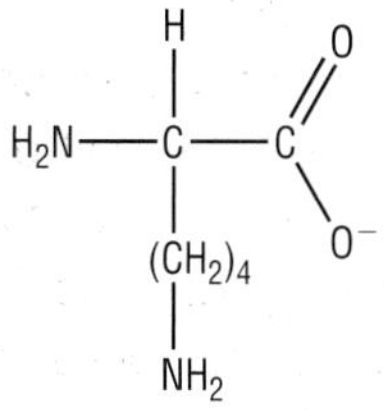

2 (a) Glutamic acid has a chiral carbon atom attached to four different atoms or groups of atoms.

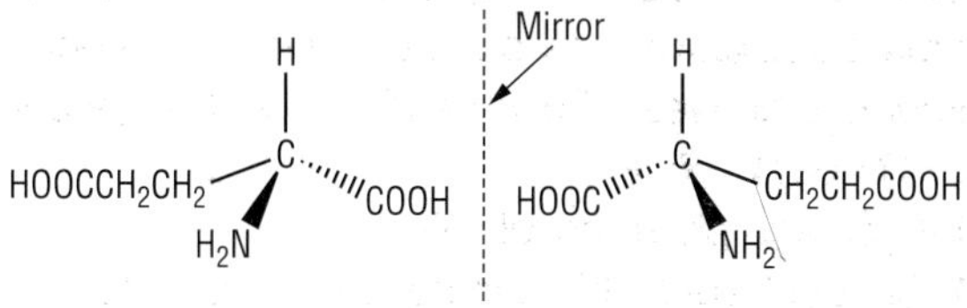

(b) (i)

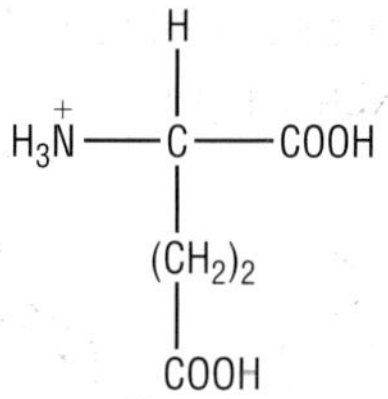

(ii)

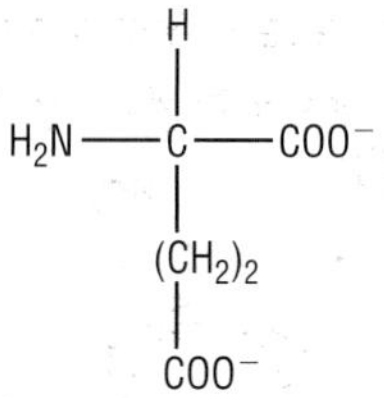

(c) (i)

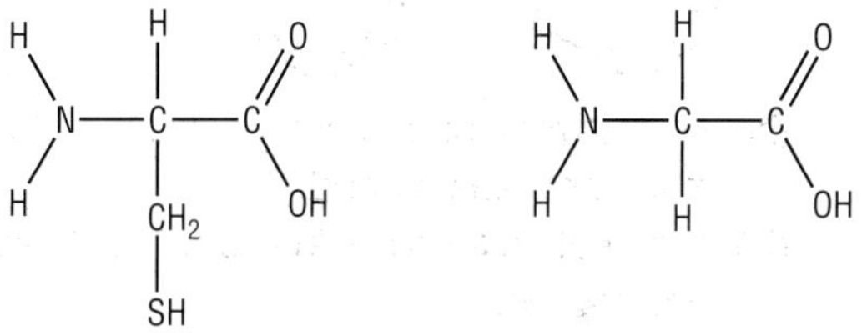

(ii) Condensation.

(iii) The reaction can be carried out by heating the polypeptide with 6 mol dm^{-3} hydrochloric acid for 24 hours.

3 (a) A chiral carbon is a carbon atom that is attached to four different atoms or groups of atoms.

(b) (i) Pentan-1-ol

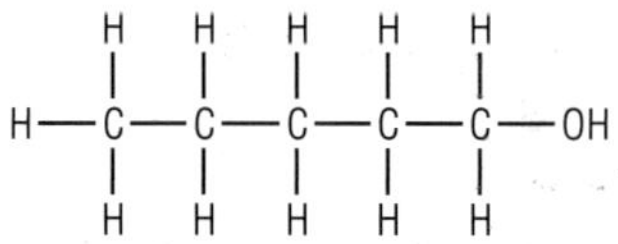

(ii) pent-2-ene

H—C—C=C—C—C—H (structure with H atoms: CH₃ on left, double bond, right side ethyl)

(iii) 2-butylamine

H—C—C*—C—C—H (with H and NH₂ groups)

(iv) 1-chloroethanol

H—C—C*—OH (with H and Cl)

(c) See on diagrams above.

4 **(a)** Condensation. **(b)** Ester.

(c) (i)

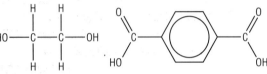

Ethane-1,2-diol   Benzene-1,4-dicarboxylic acid

(ii)

Polymer repeat unit structure (ester linkages) with subscript *n*

5 **(a)**

Polymer: —N—(CH₂)₆—N—C—(CH₂)₄—C— repeat unit with subscript *n*, two C=O and two N—H

(b) (i) Both contain the peptide linkage CONH.

(ii) The synthetic polymer has a repeat unit made up of two monomers; the protein has a sequence of repeating amino acids with different structures due to the different R– groups.

6 **(a)**

Prop-2-enoic acid (structure: C=C with COOH group)

(b)

Iodoethene (structure: C=C with I and H atoms)

(c)

Benzene-1,4-dioic acid (structure: benzene ring with two COOH groups)

(d)

Ethane-1,2-diamine (structure: N—C—C—N with H atoms)

1.3 Analysis

1 **(a)** The mobile phase is the phase that moves in chromatography; the stationary phase is the phase that does not move in chromatography.

(b) (i) In TLC, the stationary phase is a solid material, usually silica or alumina; the mobile phase is a liquid solvent.

(ii) In GC, the stationary phase is a liquid or solid held on a solid support; the mobile phase is a gas.

(c) (i) A solid stationary phase separates by adsorption; different components in a mixture bind to different degrees with the surface of the stationary phase.

(ii) A liquid stationary phase separates by relative solubility; different components in a mixture have different solubilities in the liquid of the stationary phase.

2 **(a)** Component **A** $R_f = 0.28$; component **B** $R_f = 0.61$; component **C** $R_f = 0.76$

(b) The GC separates the components in the mixture. The output is directed into a mass spectrometer; the mass spectrum of each component is analysed or matched to a database of known compounds for positive identification.

3 **(a) (i)** 5 peaks; **(ii)** 3 peaks; **(iii)** 4 peaks.

(b) Compound **A** spectrum 3; compound **B** spectrum 1; compound **C** spectrum 2.

4 **(a) (i)** 4 peaks, relative peak areas 3:2:2:1
$CH_3CH_2CH_2OH$, $\delta = 0.7$–1.6 ppm;
$CH_3\mathbf{CH_2}CH_2OH$, $\delta = 0.7$–1.6 ppm;
$CH_3CH_2\mathbf{CH_2}OH$, $\delta = 3.3$–4.3 ppm;
$CH_3CH_2CH_2O\mathbf{H}$, $\delta = 1.0$–5.5 ppm.

(ii) 3 peaks, relative peak areas 6:1:1
$CH_3CHOHCH_3$, $\delta = 0.7$–1.6 ppm;
$CH_3\mathbf{CH}OHCH_3$, $\delta = 3.3$–4.3 ppm;
$CH_3CHO\mathbf{H}CH_3$, $\delta = 1.0$–5.5 ppm.

(iii) 3 peaks, relative peak areas 3:2:3
$CH_3CH_2OCH_3$, $\delta = 0.7$–1.6 ppm;
$CH_3\mathbf{CH_2}OCH_3$, $\delta = 3.3$–4.3 ppm;
$CH_3CH_2O\mathbf{CH_3}$, $\delta = 3.3$–4.3 ppm.

(b) Compound **A** spectrum 2; compound **B** spectrum 4; compound **C** spectrum 3; compound **D** spectrum 1

5 $CDCl_3$ is used as a deuterated solvent so that it gives no signal in the NMR spectrum.

TMS is used as the standard for chemical shift measurements with a single δ value of 0 ppm.

D_2O is used to exchange with protons in OH and NH; this aids identification of OH and NH protons.

6 Compound **E** = $(CH_3)_3CCH_2COOH$.

7 Compound **F** = $CH_3CH_2CBr_2CH_3$.
Compound **G** = $(CH_3)_2CBrCH_2Br$.

8 Empirical formula = C_4H_8O; molecular formula = C_4H_8O because M_r = 72 from molecular ion peak at m/z = 72.
From NMR: CH_3CH_2 must be present from triplet/quartet splitting; the remaining protons must be present as a CH_3 with no protons on adjacent carbon.
From chemical shifts of δ = 2.4 ppm (quartet) and δ = 2.2 ppm (singlet) the protons must be on a carbon atom adjacent to C=O ($CH_2C=O$ and $CH_3C=O$).
From mass spectrum: fragment ion at m/z = 43 matches $[CH_3C=O]^+$.
Combining all the evidence, compound **H** is $CH_3CH_2COCH_3$.

2.1 Rates, equilibrium and pH

1 (a) At any instant of time, a tangent is drawn to the curve and the gradient of the tangent is measured.
  (b) $mol\ dm^{-3}\ s^{-1}$
  (c) **W** is second order; **X** is first order; **Y** is zero order
  (d) $rate = k[W]^2[X]$
  (e) Rate increases by 64 times.

2 (a) Comparing experiment 1 with experiment 2, [**R**] and [**T**] have been kept constant,
  [**S**] has increased by 2 times; the rate has also increased by 2 times; so **S** is 1st order.
  Comparing experiment 2 with experiment 3, [**S**] and [**T**] have been kept constant,
  [**R**] has decreased by 0.5 times; the rate has also decreased by 0.25 times; so **R** is 2nd order.
  Comparing experiment 3 with experiment 4, only [**R**] has been kept constant; [**S**] has increased by 3 times; but **S** is 1st order, so this change will also increase rate by **3** times; the rate has also increased by **3** times; no matter what the change in [**T**] there is no further change in rate; so **T** is zero order.
  (b) $rate = k[R]^2[S]$
  (c) From experiment 1, $k = \dfrac{4.56 \times 10^{-8}}{(0.001\,00)^2(0.002\,00)}$
  $= 22.8\ dm^6\ mol^{-2}\ s^{-1}$

3 (a) The slowest step in the reaction mechanism of a multi-step reaction.
  (b) The rate-determining step involves one molecule of **A** and one molecule of **B**.
  (c) $A + B \longrightarrow C + D$
  $D + A \longrightarrow C$

4 (a) (i) $K_c = \dfrac{[CO(g)]\,[H_2(g)]^3}{[CH_4(g)]\,[H_2O(g)]}$; units are $mol^2\ dm^{-6}$;
  (ii) K_c increases with temperature
  (b) (i) $K_c = \dfrac{[HI(g)]^2}{[H_2(g)]\,[I_2(g)]}$; no units;
  (ii) K_c decreases with temperature
  (c) (i) $K_c = \dfrac{[SO_3(g)]^2}{[SO_2(g)]^2\,[O_2(g)]}$; units are $dm^3\ mol^{-1}$;
  (ii) K_c decreases with temperature.

5 (a) $0.340\ mol\ I_2$ and $0.120\ mol\ HI$;
  (b) $K_c = 0.303$

6 (a) (i) acid 1: CH_3COOH, base 1: CH_3COO^-;
  acid 2: H_3O^+, base 2: H_2O
  (ii) acid 1: HCl, base 1: Cl^-;
  acid 2: $CH_3COOH_2^+$, base 2: CH_3COOH
  (b) (i) full: $2HCl(aq) + CaCO_3(s) \longrightarrow CaCl_2(aq) + CO_2(g) + H_2O(l)$
  ionic: $2H^+(aq) + CaCO_3(s) \longrightarrow Ca^{2+}(aq) + CO_2(g) + H_2O(l)$
  (ii) full: $2HNO_3(aq) + MgO(s) \longrightarrow Mg(NO_3)_2(aq) + H_2O(l)$
  ionic: $2H^+(aq) + MgO(s) \longrightarrow Mg^{2+}(aq) + H_2O(l)$
  (iii) full: $2CH_3COOH(aq) + Ca(OH)_2(aq) \longrightarrow (CH_3COO)_2Ca(aq) + 2H_2O(l)$
  ionic: $H^+(aq) + OH^-(aq) \longrightarrow H_2O(l)$

7 (a) (i) 2.46;   (ii) 5.32
  (b) (i) $K_a = 3.12 \times 10^{-7}\ mol\ dm^{-3}$; $pK_a = 6.51$
  (ii) $K_a = 4.48 \times 10^{-9}\ mol\ dm^{-3}$; $pK_a = 8.34$

8 (a) (i) 13.10; (ii) 10.84; (iii) 11.67 (hint: $2 \times OH^-$)
  (b) (i) $2.19 \times 10^{-3}\ mol\ dm^{-3}$;
  (ii) $0.525\ mol\ dm^{-3}$;
  (iii) $3.55\ mol\ dm^{-3}$

9 (a) The buffer equilbrium system is:
  $CH_3COOH(aq) \rightleftharpoons H^+(aq) + CH_3COO^-(aq)$
  Added acid, $H^+(aq)$, reacts with the conjugate base $CH_3COO^-(aq)$:
  $H^+(aq) + CH_3COO^-(aq) \longrightarrow CH_3COOH(aq)$
  This moves the equilibrium to the left.
  Added alkali, $OH^-(aq)$, reacts with the small H^+ concentration:
  $H^+(aq) + OH^-(aq) \longrightarrow H_2O(l)$:
  The weak acid $CH_3COOH(aq)$ dissociates in response to supply more $H^+(aq)$:
  $CH_3COOH(aq) \longrightarrow CH_3COO^-(aq) + H^+(aq)$
  This moves the equilibrium to the right.

(b) Buffer **A**: $pH = -\log(1.7 \times 10^{-5}) = 4.77$

Buffer **B**: $[H^+(aq)] = K_a \times \dfrac{[CH_3COOH(aq)]}{[CH_3COO^-(aq)]}$

$$= 1.7 \times 10^{-5} \times \dfrac{0.750}{0.250}$$

$$= 5.1 \times 10^{-5} \text{ mol dm}^{-3}$$

$pH = -\log [H^+(aq)] = -\log (5.1 \times 10^{-5}) = 4.29$

(c) $[H^+(aq)] = 10^{-5.47} = 3.39 \times 10^{-6} \text{ mol dm}^{-3}$

$$\dfrac{[CH_3COOH(aq)]}{[CH_3COO^-(aq)]} = \dfrac{[H^+]}{K_a} = \dfrac{3.39 \times 10^{-6}}{1.7 \times 10^{-5}} = 0.2 \text{ or } \dfrac{1}{5}.$$

2.2 Energy

1 (a) The enthalpy change that accompanies the formation of one mole of an ionic compound from its gaseous ions under standard conditions.

(b) (i) $K^+(g) + Cl^-(g) \longrightarrow KCl(s)$

(ii) $2Li^+(g) + O^{2-}(g) \longrightarrow Li_2O(s)$

(iii) $Al^{3+}(g) + 3F^-(g) \longrightarrow AlF_3(s)$

2 (a)

Name of enthalpy change	Equation	ΔH/kJ mol^{-1}
Enthalpy change of atomisation of potassium	$K(s) \longrightarrow K(g)$	+89
Lattice enthalpy of potassium iodide	$K^+(g) + I^-(g) \longrightarrow KI(s)$	?
Enthalpy change of formation of potassium iodide	$K(s) + \frac{1}{2}I_2(s) \longrightarrow KI(s)$	–328
Electron affinity of iodine	$I(g) + e^- \longrightarrow I^-(g)$	–295
Enthalpy change of atomisation of iodine	$\frac{1}{2}I_2(s) \longrightarrow I(g)$	+107
First ionisation energy of potassium	$K(s) \longrightarrow K^+(g) + e^-$	+419

(b)

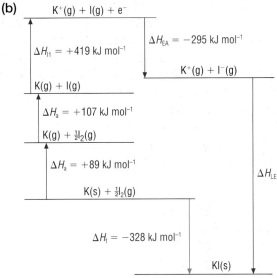

(c) $\Delta H_{LE} = -328 - (89 + 107 + 419 - 295) = -648 \text{ kJ mol}^{-1}$

(d) (i) Sodium iodide has a more exothermic lattice enthalpy; K^+ ions are larger than Na^+ ions, so electrostatic forces of attraction between Na^+ and I^- ions are stronger than between K^+ and I^- ions.

(ii) Calcium iodide has a more exothermic lattice enthalpy; Ca^{2+} ions are smaller than K^+ ions and they have a greater ionic charge. This means that electrostatic forces of attraction between Ca^{2+} and I^- ions are stronger than between K^+ and I^- ions.

3 (a) Lattice enthalpy of magnesium chloride; enthalpy change of solution of magnesium chloride; enthalpy change of hydration of chloride ions.

(b) $\Delta H_{hyd}(Mg^{2+}) = -2493 - 153 - (2 \times -363)$

$$= -1920 \text{ kJ mol}^{-1}$$

4 (a) (i) +ve;   **(ii)** –ve

(b) (i) -198 J K^{-1} mol^{-1}; $\Delta G = -33$ kJ mol^{-1}; 192 °C – above this temperature, the reaction does *not* take place spontaneously.

(ii) $+248$ J K^{-1} mol^{-1}; $\Delta G = +13$ kJ mol^{-1}; 78 °C – above this temperature, the reaction *does* take place spontaneously.

5 (a) (i) $2Ag^+(aq) + Pb(s) \longrightarrow Pb^{2+}(aq) + 2Ag(s)$

(ii) $6Fe^{2+}(aq) + Cr_2O_7^{2-}(aq) + 14H^+(aq) \longrightarrow$
$$6Fe^{3+}(aq) + Cr^{3+}(aq) + 7H_2O(l)$$

(b) (i) $3Ag(s) + NO_3^-(aq) + 4H^+(aq) \longrightarrow$
$$3Ag^+(aq) + NO(g) + 2H_2O(l)$$

(ii) $S_2O_3^{2-}(aq) + 4Cl_2(g) + 5H_2O(l) \longrightarrow$
$$2SO_4^{2-}(aq) + 8Cl^-(aq) + 10H^+(aq)$$

6 (a) The standard electrode potential of a half cell, $E^{\ominus}$, is the e.m.f. of a half cell compared with a standard hydrogen half cell, measured at 298 K with solution concentrations of 1 mol dm^{-3} and a pressure of 100 kPa (1 atmosphere).

(b) (i) 0.75 V; $5Fe^{2+}(aq) + MnO_4^-(aq) + 8H^+(aq) \longrightarrow$
$5Fe^{3+}(aq) + Mn^{2+}(aq) + 4H_2O(l)$

(ii) 2.28 V; $2\tfrac{1}{2}Zn(s) + MnO_4^-(aq) + 8H^+(aq) \longrightarrow$
$2\tfrac{1}{2}Zn^{2+}(aq) + Mn^{2+}(aq) + 4H_2O(l)$

(iii) 1.53 V; $Zn(s) + 2Fe^{3+}(aq) \longrightarrow Zn^{2+}(aq) + 2Fe^{2+}(aq)$

7 (a) $Cr^{2+}(aq)$ with $H^+(aq)$; $Cr^{2+}(aq)$ with $O_2(g)$ and $H^+(aq)$;
$H_2(g)$ with $O_2(g)$ and $H^+(aq)$

(b) $2Cr^{2+}(aq) + 2H^+(aq) \longrightarrow 2Cr^{3+}(aq) + H_2(g)$
$4Cr^{2+}(aq) + O_2(g) + 4H^+(aq) \longrightarrow 4Cr^{3+}(aq) + 2H_2O(l)$
$2H_2(g) + O_2(g) \longrightarrow 2H_2O(l)$

2.3 Transition elements

1 (a) $1s^2 2s^2 2p^6 3s^2 3p^6 4s^2 3d^6$

(b) $1s^2 2s^2 2p^6 3s^2 3p^6 3d^6$

(c) $1s^2 2s^2 2p^6 3s^2 3p^6 3d^{10}$

(d) $1s^2 2s^2 2p^6 3s^2 3p^6 4s^2 3d^7$

2 (a) $+6$;   **(b)** $+3$;   **(c)** $+3$

3 (a) $1s^2 2s^2 2p^6 3s^2 3p^6 3d^5$; iron forms an ion, Fe^{3+}, with a partially filled d-orbital – this is an essential characteristic of a transition metal.

(b) Iron is used as a catalyst in the Haber process in the manufacture of ammonia: $N_2(g) + 3H_2(g) \rightleftharpoons 2NH_3(g)$

4 (a) (i) A d-block element has a d sub-shell as the highest energy sub-shell.

(ii) A transition element forms at least one ion with an incomplete d sub-shell.

(iii) $1s^2 2s^2 2p^6 3s^2 3p^6 4s^2 3d^3$

(b)

Metal ion	Product formed with OH⁻(aq)	Colour of product
Fe^{2+}	$Fe(OH)_2(s)$	green precipitate
Fe^{3+}	$Fe(OH)_3(s)$	brown precipitate

5 (a) (i) A molecule or ion that can donate a pair of electrons to a transition metal ion.

(ii) A transition metal ion bonded to one or more ligands by coordinate bonds (dative covalent bonds).

(iii) A reaction in which one ligand in a complex ion is replaced by another ligand.

(b) (i) Aqueous ammonia, $NH_3(aq)$

(ii) $[Cu(H_2O)_6]^{2+}(aq) + 4NH_3(aq) \rightleftharpoons$
$[Cu(NH_3)_4(H_2O)_2]^{2+}(aq) + 4H_2O(l)$

(iii) $[Cu(H_2O)_6]^{2+}$ is pale blue;
$[Cu(NH_3)_4(H_2O)_2]^{2+}$ is deep blue.

(iv)

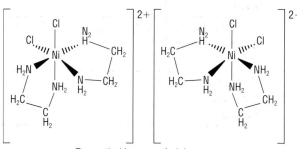

6 (a) The total number of coordinate (dative covalent) bonds formed between the central metal ion and its ligands.

(b) (i) $[Co(H_2O)_6]^{2+}$, octahedral;

(ii) $[Ni(NH_2CH_2CH_2NH_2)_3]^{2+}$, octahedral.

(c) (i)

Cis   *Trans*

(ii)

Cis   *Trans*

(d)

Trans isomer

Two optical isomers of *cis* isomer

7 (a) $MnO_4^-(aq) + 8H^+(aq) + 5e^- \longrightarrow Mn^{2+}(aq) + 4H_2O(l)$

(b) $2MnO_4^-(aq) + 6H^+(aq) + 5(COOH)_2(aq) \longrightarrow$
$2Mn^{2+}(aq) + 10CO_2(g) + 8H_2O(l)$

(c) 0.0300 mol dm^{-3}

8 (a) Colourless to pale pink.

(b) 0.0203 mol dm^{-3}

9 0.161 mol dm^{-3}

Glossary

Acid dissociation constant, K_a For an acid HA,

$$K_a = \frac{[H^+(aq)][A^-(aq)]}{[HA(aq)]}$$

$$pK_a = -\log_{10} K_a$$

$$K_a = 10^{-pK_a}.$$

Acid–base pair A pair of two species that transform into each other by gain or loss of a proton.

Activation energy The minimum energy required to start a reaction by the breaking of bonds.

Addition polymer A very long molecular chain, formed by repeated addition reactions of many unsaturated alkene molecules (monomers).

Addition reaction A reaction in which a reactant is added to an unsaturated molecule to make a saturated molecule.

Adsorption The process by which a solid holds molecules of a gas or liquid or solute as a thin film on the surface of a solid or, more rarely, a liquid.

Alicyclic hydrocarbon A hydrocarbon with carbon atoms joined together in a ring structure.

Aliphatic hydrocarbon A hydrocarbon with carbon atoms joined together in straight or branched chains.

Alkali A type of base that dissolves in water to form hydroxide ions, OH⁻(aq) ions.

Alkanes The homologous series with the general formula: C_nH_{2n+2}.

Alkyl group An alkane with a hydrogen atom removed, e.g. CH_3, C_2H_5; any alkyl group is often shown as 'R'.

Amount of substance The quantity whose unit is the mole. Chemists use 'amount of substance' as a means of counting atoms.

Anhydrous A substance that contains no water molecules.

Anion A negatively charged ion.

Atom economy

$$\text{atom economy} = \frac{\text{molecular mass of the desired product}}{\text{sum of molecular masses of all products}} \times 100\%.$$

Atomic (proton) number The number of protons in the nucleus of an atom.

Atomic orbital A region within an atom that can hold up to two electrons with opposite spins.

Average bond enthalpy The average enthalpy change that takes place when breaking by homolytic fission 1 mol of a given type of bond in the molecules of a gaseous species.

Avogadro's constant, N_A The number of atoms per mole of the carbon-12 isotope (6.02×10^{23} mol⁻¹).

Biodegradable substance A substance that is broken down naturally in the environment by other living organisms.

Biodegradable polymer A polymer that breaks down completely into carbon dioxide and water.

Boltzmann distribution A diagram showing the distribution of energies of the molecules at a particular temperature.

Bond dissociation enthalpy The enthalpy change that takes place when breaking by homolytic fission 1 mol of a given bond in the molecules of a gaseous species.

Brønsted–Lowry acid A species that is a proton, H⁺, donor.

Brønsted–Lowry base A species that is a proton, H⁺, acceptor.

Buffer solution A system that minimises pH changes on addition of small amounts of an acid or a base.

Carbanion An organic ion in which a carbon atom has a negative charge.

Carbocation An organic ion in which a carbon atom has a positive charge.

Catalyst A substance that increases the rate of a chemical reaction without being used up in the process.

Cation A positively charged ion.

Chemical shift A scale that compares the frequency of an NMR absorption with the frequency of the reference peak of TMS at $\delta = 0$ ppm.

Chiral carbon A carbon atom attached to four different atoms or groups of atoms.

Chromatogram A visible record showing the result of separation of the components of a mixture by chromatography.

***cis–trans* isomerism** A special type of *E/Z* isomerism in which each carbon of the C=C double bond carries the same atom or group: the *cis* isomer (*Z* isomer) has that group on each carbon on the same side; the *trans* isomer (*E* isomer) has that group on each carbon on different sides.

Complex ion A transition metal ion bonded to one or more ligands by coordinate bonds (dative covalent bonds).

Concentration The amount of solute, in mol, per 1 dm³ (1000 cm³) of solution.

Condensation reaction A reaction in which two small molecules react together to form a larger molecule with the elimination of a small molecule such as water.

Conjugate acid A species formed when a proton is added to a base.

Conjugate base A species formed when a proton is added to an acid.

Coordinate bond A shared pair of electrons in which the bonded pair has been provided by one of the bonding atoms only; also called a dative covalent bond.

Coordination number The total number of coordinate bonds formed between the central metal ion and any ligands.

Covalent bond A bond formed by a shared pair of electrons.

Cracking The breaking down of long-chained saturated hydrocarbons to form a mixture of shorter-chained alkanes and alkenes.

Curly arrow A symbol used in reaction mechanisms to show the movement of an electron pair during the breaking or formation of a covalent bond.

Dative covalent A shared pair of electrons in which the bonded pair has been provided by one of the bonding atoms only; also called a coordinate bond.

Degradable polymer A polymer that breaks down into smaller fragments when exposed to light, heat or moisture.

Dehydration An elimination reaction in which water is removed from a saturated molecule to make an unsaturated molecule.

Delocalised electrons Electrons that are shared between more than two atoms.

Dipole–dipole force An attractive force between permanent dipoles in neighbouring polar molecules.

Displacement reaction A reaction in which a more reactive element displaces a less reactive element from an aqueous solution of its ions.

Displayed formula A formula showing the relative positioning of all the atoms in a molecule and the bonds between them.

Disproportionation The oxidation and reduction of the same species in a redox reaction.

Dynamic equilibrium The equilibrium that exists in a closed system when the rate of the forward reaction is equal to the rate of the reverse reaction.

E/Z isomerism A type of stereoisomerism in which different groups attached to each carbon of a C=C double bond may be arranged differently in space because of the restricted rotation of the C=C bond.

(First) electron affinity The enthalpy change that accompanies the addition of one electron to each atom in one mole of gaseous atoms to form one mole of gaseous 1^- ions.

(Second) electron affinity The enthalpy change that accompanies the addition of one electron to each ion in one mole of gaseous 1^- ions to form one mole of gaseous 2^- ions.

Electron shielding The repulsion between electrons in different inner shells. Shielding reduces the net attractive force from the positive nucleus on the outer shell electrons.

Electron(ic) structure or configuration The arrangement of electrons in an atom.

Electronegativity A measure of the attraction of a bonded atom for the pair of electrons in a covalent bond.

Electrophile An atom (or group of atoms) which is attracted to an electron-rich centre or atom, where it accepts a pair of electrons to form a new covalent bond.

Electrophilic substitution A type of substitution reaction in which an electrophile is attracted to an electron-rich centre or atom, where it accepts a pair of electrons to form a new covalent bond.

Elimination reaction The removal of a molecule from a saturated molecule to make an unsaturated molecule.

Empirical formula The simplest whole-number ratio of atoms of each element present in a compound.

Enantiomers Stereoisomers that are non-superimposable mirror images of each other; also called 'optical isomers'.

End point The point in a titration at which there are equal concentrations of the weak acid and conjugate base forms of the indicator. The colour at the end point is midway between the colours of the acid and conjugate base forms.

Endothermic A reaction in which the enthalpy of the products is greater than the enthalpy of the reactants, resulting in heat being taken in from the surroundings (ΔH +ve).

(Standard) enthalpy change of atomisation The enthalpy change that takes place when one mole of gaseous atoms forms from the element in its standard state.

(Standard) enthalpy change of combustion, $\Delta H_c^{\ominus}$ The enthalpy change that takes place when one mole of a substance reacts completely with oxygen under standard conditions, all reactants and products being in their standard states.

(Standard) enthalpy change of formation, $\Delta H_f^{\ominus}$ The enthalpy change that takes place when one mole of a compound in its standard state is formed from its constituent elements in their standard states under standard conditions.

(Standard) enthalpy change of hydration The enthalpy change that takes place when one mole of isolated gaseous ions is dissolved in water, forming one mole of aqueous ions, under standard conditions.

(Standard) enthalpy change of neutralisation, $\Delta H_{neut}^{\ominus}$ The energy change that accompanies the neutralisation of an aqueous acid by an aqueous base to form one mole of $H_2O(l)$, under standard conditions.

(Standard) enthalpy change of reaction, $\Delta H_r^{\ominus}$ The enthalpy change that accompanies a reaction in the molar quantities expressed in a chemical equation under standard conditions, all reactants and products being in their standard states.

(Standard) enthalpy change of solution The enthalpy change that takes place when one mole of a compound is completely dissolved in water under standard conditions.

Enthalpy cycle A diagram showing alternative routes between reactants and products that allows the indirect determination of an enthalpy change from other known enthalpy changes using Hess' law.

Enthalpy profile diagram A diagram for a reaction to compare the enthalpy of the reactants with the enthalpy of the products.

Enthalpy, H The heat content that is stored in a chemical system.

Entropy, S The quantitative measure of the degree of disorder in a system.

(Standard) entropy change of reaction, $\Delta S^{\ominus}$ The entropy change that accompanies a reaction in the molar quantities expressed in a chemical equation under standard conditions, all reactants and products being in their standard states.

Equilibrium law For the equilibrium:
$$a\,\mathbf{A} + b\,\mathbf{B} \rightleftharpoons c\,\mathbf{C} + d\,\mathbf{D},$$
$$K_c = \frac{[\mathbf{C}]^c[\mathbf{D}]^d}{[\mathbf{A}]^a[\mathbf{B}]^b}.$$

Equivalence point The point in a titration at which the volume of one solution has reacted exactly with the volume of the second solution. This matches the stoichiometry of the reaction that is taking place.

Esterification The reaction of an alcohol with a carboxylic acid to produce an ester and water.

Exothermic A reaction in which the enthalpy of the products is smaller than the enthalpy of the reactants, resulting in heat loss to the surroundings (ΔH –ve).

Fragmentation The process in mass spectrometry that causes a positive ion to spilt into pieces, one of which is a positive fragment ion.

Free energy change, ΔG The balance between enthalpy, entropy and temperature for a process: $\Delta G = \Delta H - T\Delta S$. A process can take place spontaneously when $\Delta G < 0$.

Functional group The part of the organic molecule responsible for its chemical reactions.

General formula The simplest algebraic formula of a member of a homologous series. For example, the general formula of the alkanes is C_nH_{2n+2}.

Giant covalent lattice A three-dimensional structure of atoms, bonded together by strong covalent bonds.

Giant ionic lattice A three-dimensional structure of oppositely charged ions, bonded together by strong ionic bonds.

Giant metallic lattice A three-dimensional structure of positive ions and delocalised electrons, bonded together by strong metallic bonds.

Greenhouse effect The process in which the absorption and subsequent emission of infrared radiation by atmospheric gases warms the lower atmosphere and the planet's surface.

Group A vertical column in the Periodic Table. Elements in a group have similar chemical properties and their atoms have the same number of outer-shell electrons.

Half-life The time taken for the concentration of a reactant to reduce by half.

Hess' law If a reaction can take place by more than one route and the initial and final conditions are the same, the total enthalpy change is the same for each route.

Heterogeneous catalysis A reaction in which the catalyst has a different physical state from the reactants; frequently reactants are gases whilst the catalyst is a solid.

Heterogeneous equilibrium An equilibrium in which the species making up the reactants and products are in different physical states.

Heterolytic fission The breaking of a covalent bond with both of the bonded electrons going to one atom, forming a cation (+ ion) and an anion (– ion).

High-density lipoprotein (HDL) A type of lipoprotein that can remove cholesterol from the arteries and transport it back to the liver for excretion or re-utilisation.

Homogeneous catalysis A reaction in which the catalyst and reactants are in the same physical state, which is most frequently the aqueous or gaseous state.

Homogeneous equilibrium An equilibrium in which all the species making up the reactants and products are in the same physical state.

Homologous series A series of organic compounds with the same functional group but with each successive member differing by CH_2.

Homolytic fission The breaking of a covalent bond with one of the bonded electrons going to each atom, forming two radicals.

Hydrated A crystalline compound containing water molecules.

Hydrocarbon A compound of hydrogen and carbon only.

Hydrogen bond A strong dipole–dipole attraction between

an electron-deficient hydrogen atom ($O-H^{\delta+}$, $N-H^{\delta+}$ or $F-H^{\delta+}$) on one molecule and a lone pair of electrons on a highly electronegative atom ($H-O:^{\delta-}$, $H-N:^{\delta-}$ $H-F:^{\delta-}$) on a different molecule.

Hydrolysis   A reaction with water or hydroxide ions that breaks a chemical compound into two compounds.

Initial rate of reaction   The change in concentration of a reactant or product per unit time at the start of the reaction: $t = 0$.

Initiation   The first step in a radical substitution in which the free radicals are generated by ultraviolet radiation.

Intermediate   A species formed in one step of a multi-step reaction that is used up in a subsequent step, and is not seen as either a reactant or a product of the overall equation.

Intermolecular force   An attractive force between neighbouring molecules. Intermolecular forces can be van der Waals' forces, dipole–dipole forces or hydrogen bonding.

Ion   A positively or negatively charged atom or (covalently bonded) group of atoms (a molecular ion).

Ionic bonding   The electrostatic attraction between oppositely charged ions.

Ionic product of water, K_w $K_w = [H^+(aq)][OH^-(aq)]$ At 25°C, $K_w = 1.00 \times 10^{-14}$ mol^2 dm^{-6}.

(First) ionisation energy   The energy required to remove one electron from each atom in one mole of gaseous atoms to form one mole of gaseous 1$^+$ ions.

(Second) ionisation energy   The energy required to remove one electron from each ion in one mole of gaseous 1$^+$ ions to form one mole of gaseous 2$^+$ ions.

Isoelectric point   The pH value at which the amino acid exists as a zwitterion.

Isotopes   Atoms of the same element with different numbers of neutrons and different masses.

Lattice enthalpy   The enthalpy change that accompanies the formation of one mole of an ionic compound from its gaseous ions under standard conditions.

Le Chatelier's Principle   When a system in dynamic equilibrium is subjected to a change, the system readjusts itself to minimise the effect of the change and to restore equilibrium.

Ligand   A molecule or ion that can donate a pair of electrons with the transition metal ion to form a coordinate bond.

Ligand substitution   A reaction in which one ligand in a complex ion is replaced by another ligand.

Limiting reagent   The substance in a chemical reaction that runs out first.

Lone pair   An outer-shell pair of electrons that is not involved in chemical bonding.

Low-density lipoprotein (LDL)   A type of lipoprotein responsible for carrying cholesterol and triglycerides from the liver to the tissues.

Mass (nucleon) number   The number of particles, protons and neutrons, in the nucleus.

Mechanism   A sequence of steps, showing the path taken by electrons in a reaction.

Metallic bond   The electrostatic attraction between positive metal ions and delocalised electrons.

Mobile phase   The phase that moves in chromatography.

Molar mass, M   The mass per mole of a substance. The units of molar mass are g mol^{-1}.

Mole   The amount of any substance containing as many elementary particles as there are carbon atoms in exactly 12 g of the carbon-12 isotope.

Molecular formula   The actual number of atoms of each element in a molecule.

Molecular ion, M^+   The positive ion formed in mass spectrometry when a molecule loses an electron.

Molecule   A small group of atoms held together by covalent bonds.

Monomer   A small molecule that combines with many other monomers to form a polymer.

Neutralisation   A chemical reaction in which an acid and a base react together to produce a salt and water.

Nomenclature   A system of naming compounds.

Nucleophile   An atom (or group of atoms) which is attracted to an electron-deficient centre or atom, where it donates a pair of electrons to form a new covalent bond.

Optical isomers   Stereoisomers that are non-superimposable mirror images of each other; also called 'enantiomers'.

Order   The power to which the concentration of the reactant is raised in the rate equation.

Overall order   The sum of the individual orders: $m + n$.

Oxidation   Loss of electrons or an increase in oxidation number.

Oxidation number   A measure of the number of electrons that an atom uses to bond with atoms of another element. Oxidation numbers are derived from a set of rules.

Oxidising agent   A reagent that oxidises (takes electrons from) another species.

Peptide A compound containing amino acids linked by peptide bonds. Often the number of amino acids is indicated by the prefix, di-, tri-, tetra-:
dipeptide = 2 amino acids
tripeptide = 3 amino acids
tetrapeptide = 4 amino acids.

Percentage yield

$$\% \text{ yield} = \frac{\text{actual amount, in mol, of product}}{\text{theoretical amount, in mol, of product}} \times 100\%.$$

Period A horizontal row of elements in the Periodic Table. Elements show trends in properties across a period.

Periodicity A regular periodic variation of properties of elements with atomic number and position in the Periodic Table.

Permanent dipole A small charge difference across a bond resulting from a difference in electronegativities of the bonded atoms.

pH $\text{pH} = -\log[H^+(aq)]$
$[H^+(aq)] = 10^{-pH}.$

Pharmacological activity The beneficial or adverse effects of a drug on living matter.

Phase A physically distinctive form of a substance, such as the solid, liquid and gaseous states of ordinary matter.

pi-bond The reactive part of a double bond formed above and below the plane of the bonded atoms by sideways overlap of p-orbitals.

Polar covalent bond A bond with a permanent dipole.

Polar molecule A molecule with an overall dipole, having taken into account any dipoles across bonds.

Polymer A long molecular chain built up from monomer units.

Precipitation reaction The formation of a solid from a solution during a chemical reaction. Precipitates are often formed when two aqueous solutions are mixed together.

Principal quantum number, n A number representing the relative overall energy of each orbital, which increases with distance from the nucleus. The sets of orbitals with the same n-value are referred to as electron shells or energy levels.

Propagation The two repeated steps in radical substitution which build up the products in a chain reaction.

Radical A species with an unpaired electron.

Rate constant, k The constant that links the rate of reaction with the concentrations of the reactants raised to the powers of their orders in the rate equation.

Rate equation For a reaction: $A + B \rightarrow C$, the rate equation is given by:
$rate = k[A]^m[B]^n.$
m is the **order of reaction** with respect to **A**.
n is the **order of reaction** with respect to **B**.
$m + n =$ **overall order**.

Rate of reaction The change in concentration of a reactant or product per unit time.

Rate-determining step The slowest step in the reaction mechanism of a multi-step reaction.

Reaction mechanism A series of steps that, together, make up the overall reaction.

Redox reaction A reaction in which both reduction and oxidation take place.

Reducing agent A reagent that reduces (adds electrons to) another species.

Reduction Gain of electrons or a decrease in oxidation number.

Reflux The continuous boiling and condensing of a reaction mixture to ensure that the reaction takes place without the contents of the flask boiling dry.

Relative atomic mass, A_r The weighted mean mass of an atom of an element compared with one-twelfth of the mass of an atom of carbon-12.

Relative formula mass The weighted mean mass of the formula unit of a compound compared with one-twelfth of the mass of an atom of carbon-12.

Relative isotopic mass The mass of an atom of an isotope compared with one-twelfth of the mass of an atom of carbon-12.

Relative molecular mass, M_r The weighted mean mass of a molecule of a compound compared with one-twelfth of the mass of an atom of carbon-12.

Repeat unit A specific arrangement of atoms that occurs in the structure over and over again. Repeat units are included in brackets, outside which is the symbol n.

Retention time In gas chromatography, it is the time for a component to pass from the column inlet to the detector.

R_f value $R_f = \dfrac{\text{distance moved by component}}{\text{distance moved by solvent front}}.$

Salt A chemical compound formed from an acid, when an H^+ ion from the acid has been replaced by a metal ion or another positive ion, such as the ammonium ion, NH_4^+.

Saturated hydrocarbon A hydrocarbon with single bonds only.

Shell A group of atomic orbitals with the same principal quantum number, n. Also known as a main energy level.

Simple molecular lattice A three-dimensional structure of molecules, bonded together by weak intermolecular forces.

Skeletal formula A simplified organic formula, with hydrogen atoms removed from alkyl chains, leaving just a carbon skeleton and associated functional groups.

Specific heat capacity, c The energy required to raise the temperature of 1 g of a substance by 1 °C.

Spectator ions Ions that are present but play no part in a chemical reaction

Spin–spin coupling The interaction between spin states of non-equivalent nuclei that results in a group of peaks in an NMR spectrum.

Stability constant, K_{stab} The equilibrium constant for an equilibrium existing between a transition metal ion surrounded by water ligands and the complex formed when the same ion has undergone a ligand substitution reaction.

Standard conditions A pressure of 100 kPa (1 atmosphere), a stated temperature, usually 298 K (25 °C) and a concentration of 1 mol dm^{-3} (for reactions with aqueous solutions).

Standard electrode potential, $E^{\ominus}$ The e.m.f. of a half cell compared with a standard hydrogen half cell, measured at 298 K with solution concentrations of 1 mol dm^{-3} and a gas pressure of 100 kPa (1 atmosphere).

Stationary phase The phase that does not move in chromatography.

Stem The longest carbon chain present in an organic molecule.

Stereoisomers Species with the same structural formula but with a different arrangement of the atoms in space.

Stoichiometry The molar relationship between the relative quantities of substances taking part in a reaction.

Stratosphere The second layer of the Earth's atmosphere, containing the 'ozone layer', between about 10 km and 50 km above the Earth's surface.

Strong acid An acid that completely dissociates in solution.

Structural formula A formula showing the minimal detail for the arrangement of atoms in a molecule.

Structural isomers Molecules with the same molecular formula but with different structural arrangements of atoms.

Sub-shell A group of the same type of atomic orbitals (s, p, d or f) within a shell.

Substitution reaction A reaction in which an atom or group of atoms is replaced with a different atom or group of atoms.

Suffix The part of the name added *after* the stem.

Termination The step at the end of a radical substitution when two radicals combine to form a molecule.

Thermal decomposition The breaking up of a chemical substance with heat into at least two chemical substances.

Transition element A d-block element which forms an ion with an incomplete d sub-shell.

Troposphere The lowest layer of the Earth's atmosphere, extending from the Earth's surface up to about 7 km (above the poles) to about 20 km (above the tropics).

Unsaturated hydrocarbon A hydrocarbon containing multiple carbon-to-carbon bonds.

Valence shell The outermost shell of an atom, which contains the electrons most likely to react and bond to other atoms.

van der Waals' force An attractive force between instantaneous dipoles and induced dipoles in neighbouring molecules.

Volatility The ease with which a liquid turns into a gas. Volatility increases as boiling point decreases.

Water of crystallisation Water molecules that form an essential part of the crystalline structure of a compound.

Weak acid An acid that partially dissociates in solution.

Zwitterion A dipolar ionic form of an amino acid that is formed by the donation of a hydrogen ion from the carboxyl group to the amino group. As both charges are present there is no overall charge.

Periodic Table/Data Sheet

The Periodic Table of the Elements

Key

Relative atomic mass	
Atomic symbol	
Name	
Atomic (proton) number	

Example:
1.0
H
Hydrogen
1

1	2												3	4	5	6	7	0
																		4.0 **He** Helium 2
6.9 **Li** Lithium 3	9.0 **Be** Beryllium 4												10.8 **B** Boron 5	12.0 **C** Carbon 6	14.0 **N** Nitrogen 7	16.0 **O** Oxygen 8	19.0 **F** Fluorine 9	20.2 **Ne** Neon 10
23.0 **Na** Sodium 11	24.3 **Mg** Magnesium 12												27.0 **Al** Aluminium 13	28.1 **Si** Silicon 14	31.0 **P** Phosphorus 15	32.1 **S** Sulfur 16	35.5 **Cl** Chlorine 17	39.9 **Ar** Argon 18
39.1 **K** Potassium 19	40.1 **Ca** Calcium 20	45.0 **Sc** Scandium 21	47.9 **Ti** Titanium 22	50.9 **V** Vanadium 23	52.0 **Cr** Chromium 24	54.9 **Mn** Manganese 25	55.8 **Fe** Iron 26	58.9 **Co** Cobalt 27	58.7 **Ni** Nickel 28	63.5 **Cu** Copper 29	65.4 **Zn** Zinc 30		69.7 **Ga** Gallium 31	72.6 **Ge** Germanium 32	74.9 **As** Arsenic 33	79.0 **Se** Selenium 34	79.9 **Br** Bromine 35	83.8 **Kr** Krypton 36
85.5 **Rb** Rubidium 37	87.6 **Sr** Strontium 38	88.9 **Y** Yttrium 39	91.2 **Zr** Zirconium 40	92.9 **Nb** Niobium 41	95.9 **Mo** Molybdenum 42	(98) **Tc** Technetium 43	101.1 **Ru** Ruthenium 44	102.9 **Rh** Rhodium 45	106.4 **Pd** Palladium 46	107.9 **Ag** Silver 47	112.4 **Cd** Cadmium 48		114.8 **In** Indium 49	118.7 **Sn** Tin 50	121.8 **Sb** Antimony 51	127.6 **Te** Tellurium 52	126.9 **I** Iodine 53	131.3 **Xe** Xenon 54
132.9 **Cs** Caesium 55	137.3 **Ba** Barium 56	138.9 **La*** Lanthanum 57	178.5 **Hf** Hafnium 72	180.9 **Ta** Tantalum 73	183.8 **W** Tungsten 74	186.2 **Re** Rhenium 75	190.2 **Os** Osmium 76	192.2 **Ir** Iridium 77	195.1 **Pt** Platinum 78	197.0 **Au** Gold 79	200.6 **Hg** Mercury 80		204.4 **Tl** Thallium 81	207.2 **Pb** Lead 82	209.0 **Bi** Bismuth 83	(209) **Po** Polonium 84	(210) **At** Astatine 85	(222) **Rn** Radon 86
(223) **Fr** Francium 87	(226) **Ra** Radium 88	(227) **Ac*** Actinium 89	(261) **Rf** Rutherfordium 104	(262) **Db** Dubnium 105	(266) **Sg** Seaborgium 106	(264) **Bh** Bohrium 107	(277) **Hs** Hassium 108	(268) **Mt** Meitnerium 109	(271) **Ds** Darmstadtium 110	(272) **Rg** Roentgenium 111								

Elements with atomic numbers 112–116 have been reported but not fully authenticated

140.1 **Ce** Cerium 58	140.9 **Pr** Praseodymium 59	144.2 **Nd** Neodymium 60	144.9 **Pm** Promethium 61	150.4 **Sm** Samarium 62	152.0 **Eu** Europium 63	157.2 **Gd** Gadolinium 64	158.9 **Tb** Terbium 65	162.5 **Dy** Dysprosium 66	164.9 **Ho** Holmium 67	167.3 **Er** Erbium 68	168.9 **Tm** Thulium 69	173.0 **Yb** Ytterbium 70	175.0 **Lu** Lutetium 71
232.0 **Th** Thorium 90	(231) **Pa** Protactinium 91	238.1 **U** Uranium 92	(237) **Np** Neptunium 93	(242) **Pu** Plutonium 94	(243) **Am** Americium 95	(247) **Cm** Curium 96	(245) **Bk** Berkelium 97	(251) **Cf** Californium 98	(254) **Es** Einsteinium 99	(253) **Fm** Fermium 100	(256) **Md** Mendelevium 101	(254) **No** Nobelium 102	(257) **Lr** Lawrencium 103

Data Sheet

General information

- 1 mol of gas molecules occupies 24.0 dm^3 at room temperature and pressure, RTP.
- Avogadro constant, $N_A = 6.02 \times 10^{23}$ mol^{-1}.
- Ionic product of water, $K_W = 1.00 \times 10^{-14}$ mol^2 mol^{-6}.

^{1}H NMR chemical shifts relative to TMS

Chemical shifts are typical values and can vary slightly depending on the solvent, concentration and substitutents.

Type of proton	Chemical shift, δ/ppm	Type of proton	Chemical shift, δ/ppm
R—CH$_3$	0.7–1.6	—CH=CH—	4.5-6.0
N—H R—OH	1.0–5.5*	(amide C=O NH$_2$ / HN—)	5.0–12.0*
R—CH$_2$—R	1.2–1.4		
R$_3$CH	1.6–2.0		
H$_3$C—C(=O) , RCH$_2$—C(=O) , R$_2$CH—C(=O)	2.0–2.9	(benzene ring —H)	6.5–8.0
(benzene)—CH$_3$, (benzene)—CH$_2$R , (benzene)—CHR$_2$	2.3–2.7	(aldehyde C=O —H)	9.0–10
N—CH$_3$ N—CH$_2$R N—CHR$_2$	2.3–2.9	(carboxylic acid C=O O—H)	11.0–12.0*
O—CH$_3$ O—CH$_2$R O—CHR$_2$	3.3–4.3		
Br or Cl—CH$_3$ Br or Cl—CH$_2$R Br or Cl—CHR$_2$	3.0–4.2		
(benzene)—OH	4.5–10.0*		

* OH and NH chemical shifts are very variable (sometimes outside these limits) and are often broad. Signals are not usually seen as split peaks.

^{13}C NMR chemical shifts relative to TMS

Chemical shifts are typical values and can vary slightly depending on the solvent, concentration and substituents.

Type of carbon	Chemical shift, δ/ppm	Type of carbon	Chemical shift, δ/ppm
C—C	5–55	Carbonyl (ester, carboxylic acid, amide)	160–185
C—Cl or C—Br	30–70		
C—N (amines)	35–60		
C—O	50–70		
C=C (alkenes)	115–140	Carbonyl (aldehyde, ketone)	190–220
Aromatic	110–165		

Characteristic infrared absorptions in organic molecules

Bond	Location	Wavenumber/cm^{-1}	Bond	Location	Wavenumber/cm^{-1}
C–O	alcohols, esters, carboxylic acids	1000–1300	O–H	carboxylic acids	2500–3300 (very broad)
C=O	aldehydes, ketones, carboxylic acids, esters, amides	1640–1750	N–H	amines, amides	3200–3500
C–H	organic compound with a C–H bond	2850–3100	O–H	alcohols, phenols	3200–3550 (broad)

Index

Your Exam Café CD-ROM

In the back of this book you will find an Exam Café CD-ROM. This CD contains advice on study skills, interactive questions to test your learning, a link to our unique partnership with New Scientist, and many more useful features. Load it onto your computer to take a closer look.

Amongst the files on the CD are PDF files, for which you will need the Adobe Reader program, and editable Microsoft Word documents for you to alter and print off if you wish.

Minimum system requirements:
- Windows 2000, XP Pro or Vista
- Internet Explorer 6 or Firefox 2.0
- Flash Player 8 or higher plug-in
- Pentium III 900 MHz with 256 Mb RAM

To run your Exam Café CD, insert it into the CD drive of your computer. It should start automatically; if not, please go to My Computer (Computer on Vista), click on the CD drive and double-click on 'start.html'.

If you have difficulties running the CD, or if your copy is not there, please contact the helpdesk number given below.

Software support
For further software support between the hours of 8.30–5.00 (Mon-Fri), please contact:
Tel: 01865 888108
Fax: 01865 314091
Email: software.enquiries@pearson.com